The HISTORY of MODERN MATHEMATICS

Volume II:

INSTITUTIONS AND APPLICATIONS

Photograph of Participants, Symposium on the History of Modern Mathematics,
Vassar College, Poughkeepsie, New York, June 20–24, 1989.

The HISTORY of MODERN MATHEMATICS

Volume II: INSTITUTIONS AND APPLICATIONS

Proceedings of the Symposium on the History of Modern Mathematics
Vassar College, Poughkeepsie, New York
June 20–24, 1988

Edited by

DAVID E. ROWE

JOHN McCLEARY

ACADEMIC PRESS
Harcourt Brace Jovanovich, Publishers
Boston San Diego New York
Berkeley London Sydney
Tokyo Toronto

This book is printed on acid-free paper. ♾

ACADEMIC PRESS, INC
San Diego, California 92101

United Kingdom Edition published by
ACADEMIC PRESS LIMITED
24-28 Oval Road, London NW1 7DX

Library of Congress Cataloging-in-Publication Data

Symposium on the History of Modern Mathematics (1988 : Vassar College)
The history of modern mathematics : proceedings of the Symposium on the History of Modern Mathematics, Vassar College, Poughkeepsie, New York, 20–24 June 1988 / edited by David E. Rowe and John McCleary.
p. cm.
Includes bibliographical references.
Contents: v. 1. Ideas and their reception — v. 2. Institutions and applications.
ISBN 0-12-599661-6 (v. 1 : alk. paper). — ISBN 0-12-599662-4 (v. 2 : alk. paper)
1. Mathematics — History — Congresses. I. Rowe, David E., Date- . II. McCleary, John, Date- . III. Title.
QA21.S98 1988
510'.9—dc20 89-17766
CIP

PRINTED IN THE UNITED STATES OF AMERICA
90 91 92 93 94 9 8 7 6 5 4 3 2

Contents

Contents for Volume I

Projective and Algebraic Geometry

Abel's Theorem

Number Theory

Contributors List

The number in the parentheses refers to the pages on which the author's contribution begins.

Thomas Archibald (29)
Acadia University
Wolfville, Nova Scotia, Canada B0P 1X0

William Aspray (307)
Charles Babbage Institute
University of Minnesota
Minneapolis, MN 55455

Amy Dahan-Dalmedico (129)
CNRS
75014 Paris, France

Ivor Grattan-Guinness (109)
Middlesex Polytechnique
Queensway, Enfield
Middlesex EN 3 4SF, England

Eberhard Knobloch (251)
Technische Universität Berlin
Fachbereich 1 und 3
Wissenschaftsgeschichte-Mathematik
Strasse des 17.Juni 135
1000 Berlin 12, Federal Republic of Germany

Jesper Lützen (77)
Mathematics Institute
University of Copenhagen
2100 Copenhagen 0, Denmark

Larry Owens (287)
Department of History
University of Massachusetts
Amherst, MA 01033

Erhard Scholz (3)
Gesamthochschule Wuppertal
Fachbereich Mathematik
Gaussstrasse 20
5600 Wuppertal, Federal Republic of Germany

Gert Schubring (171)
Institut für Didaktik der Mathematik
Universität Bielefeld
4800 Bielefeld 1, Federal Republic of Germany

Dirk Struik (99)
52 Glendale Road
Belmont, MA 02170

Renate Tobies (223)
Karl-Sudhoff-Institut
Karl-Marx-Universität
Talstrasse 33
7010 Leipzig, German Democratic Republic

Preface

Within the past decade, considerable attention has been directed toward the task of studying and rewriting the history of mathematics during the 19th and early 20th centuries. While earlier generations of scholars expended most of their efforts researching the history of ancient mathematics or the mathematics of the Renaissance and Scientific Revolution, a growing number of contemporary historians have been drawn to the modern era, i.e., the period roughly spanned by the publication of Gauss's *Disquisitiones arithmeticae* in 1801 to the advent of high-speed electronic computers around 1950. This recent interest reflects a widespread recognition of the importance of an era in which mathematicians produced a vast corpus of rich new ideas that form the foundations of the subject as we know it today. It should not be overlooked, however, that this period was marked not only by revolutionary intellectual achievements but that it also witnessed the emergence of mathematics as an autonomous academic discipline. Over the course of the 19th century, mathematicians began to articulate new standards for teaching and research. The process began slowly, but by the last third of the century such issues were often debated, adopted, or promoted by various professional organizations. These mutually reinforcing intellectual and institutional developments profoundly affected the traditional role of mathematics within the exact sciences, providing it with a radically new orientation, one whose background has not received sufficient attention in the historical literature.

The twenty-four essays presented herein represent a sampling of recent efforts to redress this situation. They constitute the fruits of the Symposium on the History of Modern Mathematics held at Vassar College from June 20–24, 1988. Forty-five participants from twelve countries took part in this symposium which featured twenty-nine lectures covering a variety of topics from the era of Euler and Clairaut to the age of modern computers. In organizing this meeting, we sought to adopt a program that would accomplish three goals: (1) reflect the diverse approaches recently taken by historians of mathematics who have studied various aspects of this period; (2) stress certain characteristic themes and ideas that illustrate its development; and (3) emphasize issues and ideas that shed some light on relationships between mathematics and other scientific disciplines. By so doing, we hoped to give the symposium reasonable thematic unity while at the same time showcasing the wide spectrum of research that is now being done on the history of modern mathematics. Above all else, we wanted to avoid the pitfall of producing a hodgepodge collection of highly specialized articles. Although some of the essays below obviously presuppose a fairly sophisticated knowledge of and interest in the history of mathematics during this period, they are intended for the *general reader* who satisfies these prerequisites. These essays do not pretend to cover all major areas of research—indeed, the monumental achievements of 19th-century mathematicians in the fields of real and complex analysis receive only passing attention here—nor should they be regarded as the last word on the subjects they do address. In most cases, the authors set forth conclusions based on research that is still very much in progress. The standards by which we would like to see these studies measured, therefore, are those of cohesiveness, originality, significance, and the strength of the argumentation and the documentary evidence used to support it.

Volume I is concerned with the emergence and reception of major ideas in fields that range from foundations and set theory, algebra and invariant theory, and number theory to differential geometry, projective and algebraic geometry, line geometry, and transformation groups. The introductory essay by John McCleary raises some larger historiographical concerns regarding the reception of mathematical works *qua* literature. In one form or another these issues are echoed in many of the contributions to both volumes. For example, the complex process of assimilating and enlarging the fundamental achievements of Gauss and Riemann in differential geometry is examined from three different standpoints in the articles by Gregory Nowak, Jesper Lützen, and Dirk Struik. The essays by Walter Purkert and Harold Edwards, on the other hand, recreate the *Urwelt* of two opposing views on the foundations of mathematics—those of Cantor and Kronecker, respectively. By so doing, they reveal that both men have been badly misunderstood not only by their contemporaries but by those who have represented their respective philosophies ever since. Whereas Cantor's famous continuum hypothesis was regarded as somewhat akin to a "self-evident truth" (albeit one that required proof), Gregory Moore's analysis reveals the large number of hats it has worn and the key part it has played in the emergence of axiomatic set theory.

The reception of European educational traditions in the United States forms the larger background against which Helena Pycior describes early American algebra. By examining the British and French textbooks adopted at leading American educational institutions like Harvard and West Point, she offers some new insights about the dynamics that shaped this transmission process. Karen Parshall's essay, on the other hand, illustrates how difficult the transmission of foreign ideas can sometimes be, particularly when they are imbedded within an indigenous research tradition. J. J. Sylvester and Paul Gordan, two leading exponents of the British and German approaches to invariant theory, were both eager to understand the others work, particularly since both shared a strong desire to solve the finite basis problem. Yet despite their best efforts, neither was able to penetrate fully the other man's techniques.

Karine Chemla's essay documents the prehistory of the duality concept, which was vitally important for 19th-century projective geometry, by tracing its roots in 18th-century spherical trigonometry. The central role duality, which came into play fifty years after Gergonne called attention to it in the 1820s, is amply illustrated in the articles by Rowe, Hawkins, and Gray. David Rowe and Thomas Hawkins take up a theme that is also discussed at length in Erhard Scholz's paper, namely the work of Lie and Klein on line geometry and transformation groups during the years 1869-1873. Rowe's article stresses the background events that led up to Lie's discovery of the line-to-sphere transformation and the "Erlanger Programm," whereas Hawkins focuses his attention on the rich variety of ideas that contributed to Lie's creation of a new theory of continuous groups. Jeremy Gray's overview of major developments in algebraic geometry highlights, among other things, the fundamental significance of Abel's Theorem for the whole subject. Abel's Theorem was a major tool for two of the key players Gray considers, Alexander Brill and Max Noether. As mathematicians are sometimes wont to do, these two tried to reconstruct the source of Abel's original inspiration in their classic study *Die Entwicklung der Theorie der algebraischen Funktionen in älterer und neuerer Zeit*. Following in this grand tradition, Roger Cooke offers his own interpretation of what Abel was up to and where Brill and Noether went wrong. Volume I then concludes with a contribution to the history of number theory by Günther Frei who traces the roots of Heinrich Weber's class field theory back to Euler's ζ-function and the fundamental contributions of Dirichlet, Kronecker, and Dedekind to algebraic number theory.

Whereas Volume I is almost exclusively concerned with developments in pure mathematics, the emphasis in Volume II shifts to ideas that lie at the crossroads of mathematics and physics—crystallography and group structures, potential theory and electrodynamics, differential geometry and classical and relativistic mechanics. On the institutional front, particular attention has been focused on issues pertaining to the tensions and interplay between pure and applied mathematics as these arose in three different historical settings: early 19th-century France, Wilhelmian era Germany, and the United States during and after World War II. Erhard Scholz and Thomas Archibald offer two case studies of mathematical theories that emerged in symbiotic union with physical research. Scholz shows how crystallographers, in particular the French theorist August Bravais, were studying group structures decades before Camille Jordan, following in Bravais' footsteps, launched the abstract theory of geometrical groups. Archibald's article deals with the work of German mathematicians who sought to develop an approach to potential theory consistent with Wilhelm Weber's action-at-a-distance model for electrodynamics. This led Riemann, Carl Neumann, and others to study potentials propagated with finite velocity. As Maxwell's theory gradually displaced Weber's, however, this research program lost its initial impetus, and potential theory turned increasingly toward problems of a purely mathematical nature.

The interaction between mathematics and physics is also a major theme (though played in a minor key) in the articles by Jesper Lützen and Dirk Struik. Lützen's analysis of an unpublished paper of Liouville reveals the masterful command the Frenchman had of Gauss's intrinsic approach to surface theory. In this work Liouville derives the Hamilton-Jacobi equations of classical mechanics by treating the motion of a particle in a force field as equivalent to regarding the same particle in a force-free setting in which it moves along a geodesic curve on a certain surface. Struik's paper on the emergence of modern tensor analysis gives a first-hand account of the sudden interest this field attracted immediately after the publication of Einstein's theory of general relativity in 1915. As a final illustration of ideas that arose from a multiplicity of fields, William Aspray's essay analyzes the revolutionary impact of high-speed computers on numerical analysis by highlighting the solutions devised by John von Neumann to a variety of real-world problems.

That ubiquitous and despairingly simple question—what is applied mathematics?—crops up in a number of the essays in Volume II. Ivor Grattan-Guinness's essay shows how complex this matter had already become for Parisian mathematicians during the early decades of the 19th century. With respect to applications, he identifies two large camps of practitioners with diametrically opposite views: the theoreticians, whose prototype was Lagrange, and the engineer *savants*, typified by Lazare Carnot. Amy Dahan's paper examines how two of the theoreticians, namely Poisson and Cauchy, dealt with a problem in hydrodynamics, one which turned out to have important implications for the theory of Fourier series. Larry Owens' study reveals that a strikingly similar debate over applications took place more than 100 years later in the United States, but this time colored by the tense atmosphere of wartime. Owens identifies

Warren Weaver as the leading spokesman for "useful" applied mathematics, whereas Norbert Wiener and John von Neumann are representative examples of the theoretical applier. The debate over applications of mathematics in Wilhelmian Germany, the principal theme of Renate Tobies's article, was, if anything, even more electric owing to the institutional rift that separated the engineering schools from the universities. Her essay and the one by Gert Schubring analyze the central and oft-debated role of Felix Klein in reforming mathematics education in Germany. Both authors concur that the principal motivation behind Klein's numerous reform proposals was to ensure that mathematics instruction would not lose touch with the type of training demanded by modern engineering science. Finally, Eberhard Knobloch offers a very different approach to the study of the institutionalization of applied mathematics. His prosopographical analysis of the Berlin Technische Hochschule over the course of roughly 100 years may be regarded as a companion piece to the well-known volume by K.-R. Biermann, *Die Mathematik und ihre Dozenten an der Berliner Universität, 1810–1933*.

It goes without saying that an undertaking such as this one can only succeed through the joint effort, cooperation, and support of many people. We should note at the outset that this symposium was directly inspired by earlier meetings on the history of 19th-century mathematics held over the past eight years in Cambridge, Massachusetts, Siena, Italy, and Sandbjerg, Denmark. To the organizers of these earlier symposia, and particularly to Kirsti Andersen, go our gratitude for their efforts and inspiration. The immediate occasion for the Vassar meeting was to celebrate Walter Purkert's semester in the United States as a visiting professor at Pace University. Professor Purkert joins us in thanking Pace University, and particularly Dr. Martin Kotler, Chair of the Mathematics Department, Dr. Joseph E. Houle, Dean of Dyson College of Arts and Sciences, and Dr. Joseph M. Pastore, Jr., Provost of Pace University, for their generous support of this undertaking. We are pleased to acknowledge that major funding for the Vassar symposium was made possible through a grant from the National Science Foundation, and we are also grateful to the other organizations who helped to sponsor it: the History of Science Society, the International Commission on the History of Mathematics, and the International Union of the History and Philosophy of Science.

In editing these two volumes, we have been ably assisted by an editorial board consisting of: Joseph W. Dauben (Lehman College and the Graduate Center, CUNY), Jesper Lützen (Copenhagen University), Karen V. H. Parshall (University of Virginia), and Walter Purkert (Leipzig University). Our thanks to all four of them for their assistance, and particularly to Joseph Dauben for his advice and guidance in helping to organize the symposium and to Karen Parshall for going above and beyond the call of duty to help us edit the essays as swiftly as possible. We also wish to thank Klaus Peters of Academic Press for his enthusiatic support of this project and for his efforts in guiding it through some rough seas and safely into port.

Obviously, we could not publish these conference proceedings in a timely fashion unless the contributors themselves were willing to take time off from various other duties and projects in order to compose their articles. We were most fortunate that nearly everyone involved in this project held to the deadlines for submission of first drafts and revisions. The swift publication and general appearance of the volumes were made possible by the submission of articles as computer files and the magic of TEX, the typesetting program. We apologize for any errors that may have slipped through in the process of editing these volumes.

The highly nontrivial task of converting our authors' assorted manuscripts into the text that follows was greatly aided by the typing skills of MaryJo Santagate, Christina Mattos, Deborah Mock, and Anne Clarke. Our thanks to them, and to John Feroe for assisting us in producing several of the diagrams and tables that appear in these volumes. We greatly appreciate the friendly assistance offered to us by James Kreydt of Springer-Verlag, who arranged for us to make use of the numerous photographs reproduced here from the Springer archives. Finally, we must reserve two very special thank yous for our wives, Hilde and Carlie, those "ladies in waiting" who have patiently accompanied us through the ups and downs of this endeavor over the past four years.

David E. Rowe
John McCleary

The Crossroads of Mathematics and Physics

Camille Jordan (1838–1922)
Courtesy of Springer-Verlag Archives

Crystallographic Symmetry Concepts and Group Theory (1850-1880)

Erhard Scholz

INTRODUCTION

The first mathematician to explicitly utilize the group concept in a geometrical setting was *Camille Jordan* (1838-1922). In his *Mémoire sur les groupes de mouvements* [1869] Jordan tried to classify the subgroups of $Isom^+(\mathbb{E}^3)$, the proper isometries of Euclidean space, up to affine conjugation. At the very outset of his investigation he stated his goal in two different forms, the second of which may sound surprising or even strange at first glance. The task, according to Jordan, was: "To form all possible groups of motions ... [and] ... To form systems of molecules, which are superposable in different positions, in all different ways ...." [Jordan 1869, 232]

More light on the origin and motivation of this surprising formulation is shed by a subsequent remark in which Jordan referred directly to the sources which inspired his work: " ... It was under this second aspect that M. Bravais has studied this question: the special cases he studied and applied to crystallography in a remarkable manner are the most important ...." This reference to the crystallographic work of *Auguste Bravais* (1811–1863), a man of diverse interests and working in a broad field of disciplines as naturalist, cartographer, scientist, physicist, and theoretical crystallographer, is striking. It suggests that Bravais' work might have been a source of inspiration not only for the group-theoretic task Jordan poses in his paper, but also for the classification scheme he adopts in it ("Bravais has studied ... the most important ... special cases ... "). But in spite of this notable acknowledgement of his debt to Bravais' work on crystal structure theory, which had been written about two decades ago [Bravais 1849, 1850, 1851], Jordan felt that something more could be done along these lines from a purely mathematical point of view. And so he went on to say: " ... I nevertheless believe that today there is some interest in studying the problem in all its generality."

Jordan's assertion that there might be "some interest" in his generalization of Bravais' studies from the group theoretic point of view appears very modest, indeed. For in retrospect, his Mémoire [1869] may be considered as nothing less than one of the turning points in the development

ISBN 0-12-599662-4.

of the group concept in the 19th century. Not only was it the first publication to introduce the group concept explicitly in a geometrical context, but it also seems to have played a role in prompting the early development of the more general geometric group concept in the form of transformation groups, a development particularly associated with the names and work of Sophus Lie and Felix Klein. This will become clearer after we have had another look at their clarification of group ideas during the early 1870s in section 4 of this paper.

Before discussing this point, however, we begin with an overview of group theory and the utilization of group concepts in crystallography during the first half of the 19th century. Then, in section 2, we consider the group-theoretic elements implicit in Bravais' work on crystal structure theory. This is followed in section 3 by a discussion of Jordan's Mémoire of 1869 that attempts to assess what impact Bravais' ideas exerted on this work.

1. Group theory and crystallography in the mid-19th century

We can be fairly brief concerning the development of group-theoretic ideas during this period thanks to Hans Wussing's well-known study [1969/1980]. Not so well known is the development of symmetry concepts in geometric crystallography, which has only very recently been the object of historical investigations [Burckhardt 1988; Scholz 1989a, 1989b]. Around 1850 the *explicit use of the group concept was extensionally restricted to permutation groups mainly in the algebraic theory of equations*, by consequence, it was intensionally restricted to finite systems of substitutions. Let us just recall that Galois' work on the subject, written between 1830 and 1832[1], had not been accessible before its publication in 1846. On the other hand, just before the appearance of Galois' papers, Cauchy published his long essay on permutation groups from a point of view that was completely compatible with Galois' own [Cauchy 1844].[2] After a delay of several more years there then began the reception and extension of Galois' programmatic ideas in Italy [Betti 1852], Germany [Kronecker 1853, 1856; Dedekind 1856], and France (Liouville, Serret, Hermite, and Bertrand).

During this time there was also a growing *awareness of group structures in number theory*—particularly in the theory of residue classes and in Gauss' theory of binary quadratic forms—even if it was not yet worked out explicitly. Kummer realized that there was a structural analogy between the composition of equivalence classes of binary quadratic forms and those of ideal classes in quadratic number fields [Kummer 1845]. And when Dedekind began to work out the Galois theory of algebraic solutions of equations in his lectures given at Göttingen as a young "Privatdozent," he developed the theory of permutation groups systematically as a first step to his theory. At the end of the chapter dealing with this question, he remarked that all he had done could in principle be derived from the general laws of composition of the permutations, which thus could have served as logical starting point, making the theory in principle applicable to the Gaussian theory of composition of binary forms [Dedekind 1856, 63]. Dedekind therefore saw this relationship, but he did not work out its implications.

In geometry group ideas were still completely implicit, even if partially quite well developed. They grew in *projective geometry,* at least in its second phase of development, when it was taken up by the German school. In his *Barycentrische Calcul* [Möbius 1827], A.F. Möbius discussed not only projective collineations as leading to a meaningful (i.e. transitively closed) system of geometrical equivalence classes, but moreover he considered a whole hierarchy of such "geometrical relationships" (*geometrische Verwandtschaften*). These, looked upon from an operational point of view, would lead to a characterization of a methodological level of abstraction in geometry similar to the one later advanced by Felix Klein in his *Erlanger Programm.* J. Steiner characterized general projective collineations in [Steiner 1832], whereas Poncelet and others had still restricted their attention to special collineations, i.e., projections ("homothéties") in the restricted sense, which, strangely enough, were not even closed under composition.

Cinematical considerations led to a deeper understanding of compositions of *rotations in Euclidean space* $\mathbb{E}^3$ [Rodrigues 1840][3] and of general *screw motions* [Poinsot 1851]. These investigations revealed implicit aspects of the semidirect product structure of the Euclidean isometries

$$Isom^+(\mathbb{E}^3) = \mathbb{R}^3 * SO(3,\mathbb{R}).$$

Composition in the Euclidean rotation group, traditionally computed by use of Eulerian angles, was looked upon from the point of view of the quaternion calculus by Cayley [1845] and Hamilton [1847]. Their application of unitary quaternions operating on Euclidean, space embedded as the purely imaginary quantities in the quaternion field, by multiplicative conjugation was a first implicit spin representation of $SO(3,\mathbb{R})$.

Of course the whole field of *linear algebra* was full of even more implicit group structures. They were usually interwoven, however, with additional structures, such as vector spaces [Grassmann 1844 etc.] or associative algebras (in particular quaternions or matrices)[4], and were therefore not looked upon from a group-theoretic point of view. Thus through a wide variety of mathematical investigations, the relationships between a number of implicit group structures became more apparent by the middle of the 19th century. These relationships, however, were not sufficiently impressed on the minds of working mathematicians that Arthur Cayley's attempt [Cayley 1854] to give a *general definition* of a group as a (countable) "set of symbols" with an associative rule of composition[5] received more than passing notice.

Another field in which detailed knowledge of the implicit structure of the finite orthogonal groups in Euclidean space had arisen, and received next to no attention from the mathematical community was *geometric crystallography* with its variant usages of symmetry concepts. This line of research was initiated by *R.J. Haüy* (1743–1822), creator of an atomistic crystal structure theory based on detailed phenomenological knowledge gained in the 18th century. Haüy proposed a derivation of the admissible crystal forms that obeyed a specific "law of symmetry" [Haüy 1815]. Initially he had six polyhedrical primitive forms, but he later adopted eighteen such structures. On these "nucleus polyhedra" he studied arrangements of molecules in layers shaped like parallelepipeds. These could fall back along the edges, summits, or face diagonals according to certain rules which required that new plane faces inclined to the old ones arose. Haüy then postulated in his *"law of symmetry"* that the "falling-back" behaviour was the same for all those parts of the primitive form (edges, summits, and face diagonals), which are *indistinguishable to the eye*.

By following this morphological scheme, Haüy was able to characterize point symmetry systems with respect to the primitive forms. This gave him an implicit representation of the 8 finite orthogonal groups (crystall classes) O^*, T_d, D_{6h}, D_{3d}, D_{4h}, D_{2h}, C_{2h}, C_i, written in Schoenflies notation. Haüy's characterization was a great step forward when compared with the elementary symmetry concepts then in use in mathematical geometry, which did not go beyond simple mirror symmetry [Legendre 1794]; it was, however, still a bit clumsy due to its rigid morphological adherance to the polyhedrical primitive forms.

This latter problem was overcome by developments that took place within the dynamistic tradition of crystallography, which flourished during the first half of the 19th century was strongly influenced by the dialectical philosophy of nature then in vogue throughout much of the German-speaking world. Inspired by their dynamistic convictions, the German theoreticians[6] looked for "axes" of the crystall forms that dominated and regulated these figures. This led them to analyze "axes" as symmetry elements and to consider their possible combinations. The result was a more direct, and even partially algebraicized characterization of the 7 holoedric crystal classes by C.S. Weiss (in 1815) and J.G. Grassmann (1829), the 32 crystal classes (Frankenheim 1826), and even a full list of all finite point symmetry systems, i.e. implicitly the finite orthogonal groups [Hessel 1830]. This story cannot be dealt with here, nor need it be, as it has recently been discussed in detail elsewhere.[7] One remark, however, ought to be made here. Most of the (German) dynamistic theoreticians knew the (French) atomistic theory quite well, and learned from it, although they did not accept the atomists ontological views regarding the formation of matter. The majority of the atomists, on the other hand, did not seriously study the dynamists, and even thought of them as weird and obscure. This was not the case, however, with A. Bravais, who knew the dynamistic theoreticians well and cited their work positively[8]; he may have even sympathized with their philosophy.[9]

2. Bravais' theory of crystal structure

From the 1820s onwards Haüy's hypothesis of polyhedral molecules lying side by side in bricklike packages became less and less acceptable from the chemical point of view. For a variety of reasons it became fashionable to think of molecules as spatial constellations of atoms, perhaps spherical, with nonzero interatomic distances. Thus, in the early 1840s, proposals were made to adapt Haüy's concept to this new point of view. *Gabriel Delafosse* (1796–1878) proposed a model for molecules having a polyhedral skeletal form with only the summits occupied by spherical atoms. He considered these as distributed in a three-fold periodic pattern throughout space with the molecule centers forming the points of a spatial lattice.[10] The regular distribution along the lattice points was conceived as a substitute for Haüy's parallelepiped structure for the molecules, and the shape of the polyhedral molecules was assumed to somehow determine the point symmetry type of the morphology and physical behaviour of the crystal.

This set the framework for a modification of Haüy's program. To pursue it further would presuppose, of course, a subtler analysis of the symmetry types of the molecules, a symmetry analysis of spatial point lattices, and the formulation and evaluation of principles according to which the molecule and lattice symmetries are coordinated.

All these problems were analyzed and solved to a certain degree by Bravais in three consecutive publications [1849, 1850, 1851]. I shall discuss the first of these, *Mémoire sur les polyèdres symétriques* [1849], only very briefly, although it contains very interesting considerations about finite point symmetry systems, or "symmetric polyhedra" (*polyèdres symétriques*) as Bravais called them. He began with an enumeration of elementary symmetry relations and with associated geometrical symmetry elements and their composition:

- line (*ligne*) L^q for an axis of rotation of order q,
- plane (*plan*) P^q for a mirror plane orthogonal to an associated L^q,
- inversion centre (*centre*) C,
- and Λ^q or Π^q for a principal axis, respectively principal plane.[11]

Then he went on to consider all combinations of such symmetry elements and to derive a nearly complete list of point symmetry systems ("symmetric polyhedra"), the only incompleteness resulting from the fact that he did not consider mirror reflections as symmetry elements of their own. The result was that from the seven infinite series of finite orthogonal groups C_n, $C_{n,v}$, $C_{n,h}$, D_n, $D_{n,h}$, $D_{n,d}$, $\overline{C}_{2n}$, Bravais overlooked those $\overline{C}_{2n}$ with n even, as these cannot be obtained by the symmetry elements that he considered.[12]

For each symmetry type Bravais wrote down a "symmetry symbol" listing type, order, and number of conjugate symmetry elements, e.g.

$$D_{2q,h} = [\Lambda^{2q}, qL^2, qL'^2, C, \Pi, qP^2, qP'^2],$$
$$C_{2q,h} = [\Lambda^{2q}, C, \Pi], \quad C_{2q+1,h} = [\Lambda^{2q+1}, \Pi] \quad etc.$$

The last example shows how Bravais' mode of symbolical representation makes the Schoenflies series split into two parts. Together with some singular cases he thus derived a list of 23 "classes of polyhedra" (*classes des polyèdres*) characterizing all the finite orthogonal groups with the exception of $\overline{C}_{4n}$. As there is no reference to [Hessel 1830], whereas there are many in his later work to Weiss, Frankenheim and other dynamistic crystallographers, and as Bravais would probably not have overlooked the symmetries of type $\overline{C}_{4n}$ if he had read Hessel more than just superficially, we may conclude that Bravais derived this list completely anew and independently of Hessel. This investigation then prepared the way for his subsequent symmetry studies.

In his next paper, *Mémoire sur les points distribuées régulièrement sur un plan ou dans l'éspace* [Bravais 1850], he investigated plane and spatial point lattices from the point of view of their symmetry properties. This new approach brought with it two consequences right from the start. The first was that Bravais now had to consider superpositions of a lattice onto itself by translations, rotations, reflections, or inversions. Thus he considered geometrically defined *operations on infinite unbounded point sets* in space. And although he did not explicitly work with mappings of the whole space, the idea of the latter was already potentially very close, even when realized explicitly only on the grid given by the lattice points.

The second consequence of Bravais' symmetry studies of lattices was that he now had to work with *infinite systems of symmetries*, thus implicitly with infinite subgroups of the Euclidean isometries. Bravais touched upon the idea of the totality of all symmetry operations of a given lattice Γ,

$$Isom\ \Gamma \cong \Gamma * H_\Gamma,$$

H_Γ orthogonal (holoedric) symmetries of Γ, and remarked that to each symmetry element there are infinitely many "of the same species" (conjugate in $Isom\ \Gamma$), and to each symmetry element lying anywhere in the lattice (e.g. intermediate between the lattice points) there is one of the "same type and parallel" (conjugate in $Isom(\mathbb{E}^3)$) through any lattice point.

This last observation allowed him to reduce the complexity of the infinite symmetry systems and consider only those symmetry elements meeting one of the lattice points. The latter were obviously finitely many, and so he was back to the situation he had analyzed in his Mémoire [1849]. Looked upon from the point of view of modern group theory, this was a geometrical version of the transition from $Isom\ \Gamma$ to its *orthogonal part*, according to the natural homomorphism ρ resulting from the semidirect product structure of $Isom(\mathbb{E}^3)$)

$$0 \longrightarrow \mathbb{R}^3 \longrightarrow Isom(\mathbb{E}^3) \xrightarrow{\rho} O(3,\mathbb{R}) \longrightarrow 1$$

This made it possible for Bravais to associate a "symmetrical polyhedron" (*polyèdre symetrique*) together with its "symbol of symmetry" (*symbol de symétrie*) to any point lattice and to derive as his first main result:

THEOREM 1 [BRAVAIS 1850, 65FF.]. *A spatial lattice Γ possesses one of the following 7 point symmetry systems (in Schoenflies notation): O^*, D_{6h}, D_{3d}, D_{4h}, D_{2h}, C_{2h}, C_i (the holoedric crystal classes).*

According to their maximal point symmetry, the spatial lattices therefore subdivide into seven (Bravais) lattice systems[13].

Bravais did not end his investigation here, but went on to analyze when two lattices Γ, Γ' of the same maximal point symmetry H could be transformed into one another by continuous deformation without ever losing the symmetry H. In so doing, he was led to a more refined notion of lattice types in the following sense:

DEFINITION:. *Two lattices Γ, Γ' of the same holoedry (maximal point symmetry type H) are said to be of the same lattice type, if there exists a continuous 1-parameter deformation of lattices, $\Gamma(t)$, $0 \le t \le 1$, $\Gamma(0) = \Gamma$, $\Gamma(1) = \Gamma'$, without break of symmetry, i.e. the maximal point symmetry of $\Gamma(t)$ is H for all $0 \le t \le 1$.*

On the basis of this finer equivalence relation, Bravais went on to prove his second main result:

THEOREM 2 [BRAVAIS 1850, 65FF.]. *There are exactly 14 different lattice types in* $\mathbb{E}^3$.

Referring back to his consideration of the totality of symmetry elements associated with a lattice, this result implied ***an implicit characterization of 14 infinite spatial symmetry systems*** which,from the group-theoretic point of view, represent the 14 crystallographic space group types arising as semidirect products of translation lattices with their holoedric (maximal) orthogonal symmetry groups

$G = Isom\ \Gamma \cong \Gamma * H_\Gamma$, where

Γ representative of one of the 14 lattice types in $\mathbb{E}^3$, and

H_Γ is the orthogonal symmetry group (holoedry) of Γ.

All this was done, to say it once more, without any explicit use of the group concept and *a fortiori* without semidirect products of groups. The geometric context and problem structure suggested this semidirect product type for symmetry systems, so to speak, completely naturally and by its own intrinsic logic.

This last point is precisely what makes Bravais' work *Etudes crystallographiques* [Bravais 1851] even more interesting than the papers that preceded it. In this last contribution to his crystallographic "trilogy," he combined the mathematical results of the first two parts to give a detailed exposition of how this approach could be used to modernize the atomistic theory of crystal structure and give it the flexibility needed to accomodate new developments in the field. Changes were necessary with respect to the conception of molecular structure, mainly an offspring of chemistry, and the knowledge of symmetries, which were largely derived as part of the dynamistic program during the preceding decades.

Bravais started from the same basic hypotheses as Delafosse, but he could characterize molecular symmetry more precisely by the use of one of his "symbols of symmetry" S_m ("m" for *molécule*) derived in his [1849]. Here arose the question: in which lattice structure Γ may such molecules of symmetry S_m crystallize? Denoting the orthogonal (holoedric) symmetry of Γ by S_a ("a" for *assemblage*—lattice), Bravais had to deal with the problem that the two symmetries, S_m and S_a, generally are not identical and may not even "fit together" neatly (in the sense that S_m may not be contained in S_a). And, of course, there was the question of establishing which principles regulate the coordination of S_m and the lattice system (characterized by S_a).

The first problem was not difficult to solve. Bravais observed that by the superposition of molecule and lattice symmetry, S_m and S_a, the global behaviour of the whole configuration, and thus of the crystal itself—morphology, physical properties like rigidity, elasticity, optical refraction and polarization, pyroelectricity, etc.—is determined by the intersection

$$S_c = S_m \cap S_a$$

("*c*" for *commun*).[14] Next he proposed two principles for regulating the selection of the Bravais systems of lattices Γ in which a certain substance with given molecule symmetry S_m may crystallize.

POSTULATE 1 [BRAVAIS 1851; 206]:. *Molecules of symmetry S_m crystallize in a lattice Γ of symmetry S_a such that $S_c = S_m \cap S_a$ is as large as possible when compared with other choices of Γ.*

This principle of maximal intersection symmetry, although in itself rather plausible *a priori*, did not suffice, of course. The molecules, if regulated only by this principle, could always choose lattices, for example, with "large" symmetries[15] and more or less forget about lattices with smaller ones. Bravais overcame this difficulty by employing a complementary assumption regarding the generic type of lattice that could occur within the range of possibilities controlled by postulate 1.

POSTULATE 2 [BRAVAIS 1851, 208FF.]:. *Among the range of choices for lattices Γ allowable by Postulate 1, those with the largest possible number of free metrical parameters are selected by a crystallizing substance with a given molecule symmetry S_m.*

As the number of free lattice parameters increases with reduction of symmetry, this postulate excludes blind symmetries of the lattice with respect to molecule symmetry, that is symmetries that are without influence on the common symmetry S_c. Thus this genericity principle can also be read as a minimality condition complementary to the maximality assumed in postulate 1 (and restricted to the range of validity of the latter).

In utilizing these principles, Bravais discussed the necessary reduction of molecule symmetries S_m to S_c by intersecting with lattice symmetries all of the 23 "classes" derived in [Bravais 1849], the greater part of which contained infinite series of "symbols of symmetry". That left no more than 31 "symbols of symmetry" characterizing the crystallographic point symmetry systems, the crystal classes. As a result of his omission of the Schoenflies series $\overline{C}_{4n}$, Bravais was missing just one of the 32 crystal classes, the spheroidal class $S_4 = \overline{C}_4$ in Schoenflies notation.

Moreover, he derived a list of admissible combinations of reduced symmetry types S_c (crystal classes) with lattices, according to his postulates 1 and 2. This list is most interesting from the point of view of implicit group considerations, as it contains in its geometrico-crystallographic form an implicit, but nevertheless nearly complete, determination of those lattices that can be used for semidirect extensions of orthogonal groups of the type of crystal classes:

$$0 \longrightarrow \Gamma \longrightarrow G \cong \Gamma * K \xrightarrow{\rho} K \longrightarrow 1$$

$K < O(3, \mathbf{R})$ crystal class, Γ euclidean lattice with $K < \mathrm{Aut}\Gamma$, K operating on Γ by conjugation in $Isom(\mathrm{E}^3)$.

It ought to be said, however, that Bravais' viewpoint was not particularly attuned to the group-theoretic implications of his argumentation, but was led rather by considerations of crystal structure theory. This is underscored by the fact that in this study he only listed the admissible combinations of crystal classes with lattice *systems* (not lattice types, which were of course implied by this information, but were not considered particularly relevant by Bravais). Nor did Bravais take up his earlier investigations of infinite systems of symmetry elements with respect to a lattice distributed throughout all of space space [Bravais 1850].

Still, this context was explicitly contained in his work, and so it is legitimate to say that in his list of admissible combinations of crystal classes with lattice systems Bravais implied a survey of possible combinations of crystal classes with lattice types. Moreover, read against the background of his descriptions of infinite symmetry systems of lattices, he gave an implicit characterization of the crystallographic space group types arising as semidirect products of lattices with crystal classes (called *symmorphic space groups* in the crystallographic literature). Allowing for Bravais' omission of the crystal class $\overline{C}_4$, we thus can say *that Bravais implicitly characterized 71 of the 73 symmorphic space groups. We could, therefore, just as well call them (the symmorphic groups) "Bravais groups".*

This was Bravais' most important contribution to crystallographic symmetry. It was slightly more hidden by crystal structure considerations than his more directly geometrical contributions to point symmetry systems and lattices, but nevertheless it seems not to have been without influence on the further development of the geometrical group concept.

3. Jordan's first explicit use of the group concept in geometry

After this excursion into the most interesting development of symmetry ideas in geometrical crystallography in the first half of the 19th century, we return to our starting point, Camille Jordan's *Mémoire sur les groupes de mouvements* [Jordan 1869]. During the 1860s, Jordan had studied Galois' work intensely and had written noteworthy commentaries on Galois along with important research results of his own. This work gained him considerable attention within Parisian circles as a talented young mathematician. Jordan's intimate knowledge of Galois theory enabled him to read Bravais' papers with different eyes than those of his predecessors. The group structures implicit in these works, some apparent, others more hidden, helped focus his interest on groups of motion in Euclidean geometry, the starting point of his Mémoire.

My object here is not to discuss Jordan's Mémoire in all details, but rather to concentrate on questions directly relating to Bravais' work. The latter served not only as a programmatic animation for Jordan, but also as a source of technically relevant information in some parts of his investigation, particularly in connection with the symmetry classification of lattices and the relationship between crystal classes and lattices. But let us first discuss the general outline of Jordan's paper.

Jordan defined a *motion group (groupe de mouvements)* to be a system of Euclidean motions in $\mathbb{E}^3$ that is closed under composition [1869, 231]. He also presupposed, without further comment, the existence of an identity and of inverse elements (*mouvements égales et contraires*). This is hardly surprising in view of the fact that it still took over a decade to delimit more precisely the concepts of group and semigroup and to give a proper axiomatic formulation of the general group concept [Wussing 1969, 166ff.; Hofmann 1985]. Secondly, he presupposed a *strict dichotomy between discrete and continuous motion groups.* He excluded without any further deliberation the possibility of a case that was neither continuous nor discrete[16], apparently because he was unaware of the existence of such intermediate phenomena.

In describing the motions and their composition, Jordan built upon Poinsot's characterization of screw motions, denoting by $A_{\alpha,t}$ such a motion of angle α about the (oriented) axis A combined with a translation of amount t along A. After choosing a base point P, he represented the rotational part of the screw motions $A_{\alpha,t}$, $A'_{\alpha',t'}$ etc. by pure rotations B_α, $B_{\alpha'}$ etc. about P. He then regarded the screw motions themselves as compositions of the translational parts $t+t_1$, $t'+t'_1$, etc. (t_1, t'_1 orthogonal to t, t') with the rotational parts

$$A_{\alpha,t} = (t+t_1)B_\alpha, \qquad A'_{\alpha',t'} = (t'+t'_1)B_{\alpha'}.$$

This enabled him to calculate the composition of two such screw motions in the usual way, which clearly reflects the semidirect product structure of the group $Isom(\mathbb{E}^3)$.[17]

These screw calculations suggested quite naturally the problem of associating to each group of motions $G < Isom^+(\mathbb{E}^3)$ the respective group of purely rotational parts G_0. This led Jordan to a first rough scheme with six "categories" for the classification of rotation groups:

- G_0 generated by a (finite or "infinitely small") rotation (finite or infinite continuous cyclic group),
- G_0 as above but with an additional orthogonal binary rotation (dihedral groups),
- $G_0 = T, O, I$ (groups of the regular polyhedra - three different "categories"),
- $G_0 = SO(3, \mathbb{R})$.

Following this scheme, Jordan began to discuss *"species"(espèces)* of groups within the different categories. This combined the intuitive idea of equivalence under affine conjugation with certain types of series formation (inside one "species" the affine conjugation class may depend on a discret parameter $n \in \mathbb{N}$). All in all he discussed 174 "species;" these were not all different from one another, however.

Jordan's analysis went beyond Bravais' in two respects. First, he used group ideas and terminology *explicitly*. Secondly, he gave a much more subtle analysis of the question of how orthogonal groups can be extended by translation groups. Bravais' approach in his crystal structure theory dealt with this question implicitly by "superposition" of the molecule and lattice symmetries, but this did not allow him to go beyond semidirect extensions. Jordan, by contrast, analyzed the additional possibilities for group extensions

$$0 \longrightarrow \Gamma \longrightarrow G \xrightarrow{\rho} K \longrightarrow 1$$

(Γ translation lattice in $\mathbb{E}^3$, $K < \mathrm{Aut}_{orth}(\Gamma)$). These extensions arise via the representation of elements of an orthogonal group K compatible with the orthogonal lattice symmetries $\mathrm{Aut}_{orth}(\Gamma)$ by screw motions

$$(t, \alpha) \in Isom^+(\mathbb{E}^3) \cong \mathbb{R}^3 * SO(3, \mathbb{R}), \qquad \alpha(t) = t$$

(the *centered screw*) or

$$(t', 1)(t, \alpha)(-t', 1) = ((1 - \alpha)t' + t, \alpha), \qquad \alpha(t) = t, t' \perp t$$

(the *decentered screw*).

Although he did not explicitly discuss the conditions under which a choice of representatives $\tilde{\alpha}$ for elements $\alpha \in K$ would give rise to a system of screw motions consistent with the translation lattice Γ, Jordan's approach implicitly made use of the following principle:

CRITERION OF COMPATIBILITY:. *If $\alpha_1, \ldots, \alpha_r \in K$ satisfy a relation $R(\alpha_1, \ldots, \alpha_r) = 1$ and if $\tilde{\alpha}_1, \ldots, \tilde{\alpha}_r$ are specified representatives for $\rho^{-1}(\tilde{\alpha}_1), \ldots, \rho^{-1}(\tilde{\alpha}_r)$ in G, then the relation $R(\tilde{\alpha}_1, \ldots, \tilde{\alpha}_r) \in \Gamma$ holds.*[18]

Of course, it is sufficient to check this condition for generators of K and their relations; and, in fact, Jordan wrote down the screw motions only for generators of the group G.

Thus by combining ideas developed earlier by Bravais, Poinsot, and Galois, Jordan was able to characterize satisfactorily, if not absolutely completely, the possible extensions of orthogonal groups by translation lattices. He found 59 out of 65 such extensions in all, if only proper motions of $\mathbb{E}^3$ are taken into consideration. This was a remarkable step forward that transcended the range of symmetry constellations open to Bravais' approach. We must not overlook, however, that Jordan never addressed the question of why it was necessary and sufficient to consider exactly those lattice types that he employed. These were, in fact, identical with the ones classified earlier by Bravais. And so, if we take his reference to Bravais cited at the beginning of this article into account, *we may be quite sure that Bravais' analysis of lattice systems and lattice types was indeed an important source of technical information for Jordan's own investigations.*

To conclude this discussion of Jordan's Mémoire and its relation to Bravais' theory, it should be pointed out that not all of the groups considered by Jordan were discrete or even crystallographic. Among the 174 "species of groups" he studied, 5 were finite, 59 spatial crystallographic, 16 were crystallographic plane group types, and more than 50 comprised "species" of continuous groups. The latter contained, of course, several species that were isomorphic to one another as abstract Lie groups, but different with respect to affine conjugation. These results make Jordan's paper much richer in content than is suggested by our analysis of Bravais' influence. But since that is what interests us at the moment, it would be neither plausible nor useful to try to give a full discussion of Jordan's work here.

4. The influence of Jordan's Mémoire (1869) on the development of Lie's and Klein's ideas regarding transformation groups

Now that we have examined how Bravais' work became important for Jordans' *Mémoire sur les groupes de mouvements*, it remains to consider the question of how important Jordan's Mémoire may have been for *Sophus Lie* (1842–1899) and *Felix Klein* (1849–1925) during the early 1870s. During these years they formulated their concept of transformation groups as well as their respective scientific programs, Klein's *Erlanger Programm* and Lie's classification program for finding the Lie subgroups of the projective group.

To avoid any misunderstanding, it shall not be implied in what follows that the primary or only source of motivation for Lie and Klein's formulation and use of the transformation group concept was Jordan's Mémoire. Thus the generally accepted view that their prime motivation came from projective geometry shall not be challenged. The question remains, however, whether or not Jordan's Mémoire played some role in clarifying their ideas, and if so, what that role might have been. A precise reading of their early publications on transformation groups shows, in fact, that Jordan's investigations of Euclidean motion groups influenced their thought during a decisive phase of its development.

In the course of their research on projective geometry, Lie and Klein, like other mathematicians of their time, came across objects which were—implicitly—subgroups of $PSL(n, \mathbb{C})$. This occured shortly after they first met one another in Berlin in 1869. Lie's first, but already quite clear, use of transformation groups can be found in his investigations of "Reye line complexes" [Lie 1870]. Klein was very familiar with Lie's work in this field, and it was he who aided Lie in the final German formulation.[19] Lie considered a specific quadric submanifold R, the "Reye line complex," or "tetrahedral complex," in the Plücker manifold M_4 of lines of complex projective 3-space (M_4 itself being represented as a four-dimensional submanifold of $P(5, \mathbb{C})$), together with the system (group) G_R of collineations in $P(5, \mathbb{C})$ which leave R invariant. He realized that G_R consists of a "threefold infinite number of homographic [i.e., projective] transformations" operating transitively and freely on R. Klein later realized that there exists a class of curves in R which again are left invariant under a "simply infinite number of transformations" of G_R. These curves ("W-curves" in their terminology) and their transformations were the subject of Lie's and Klein's first common article written in June 1870 during their common sojourn in Paris [Klein/Lie 1870].[20]

In these early studies it was the *continuous, i.e. manifold, character of these sets of transformations* that seemed the most striking feature to them. This is reflected quite clearly in their choice of terminology, where they spoke of "a simply , twofold, ... infinite family of transformations (*eine einfach, zweifach, ... unendliche Schar von Transformationen*)."[21] Of course they used the fact that these systems of transformations were closed under composition, but this property remained in the background and received little attention. Thus it was only in a footnote to their joint article [Klein/Lie 1871, 430], that they noted that there was a sort of kinship between their investigations of plane curves " ... which map into themselves under a closed system of one-fold infinitely many commuting linear transformations ... " and the substitution groups studied in Galois theory.

Their terminology was still vague and wavering. Not yet ripe to be considered as a well-formed mathematical notion, Lie and Klein discussed transformation groups by using a variety of fluently interchangeable terms—"family, cycle, or system of transformations"(Schar, Zyklus, System von Transformationen)—that mainly emphasized the continuous character of the sets of transformations involved. *We should therefore look more carefully at those places in their work where they consciously and explicitly used the group idea to describe these infinite systems of transformations.*

S. Lie took up the terminology and concept of transformation groups for the first time in a note appended to the second part of his doctoral dissertation, written in the Winter 1870/71. There he wrote:

> We can pose the problem of listing all groups of linear transformations. In this connection, I say that a continuous or discontinuous family of transformations forms a group, if in each case the combination of several of these transformations is equivalent to a transformation of the family. Mr. Jordan has in particular determined all groups of motions. [Lie 1871b, 208 ff.; my translation]

Lie continued to explain that the task he had in mind was far more general than the one attacked by Jordan and included the latter as a special case.

It should be said, however, that the program Lie so concisely posed here had almost nothing to do with the rest of his dissertation except for the fact that "families of transformations" cropped up in it, just as they did in other works of Lie and Klein at this time. Thus the idea behind this program was not derived from or used in other parts of Lie's dissertation. It appeared instead as a kind of amendment to the finished work, one that contrasted sharply in terminology, conceptual content, and style with the main body of the text.

This impression of an amendment, a sort of extended footnote that Lie added after reading an investigation that was only indirectly related to the work he had just done, becomes even more striking when one looks at the course of Lie's subsequent work. At the time he gave this initial formulation of the group classification problem its solution lay outside the range of methods available to Lie (and Klein) for handling it. It even appeared unclear to Lie how much sense it would make to pose such a generalized problem. Jordan's *Mémoire* seems to have helped him organize his thoughts on a subject of considerable uncertainty. The fact that Lie used essentially the same wording, including the same pointed reference to Jordan, when he next referred to this problem in [Lie 1873, 27], supports our interpretation that Jordan's *Mémoire* apparently guided him in formulating this programmatic conception, though in a heuristic rather than a technical sense. It was not until 1874 that he first began to attack the problem stated in the above passage from his dissertation. Only after he had taken his first steps into the theory of differential equations and formulated his seminal idea of using symmetry groups for integration theory in a manner similar to the way Galois groups are used in the theory of algebraic equations, did Lie return to the classification program of finding all continuous subgroups of the projective group [Lie 1874].[22] This now became the starting point for a whole series of works devoted to this problem [Lie 1876a, b, 1878a, b, 1880 etc.]. Furthermore, it also marked the final changeover to Lie's use of "transformation group" as a technical term in his mathematical vocabulary.[23]

I conclude from this evidence that Jordan's classification of subgroups of the group of proper Euclidean motions acted as a latent inducement on Lie to take up the generalized group classification problem and thus to clarify the more general concept of transformation group. This notion remained for some years an isolated idea, but when the time became ripe and a substantial body of problems had arisen in his own work it emerged as a crystallizing centre in the organization of Lie's ideas.

The question we must next consider is whether we find similar traces of Jordan's Mémoire imprinted on Klein's work during these years when he was thinking about transformation groups. What we find in Klein's publications is that, again, there is no reference to Jordan's [1869] before a certain date. Moreover, just as with Lie, the first reference to Jordan's Mémoire goes hand in hand with a transformation of terminology and a clarification of Klein's programmatic ideas. In Klein's case, this transition first occurs in his *"Zweiter Aufsatz über die sogenannte Nicht-Euklidische Geometrie (Second Essay on the so-called non-Euclidean Geometry)"* [Klein 1873], written in May/June 1872, a few months before his *Erlanger Programm* (September/October 1872), but only published in 1873 in *Mathematische Annalen.*

In this essay, Klein stated for the first time—albeit restricted to the range of the classical metrical (Euclidean, hyperbolic and elliptic non-Euclidean) geometries—the central ideas underlying the methodological unification of different geometrical fields that he soon thereafter formulated in greater generality in his *Erlanger Programm.* As is well known, Klein's *Erlanger Programm* was a generalization of his successful attempt to understand the classical metrical geometries as parts of projective space by specifying a certain subgroup of the projective group and an invariant metric defined on a chosen submanifold of projective space. Klein sought to bring order into the flowering garden of geometric research programs and subdisciplines by utilizing in each case an underlying manifold endowed with an appropriate geometric structure group (*Hauptgruppe*) and studying the class of objects of the manifold that remained invariant under the transformations of the group.

We therefore observe that *Klein's "Second essay on non-Euclidean geometry" contains his first use of the terminology of transformation groups combined with a clear methodological idea of the group concept. Furthermore, it also contains his first reference to Jordan's Mémoire [1869]*, even if in a more subordinate role than in Lie's case.

Klein consistently used group terminology from this point onward. This is quite understandable, considering that, for Klein, the programmatic context of the new concept was central to his thought from the beginning, whereas, for Lie, it only took on this larger significance several years later. *Thus we can say that Klein took notice of Jordan's Mémoire in early 1872 and could directly adapt its message to his context.*

The question remains, what exactly was this message for Klein in 1872? Was it only the explicit use of the group concept in a geometrical context, and thus the generalization, still vague, to infinite systems of transformations? That may have been so, but there seems to have been more to it than just this.

Let us remember that Jordan's Mémoire began with a clear and surprising reference to Bravais' theory. Moreover, the reference alluded to a notion that must have been quite dear to Klein's heart in early 1872, namely the study of the molecular structure of crystalline matter by characterizing its associated *groupe de mouvements*. This idea bears a striking resemblance to Klein's approach in characterizing the structure of the Euclidean and non-Euclidean geometries. Moreover, Klein came across this reference at a time when he was trying to formulate this idea explicitly and generalize it to the whole range of geometry.

Presumably such a striking methodological parallel turning up elsewhere in a remotely related field would not go unnoticed, and could easily serve as a catalyst for the maturation of nascent mathematical or methodological ideas. *We therefore conclude that it is very likely, though not completely clear on the basis of our source material, that Jordan's "Mémoire sur les groupes de mouvements" played a heuristic role for Klein by serving as a sort of signpost that helped point him along the road to the "Erlanger Programm".*

If this is correct, then we have found an indirect, but nevertheless remarkable influence of the geometrical theory of crystal structure on the formation of one of the most important methodological programs for geometry at the end of the 19th century. In any case, we may be sure that via Jordan's Mémoire and its reception by Lie and Klein there was a certain influence of crystallography on the formation of the geometric group concept, even if only indirectly. This fertilization process came full circle two decades later as a result of the more involved crystallographic space symmetry systems studied by Fedorov and Schoenflies. Indeed, Schoenflies was able to successfully introduce these group-theoretic aspects into the field of crystallography, where they now hold a place of central importance. But this is another story that cannot be be dealt with here.[24]

NOTES

1. In particular, Galois' Mémoire [1831/1846].
2. Cf. [Dahan-Dalmedico 1980/81].
3. Cf. [Gray 1980].
4. [Hamilton 1844a, 1844 b, 1847; [Cayley 1858]; cf. [van der Waerden 1985].
5. Invertibility was implicitly assumed in the postulate of bijectivity of the multiplication map with a given "symbol." Cf. [Wussing 1969/1980].
6. C.S. Weiss (1780–1856), M.L. Frankenheim (1801–1869), J.F.C. Hessel (1796–1872), J.G. Grassmann (1779–1852), et al..
7. In particular Frankenheim and Weiss; Bravais apparently did not know [Hessel 1830].
8. Auguste Bravais, together with his brother Louis, wrote a detailed study about morphology akin to the aspirations of the dynamistic botanists of the time [Bravais, A.; Bravais, L. 1837a,b,c].
9. Cf. [Burckhardt 1988] and [Scholz 1989a; 1989b].
10. This idea was not completely new, it had already been proposed by Gauss' student Seeber [1824], by Gauss himself [1831], and at about the same time as Delafosse it was taken up by M.L. Frankenheim [1835].
11. "Principal axis" or "principal plane" are symmetry elements that are mapped onto themselves under any symmetry operation of the group (the subgroup associated with a "principal" element is normal).
12. For n odd, $\overline{C}_{2n}$ contains an inversion i and can be generated as, $\langle C_{4n}, i\rangle$. Here $\overline{C}_{2n}$ is the finite orthogonal group generated by a rotoreflexion of order $2n$.
13. Four in the plane, with the holoedries D_3, D_4, D_2, C_2 [Bravais 1850, 29ff.].
14. Intersection may be taken literally for Bravais' "symbols of symmetry." If one reads S_m etc. as orthogonal groups, however, one must take care of the "correct" choice of the representatives in the conjugation class in $O(3,\mathbb{R})$ (i.e. so that the intersection becomes maximal).

15. To make this more precise: Take the maximal elements in the inclusion graph of the holoedries considered as finite orthogonal groups. This leads to O^* and D_{6n} and the corresponding regular and hexagonal Bravais lattice systems.
16. Such as, e.g. the translations of a rational vectorspace in $\mathbf{E}^3$, or the rotation groups generated by a rotation not commensurable with 2π, etc.
17. $A_{\alpha,t} \cdot A_{\alpha',t'} = (t+t_1)B_\alpha \cdot (t'+t_1')B'_\alpha = (t+t_1+B_\alpha(t'+t"_1)) \cdot B'_\alpha B_{\alpha'} = (t''+t_1'')B''_{\alpha''} = A_{\alpha'',t''}$ with t'' along the axis of $B_{\alpha''}$ and t_1'' orthogonal to it.
18. It is easily verified that this "criterion of compatibility" implies the "Frobenius congruencies" for the translational parts t_α of the $\alpha \in K$ $\alpha t_n - t_{\alpha\beta} + t_\beta \equiv 0 \pmod{\Gamma}$ for all $\alpha, \beta \in K$. This is nothing other than the special version of the condition that $f_{\alpha,\beta} = (\delta t)_{\alpha,\beta} \in \Gamma$ defines a factor system in the sense of cohomological group extension theory (Frobenius 1911; Eilenberg/Mac Lane 1946).
19. According to Klein [1892]. For more details on this phase in Lie's and Klein's work, see the articles by D. Rowe and T. Hawkins in these proceedings.
20. For further details, see the articles by D. Rowe and T. Hawkins in these proceedings. Regarding this work on W-curves, I essentially share Rowe's viewpoint that the group idea was "lurking in the background," that is, that it was still only present implicitly. The remarks in this section deal with the question of how this implicit use of the group notion gradually transformed into an explicit programmatic formulation of their role in geometry.
21. In their French publication [Klein/Lie 1871] they spoke of "une infinité, un nombre doublement infini de transformations," etc..
22. For a discussion of the content of Lie's dissertation, see the articles by D. Rowe and T. Hawkins.
23. These steps are carefully analyzed in Hawkins' article.
24. The time lag between Lie's first use of explicit group terminology, including the reference to Jordans's [1869], and the effective assimilation of the latter has already been noted by Wussing [1969, 151].
25. Cf. for example [Scholz 1989a, section 5].

Literature and Sources

Betti, Enrico

1852. *Sulla resoluzione delle equazioni algebriche.* Annali di Scienze Matematiche e Fisiche, 3. Opere matematiche, vol. 1. Milano 1903, 31–80.

BRAVAIS, AUGUSTE

1849. *Mémoire sur les polyèdres de forme symétrique.* Journal de Mathematique, **14**, 141–180. In: [1866, IXX - LXII].

1850. *Mémoire sur les systèmes des points distribuées régulièrement sur un plan ou dans l'espace.* Journal Ecole Polytechnique, **19**, 1–128.

1850. *On the systems formed by points regularly distributed on a plane or in space.* English translation by A. J. Shaler. Michigan 1949.

1851. *Etudes cristallographiques.* Journal Ecole Polytechnique, **20**, 101–278.

1866. *Etudes cristallographiques.* Paris. (Book edition of 1849a, 1849b, 1850, 1851).

BRAVAIS, AUGUSTE; BRAVAIS, LOUIS

1837a. *Essai sur les dispositions des feuilles curvisériées.* Annales des Sciences Naturelles, 2nd ser. Botanique, **7**, 42–110.

1837b. *Essai sur la disposition symétrique des inflorescences.* Annales des Sciences Naturelles, 2nd. ser. Botanique, **7**, 193–221, 291–348; **8**, 11–42; **7**, Pl II, III, VII–XI.

1837c. *Resumé des travaux de MM. Schimper et Braun sur la disposition spirale des organes appendiculaires.* Annales des Sciences Naturelles, 2nd. ser. Botanique, **8**, 161–183.

BURCKHARDT, JOHANN JAKOB

1988. *Die Symmetrie der Kristalle. Von René Just Haüy zur kristallographischen Schule in Zürich. Mit einem Beitrag von Erhard Scholz.* Basel.

CAUCHY, AUGUSTE L.

1844. *Mémoire sur les arangements que l'on peut former avec des lettres donnes ... .* Exercises d'Analyse et de Physique Mathématiques, **3**, Paris 151-252. Oeuvres (2) vol. 13, Paris, 171-282.

CAYLEY, ARTHUR

1854. *On the theory of groups, as depending on the symbolic equation $\theta^n = 1$.* Philosophical Magazine, **7**. In: [1889, vol. 2, 123–132].

1858. *A memoir on the theory of matrices.* Philosophical Transactions Royal Society London, **148**. In: [1889, vol. 2, 475–496].

1845. *On certain results relating to quaternions.* Philosophical Magazine, **26**, 141-145. In: [1889, vol. 1, 123–126].

1889. Collected Mathematical Papers, vols. 1,2. Cambridge, 1889.

DAHAN-DALMEDICO, AMY

1980 *Les travaux de Cauchy sur les substitutions. Etude de son approche du concept du groupe.* Archive for History of Exact Sciences, **23**, 279–319.

DEDEKIND, RICHARD

1856. *Eine Vorlesung über Algebra.* In: [Scharlau 1981, 59-100].

EILENBERG, SAMUEL; MAC LANE, SAUNDERS

1946. *Cohomology theory in abstract groups I, II.* Annals of Mathematics, **48**, 51–78, 326–341.

FRANKENHEIM, MORITZ LUDWIG

1835. *Die Lehre von der Cohäsion, umfassend die Elasticität der Gase, die Elasticität und Cohärenz der flüssigen und festen Körper und die Krystallkunde.* Breslau.

FROBENIUS, GEORG

1911. *Über die unzerlegbaren diskreten Bewegungsgruppen.* Sitzungsberichte der Preussische Akademie der Wissenschaften, 654-655.

GALOIS, EVARISTE

1830. *Analyse d'un Mémoire sur la résolution algébrique des équations.* Bulletin des Sciences Mathématiques, Physique et Chimique, **13**, 271-271; Oeuvres, Paris 1897, 11-12. In: [1962, 163-165].

1831. *Mémoire sur les conditions de résolubilité des équations par radicaux.* Journal de Mathematique, **11**. In: [1962, 43–72].

1832. *Lettre à Auguste Chévalier (29.5.1832).* Révue Encyclopédique. In: [1962, 173-185].

1962. *Ecrits et mémoires mathématiques.* Paris.

GAUSS, CARL FRIEDRICH

1831. *Recension der "Untersuchungen über die Eigenschaften der positiven ternären quadratischen Formen" von Ludwig August Seeber* (Seeber 1831). Göttingische gelehrte Anzeigen 108. Wiederabdruck, Journal für reine und angewandte Mathematik, **20** (1840), 312-320. Werke, vol. 2, Göttingen 1863, reprint, Hildesheim: Olms, 1973, 188-196.

GRASSMANN, HERMANN GÜNTHER

1844. *Die lineale Ausdehnungslehre, ein neuer Zweig der Mathematik, dargestellt und durch neue Anwendungen auf die übrigen Zweige der Mathematik, die Lehre vom Magnetismus und die Krystallonomie erläutert.* Leipzig: Teubner. Werke, vol. 1(1), Leipzig: Teubner, 1894.

1862. *Die Ausdehnungslehre, vollständig und in neuer Form bearbeitet.* Berlin. Werke, vol. 1(2), Leipzig: Teubner, 1894.

1894 *Gesammelte mathematische und physikalische Werke,* 3 vols. in 6. Leipzig: Teubner, 1894-1911.

GRAY, JEREMY J.

1980. *Olinde Rodrigues' paper of 1840 on transformation groups.* Archive for History of Exact Sciences, **21**, 375–385.

Hamilton, William Rowan

1844a. *Letter to Graves, 17.10.1843.* Philosophical Magazine (3), **25**, 490-495. Collected Mathematical Works, vol. 3. Cambridge 1967, 106ff.

1844b. *On quaternions or a new system of imaginaries in algebra.* Philosophical Magazine (3), **25**, 10–13, 241–246, 489–495.

1847. *On quaternions, or on a new system of imaginaries in algebra; with some geometrical illustrations.* Proceedings Royal Irish Academy of Science (3), **25** 1–16. Collected Mathematical Works, vol. 3, Cambridge 1967, 355–362.

Haüy, René-Just

1815. *Mémoire sur une loi de cristallisation, appelée loi de symétrie.* Mémoire du Muséum d'Histoire Naturelle, **1**, 81–101, 206–225, 273–298, 341–352. Also in: Journal des Mines, **37** (1815), 215–235, 347–369; **38** 5–34, 161–174.

Hessel, Johann Friedrich Christian

1830. *Krystallonometrie oder Krystallometrie und Krystallographie, auf eigenthümliche Weise und mit Zugrundelegung neuer allgemeiner Lehren der reinen Gestaltenkunde, sowie mit vollständiger Berücksichtigung der wichtigsten Arbeiten und Methoden anderer Krystallographenn bearbeitet.* In: H. W. Brandes e.a. (Hrsg.), Johann Samuel Traugott Gehlers Physikalisches Wörterbuch, vol. 8, Leipzig. Separate printing, Leipzig 1831. Reprint Ostwalds Klassiker, **88/89**. Leipzig 1897.

Hofmann, Karl H.

1985. *Semigroups in the 19th century? A historical note.* Preprint No. **874**, Fachbereich Mathematik, Technische Hochschule Darmstadt.

Jordan, Camille

1869. *Mémoire sur les groupes de mouvements.* Annali di Matematica, **2**, 167-215, 322-345. Oeuvres, vol. 4, Paris 1964, 231-302.

1870. *Traité des substitutions et des équations algébriques.* Paris.

Klein, Felix

1871. *Über die sogenannte Nicht-Euklidische Geometrie.* Mathematische Annalen, **4**, 573-625. In: [1921. 244–305].

1872a. *Über Liniengeometrie und metrische Geometrie.* Mathematische Annalen, **5**, 257-277. In: [1921, 106–126].

1872b. *Vergleichende Betrachtungen über neuere geometrische Forschungen.* Erlangen: Deichert. Auch in Mathematische Annalen, **43**, 63–100. In: [1921, 460–497].

1873. *Über die sogenannte Nicht-Euklidische Geometrie, 2. Aufsatz.* Mathematische Annalen, **6**, 112–145. In: [1921, 311–343].

1892. *Über Lies und meine ältere geometrische Arbeiten.* Erster Borkumer Entwurf vom August 1892. Niedersächsische Staats- und Universitätsbibliothek Göttingen, Handschriftenabteilung. Codex Ms. Klein, XXIIG. Typewritten transcription courtesy of D. Rowe.

1921. *Gesammelte Mathematische Abhandlungen,* vol. 1. Berlin. Reprint Berlin-Heidelberg-New York 1973.

KLEIN, FELIX; LIE, SOPHUS

1870. *Deux notes sur une famille de courbes et surfaces.* Comptes Rendus, **70**. In: [Lie 1934, 78–85; Klein 1921, 416-423]. KLEIN, FELIX; LIE, SOPHUS

1871. *Über diejenigen ebenen Kurven, welche durch ein geschlossenes System von einfach unendlich vielen vertauschbaren linearen Transformationen in sich übergehen.* Mathematische Annalen, **4**, 50–84. In: [Lie 1934, 229–266; Klein 1921, 424–459].

KRONECKER, LEOPOLD

1853. *Über die algebraisch auflösbaren Gleichungen, Teil I.* Monatsberichte Akademie der Wissenschaften Berlin. In: Mathematische Werke, vol. 4, Berlin 1929, 1–11.

1856. *Über die algebraisch auflösbaren Gleichungen, Teil II.* Monatsberichte Akademie der Wissenschaften Berlin. In: Mathematische Werke, vol. **4**, Berlin 1929, 25–37.

KUMMER, ERNST E.

1845. *Zur Theorie der complexen Zahlen.* Monatsberichte der Bayerische Akademie der Wissenschaften. Journal für reine und angewandte Mathematik, **35** (1847), 319-326. Collected Papers, vol. 1, Berlin/Heidelberg/New York: Springer, 1975, 203–210.

LEGENDRE, ADRIEN MARIE

1794. *Elements de géométrie.* Paris 1813.

LIE, SOPHUS

1870. *Über die Reciprocitätsverhältnisse des Reyeschen Komplexes.* Nachrichten der Gesellschaft der Wissenschaften Göttingen, **4**, 53-66. In: [1934, 68–77].

1871a. *Over en classe geometriske transformationer.* Christiania Forhandlinger i Vidsenskabs-Selskabet Aar 1871 (1872), 67–109. In: [1934, 105–152, German].

1871b. *Über eine Klasse geometrischer Transformationen* (Fortsetzung). Christiania Forhandlinger i Vidsenskabs- Selskabet Aar 1871 (1872), 182–245. In: [1934, 153–210].

1871c. *Selbstanzeige von 1871a, b.* Jahrbuch für Fortschritte der Mathematik, **3**, Jahrgang 1871 (1874). In: [1934, 212–214].

1872. "*Uber Komplexe, insbesondere Linien- und Kugelkomplexe, mit Anwendungen auf die Theorie partieller Differentialgleichungen*. Mathematische Annalen, **5**, 145-256. In [1935, 1-121].

1874. *Über Gruppen von Transformationen*. Nachrichten Gesellschaft der Wissenschaften Göttingen. In: [1924, 1–8].

1876. *Theorie der Transformationsgruppen, Abhandlung I, II*. Archiv für Mathematik, **1**, 19–57, 152–193. In: [1924, 9–41, 42–75].

1878. *Theorie der Transformationsgruppen, Abhandlung III, IV*. Archiv für Mathematik, **3**, 93–165. 375–460. In: [1924, 78–132, 136–196].

1880. *Theorie der Transformationsgruppen I*. Mathematische Annalen **1** **441**–528. In: [1927, 1–94].

1924. *Gesammelte Abhandlungen*, vol. 5. Christiania/Leipzig.

1927. *Gesammelte Abhandlungen*, vol. 6. Leipzig/Oslo.

1934. *Gesammelte Abhandlungen*, vol. 1. Leipzig/Oslo.

1935. *Gesammelte Abhandlungen*, vol. 2. Leipzig/Oslo.

MÖBIUS, AUGUST FERDINAND

1827. *Der barycentrische Calcul*. Leipzig: Teubner. In: [1885, 1-388].

1885. *Gesammelte Werke*, vol. 1. Leipzig: Teubner. Reprint Wiesbaden: Sändig 1967.

POINSOT, LOUIS

1851. *Théorie nouvelle de la rotation des corps*. Journal de Mathematique, **16**, 9–129, 289–336.

RODRIGUES, OLINDE

1840. *Des lois géométriques qui régissent les déplacements d'un système solide dans l'espace*. Journal de Mathematique, **5**, 380–440.

SCHARLAU, WINFRIED (Hrsg.)

1981. *Richard Dedekind 1831/1981*. Braunschweig: Vieweg.

SCHOLZ, ERHARD

1989a. *Symmetrie - Gruppe - Dualität. Zur Beziehung zwischen theoretischer Mathematik und Anwendungen in Kristallographie und Baustatik des 19. Jahrhunderts*. Science Networks - Historical Studies **1**. Basel/Berlin/Boston: Birhäuser.

1989b. *The rise of symmetry concepts in the atomistic and dynamistic schools of crystallography (1815–1830)*. To appear in the Proceedings of the conference "La Mathématisation, 1800–1830" (Ed. A. Dahan): Révue d'Histoire des Sciences.

SEEBER, LUDWIG

1824. *Versuch einer Erklärung des inneren Baus der festen Körper*. Annalen der Physik, **76**, 229–248, 349–372.

1831. *Untersuchungen über die Eigenschaften der positiven ternären quadratischen Formen*. Freiburg.

SERRET, JOSEPH ALFRED
1866. *Cours d'algèbre* 3rd ed., 2 vol. Paris.

STEINER, JACOB
1832. *Systematische Entwicklung der Abhängigkeiten geometrischer Gestalten voneinander*. Berlin. Gesammelte Werke, vol. 1. Berlin 1881, 229–458.

VAN DER WAERDEN, BAARTEL L.
1985. *A history of algebra. From al-Khwārizmi to Emmy Noether*. Berlin/Heidelberg/New York: Springer.

WUSSING, HANS
1969. *Die Genesis des abstrakten Gruppenbegriffs*. Berlin: VEB Deutscher Verlag der Wissenschaften.
1984. *The genesis of the abstract group concept: A contribution to the origin of abstract group theory*. Translated by A. Shenitzer and H. Grant. Cambridge, Mass.: MIT Press.

A contemporary illustration showing Gauss
and Weber with their electromagnetic telegraph.
Courtesy of Fremdenverkehrsverein, Göttingen.

Physics as a Constraint on Mathematical Research: The Case of Potential Theory and Electrodynamics

Thomas Archibald

1. Introduction

The complicated interaction between developments in mathematics and those in physical theory is a central feature, perhaps *the* central feature, of the history of mathematics since the seventeenth century. The profound effect that physical research exercises on the history of mathematics is particularly well-illustrated by the relationship between electromagnetic theory and mathematical potential theory in Germany between approximately 1840 and 1880. This interaction demonstrates several different modes of mutual influence between the two research areas: physical questions served not only as a stimulus to mathematical research, but also acted to channel it in certain directions. Indeed, it is not always possible, nor perhaps would it be desirable, to entirely untangle the mathematical and physical aspects of this research, in part because both aspects are frequently to be observed in the work of a single individual.

It is the purpose of this paper to outline some key features of historical developments associated with the notion of potential during this period, especially in the context of electrodynamics. By electrodynamics is meant that portion of electromagnetic theory having to do with the behaviour and interaction of electric currents, namely the study of electromagnetism, electromagnetic induction, and electric conduction: that is how the term was used in the mid-nineteenth century by continental writers. During the period between the publication of Gauss' *Allgemeine Lehrsätze* in 1840 and the effective adoption by German writers of Clerk Maxwell's views on electromagnetic theory the notion of potential was a key tool for electrodynamics. In fact, I shall argue that because of then-prevalent notions about electrodynamics we may conveniently regard potential-theoretic studies in Germany during this period as part of a unified research program, one in which most writers shared assumptions concerning the appropriate fundamental hypotheses of the theory, and about what were the most important open problems. In addition, the period encompasses a series of events associated with the maturation of potential as a physical notion; in 1840, a potential was only a mathematical tool employed in physics, but it gradually acquired an interpretation

ISBN 0-12-599662-4.

as potential energy during the 1850s and 1860s. This latter development also plays an important role in delimiting the period and the research program I discuss below.

In order to grasp these developments, we must bear in mind some basic features of the continental electrodynamics of the time. While there was no single universally accepted theory, almost all writers used an action-at-a-distance model along the lines of that employed by Laplace, Poisson and others for the theories of gravitation, electrostatics, and magnetostatics. Because of the non-central character of the macroscopic interaction between filamentary electric currents, Ampère had found it necessary during the 1820s to introduce a dependence of the electromagnetic force on the *direction* of hypothetical line elements which he viewed as constituting the electric circuit. The resulting law expressing the force between infinitesimal circuit elements was fundamental to later work, whether it employed a microscopic model for the electric current (as did the work of Gauss' collaborator Wilhelm Weber) or whether it was phenomenological in character (as, for example, the induction law due to Franz Neumann).[1]

The theories of Weber and Neumann, elaborated during the years from about 1843 to 1850, were the principal electrodynamic theories of the period 1840-1870. They were not competing theories, but rather arrived at much the same end via different routes, and coexisted in a sort of synthesis which was widely known and highly regarded in the mathematico-physical community. Neumann's theory, which I will deal with in more detail later, derived the Ampère electromagnetic force law and the law describing induced currents from a single potential function. Weber's, on the other hand, posited an atomistic model of electric current in which a current consists of a double stream of oppositely-directed, oppositely-charged particles. In addition to the usual attractive and repulsive electrostatic forces, Weber postulated a term in the force law which depended on the relative speed and acceleration of the particles. Using this velocity-dependent force law, Weber was able to derive the Ampère law and Neumann's induction law. Weber's model also accounted qualitatively for a number of other phenomena, among them diamagnetism, and in part because of this perceived completeness was the main reference point for further work; for example, Kirchhoff's conduction theory was based on it.

Potential theory was important in electrodynamic research both because of its application to problems posed in the context of the Weber-Neumann theory, and because it was employed as a tool in attempts to supersede that theory. I shall outline in what follows some of the major points of contact between electrodynamics and potential theory; but first a few comments are called for about the laying of the foundations of potential theory by Gauss.

2. Gauss and the founding of potential theory as a research specialty

a) The Allgemeine Lehrsätze and its physical background

The notion of potential arose in the eighteenth century, though the name was not used at the time, and found significant application in the works of Laplace and Poisson on gravitation and electrostatics. By a potential for a given force, we mean a function the rate of change of which gives the force. More precisely, a function V is a potential for a force F if the partial derivatives of V at a given point give the components of the force at that point:

$$F = (\partial V/\partial x, \partial V/\partial y, \partial V/\partial z).$$

This is useful for a multitude of reasons, not the least of which is that when we deal with potentials we deal with scalars, that is, with only a single component, which thereby gives information about all three components of the force function.

Gauss had employed these functions in several of his works, but undertook a detailed study of their properties in his 1839 paper, *General theorems in relation to attractive and repulsive forces which act in inverse proportion to the square of the distance*[2]. I shall refer to this paper as the *Allgemeine Lehrsätze*, following the German title.

Here Gauss employed the term potential as I have done earlier, thus giving the name 'potential theory' to the area of research which he founded with this paper. The *Allgemeine Lehrsätze* not only provided this theory with a set of fundamental results on which later researchers were to build; it illustrated the physical and mathematical importance of the subject by powerful examples.

Defining the potential as the sum of all acting mass points, each divided by its distance from the given point (x, y, z), Gauss states

> Research into the properties of this function is the key to the theory of attractive and repulsive forces.[3]

Gauss terms this function the potential of the masses, noting that this notion could be generalized to non-inverse-square forces by the same means—i.e. by finding a function with partial derivatives equal to the force components. The approach Gauss employs, as he acknowledges, works only for the central forces of gravitation, electrostatics and magnetostatics. Indeed, the problem of extending this method to non-central electromagnetic forces was specifically stated by Gauss as an open problem (one later taken up by F. Neumann, Riemann and others).

The *Allgemeine Lehrsätze* was a very natural outgrowth of an area with which Gauss had been much concerned in the past few years, namely the theory of terrestial magnetism. Both the 'magnetic crusade' , and Gauss' part in it have received considerable attention from historians in recent years.[4] In fact, Gauss portrayed the general aim of the treatise as an attempt to clear up what he saw as a confusion in the foundations of the theory of magnetism. Physical writers of Gauss' time viewed magnetization as the result of a separation of two magnetic fluids, and spoke as though all ' north' magnetic fluid migrated to one end of a magnetized iron bar. Yet the experimental existence of magnetic monopoles, or 'free magnetic fluid' was not established, so that the theory depended on a quantity that was not measurable, or even detectable. As Gauss portrayed it, the central result of his paper cleared this up:

> In the place of an arbitrary distribution of magnetic fluids inside a bounded region of space may be substituted an ideal distribution on the surface of the region, with the result that exactly the same action will be exerted on any point of the external space as that which is actually exerted.[5]

In the case of terrestrial magnetism, this theorem is comforting, since the surface distribution can be mapped while that of the interior cannot.

Gauss devoted the *Allgemeine Lehrsätze* to the chain of connected results which lead to this theorem. The theorems so obtained, however, went well beyond the confines of magnetic theory, and in fact touched upon a number of fundamental issues, even foundational issues, confronting the analysis of Gauss' time. These theorems constituted the basic repertoire of potential theory, studied by the next generations of students, and built upon by researchers.

The theorems derived by Gauss are for the most part quite familiar, and fall into several general classes. First, he discusses the existence and continuity of the potentials and their related forces, showing that these potentials satisfy the Laplace equation (respectively Poisson equation) at points outside (respectively inside) the force-exerting mass. He then proceeds to derive a number of theorems which we now class as standard vector analysis, providing techniques for transforming integrals over bounded volumes into integrals of related functions over the bounding surface. A further class of theorems are mean-value theorems stating, for example, that surface distributions over closed surfaces will exert the same potential as their average mass concentrated at a centre of mass in the interior of the surface. Together with the integral theorems, the mean value theorems serve to prove the key result stating that under appropriate hypotheses concerning continuity and differentiability of the densities and potentials it is possible to treat any volume distribution as a surface distribution for the purposes of calculating its effects.

b) The hypotheses of potential theory in Gauss' work

From the point of view of those who sought to apply Gauss' theory to electrodynamics, the fact that the hypotheses of these theorems were not always as general as possible, or as clear as desirable, makes little difference. As with physicists today, the assumption that the functions they study are continous (or differentiable), that surfaces can be described by continuous functions, or that differential equations have solutions frequently followed from physical hypotheses and experience. Thus the assumptions employed by Gauss, and the justifications for them, were either explicitly or tacitly assumed by most writers on electromagnetic theory employing Gauss' results. The character of these assumptions is illustrated in Gauss' proof that the potential at points interior to a force-exerting mass satisfies Poisson's equation. The potential V at a point inside a connected body containing a mass distribution of density ρ is defined as

$$V = \int \frac{\rho dT}{r} \tag{2.1a}$$

where r is the distance from the volume element dT with coordinates (a, b, c) to the point at (x, y, z) at which the potential is being evaluated. For the force component, this yields

$$F_x = \frac{\partial V}{\partial x} = \int \frac{\rho(a-x)}{r^3}\, dT. \tag{2.1b}$$

As r approaches zero, the integral will become unbounded (*unendlich gross*); Gauss argued, however, that a proper integration is implied (*eine wahre Integration*), since a change of coordinate systems (to spherical coordinates) removes the singularity. This procedure yields

$$V = \int k\rho r \sin\phi d\phi d\vartheta dr$$

as the potential for points inside the force-exerting mass. Likewise, first derivatives of the potential, which are the force components, are finite inside the region, by the same transformation:

> [these components] have a definite finite value, which varies continuously [*nach der Stetigkeit*], because every

> element lying infinitesimally near the point 0 [where $r = 0$] only contributes an infinitesimal amount.[6]

To show that Poisson's equation $\nabla^2 V = 4\pi\rho$ holds at interior points of the mass, involves the second derivatives of the potential, which cannot be salvaged by the coordinate transformation just described.

Notice that this change to polar coordinates is not simply a means of evading the problem. The essential notion is that arbitrarily small contributions to dT around 0 give arbitrarily small contributions to the integral, because r^3 and dT are of the same order. The fact that we can show this directly by switching to polar coordinates simply assures us that nothing goes wrong. On the other hand, for higher order derivatives we get a new contribution of r^2 to the denominator each time we differentiate. This renders the order of the denominator too great, and in such a case another means must be found to establish the correctness of Poisson's equation. To see the difficulty Gauss encountered, notice that for example

$$\frac{\partial^2 V}{\partial x^2} = \int \rho dT \, \frac{\partial}{\partial x}\left(\frac{a-x}{r^3}\right)$$

$$= \int \rho \left(\frac{3(a-x)^2 - r^2}{r^5}\right) dT$$

and transforming to spherical coordinates is unhelpful. Gauss said

> Strictly speaking this expression is only a sign without a clear, definite meaning. For since it may be shown that, in fact, within an arbitrarily small portion of T containing the point there is a [finite] volume over which the integral will pass any previously chosen value ... thus the condition fails here according to which alone the entire integral can be dealt with, namely the applicability of the method of exhaustion.[7]

To resolve this difficulty Gauss argued (from a definition of the derivative as the limit of a difference quotient) that one could write the second partial derivatives as a sum of two terms:

$$\frac{\partial^2 V}{\partial x^2} = \int \frac{\frac{\partial \rho}{\partial a}(a-x)}{r^3} - \int \frac{k(a-x)\cos\alpha ds}{r^3}$$

where ρ is the density, assumed to be a continuous function of position (a, b, c) and α is the angle between the surface element ds and the outward normal.[8] The assumption of a continuous, (indeed differentiable) density was enunciated by Gauss in the context of a specific example, showing that the potential satisfies Poisson's equation at interior points of a spherical mass. He described this assumption as "the most satisfactory way to ground this important theorem,"[9] and retained it in the more general theorems which follow.

From this brief example, we grasp in a general way how Gauss employed his assumption regarding the density function in the *Allgemeine Lehrsätze*. The resulting body of theorems provided basic tools. The now-familiar concept of equipotential surfaces, the notion that the integral of a central inverse-square force along a path in space was given by a potential, and a collection of results linking volume and surface distributions while elucidating their dependence on continuity and differentiability conditions – all these provided models to pure mathematicians, and useful results to those studying such forces. The first to publish such research in the context of electrodynamics was Franz Neumann (1793-1892) of Königsberg, though Gauss himself had made earlier unpublished efforts along the same lines.

3. The introduction of potentials in electrodynamics: Franz Neumann

In 1845 Franz Neumann published a derivation of a law which expressed the induced electromotive force acting on a circuit (of fixed form) due to a nearby moving filamentary circuit or magnet. Note that this is a force which acts to create a current, measured electromagnetically, and which therefore acts to "move electricity". Such electromotive forces were commonly distinguished from ponderomotive forces (such as gravitation and electromagnetism) which accelerate substances which possess inertial mass. This distinction is a property of the class of models which assigns different forces to different types of substance. Electric and magnetic "fluids" appeared to be massless, thus posing a variety of problems for theoretical mechanics. Neumann expressed the induced electromotive force as a function of the velocity of the moving circuit and of the ponderomotive force (as given by Ampère's law) exerted on the moving circuit by an adjacent closed circuit. Neumann's work thus provided a basis for the examination of induced currents in a fashion which was linked to Ampère's theory. More important, Neumann treated induction as the result of a change in the potential between two circuits. This gave his theory the possibility of eventual assimilation to the point of view of energy conservation.

It is probable that Neumann's association with Carl Gustav Jacobi was influential in turning his interest to questions about induced currents and in using potentials in their study. The brother of Carl Jacobi was Moritz Jacobi, then at Petersburg, who was the collaborator of Emil Lenz in research on induced currents. The brothers corresponded regularly, although I have been unable to determine whether M. Jacobi actually visited Königsberg in the period between 1834 and 1843. Moritz Jacobi was also among those who recognized early the correctness of Ohm's law, an important starting point for Neumann's work.[10]

a) Neumann's Treatise of 1845

Neumann's first major theoretical investigation of induced currents began in 1844, and was completed by August or September of 1845. Neumann submitted the paper to the Berlin Academy by way of his former colleague at Königsberg, C. Jacobi, at Berlin since 1843. Jacobi in turn gave it to Poggendorff as the man in Berlin best acquainted with Ampère's work.[11] However, as Jacobi told Neumann:

> Indeed your work seems to him [Poggendorff] to be in Chinese. He didn't understand a single symbol in it."[12]

Jacobi suggested to Neumann that a general summary of some kind would be necessary for the Berlin physicists, and reported that none of them had obtained the least idea of the contents from Poggendorff's report.[13] In fact, Jacobi at this point suggested that if Neumann himself could not come to Berlin to explain the content of his work, then one of his students—ideally Kirchhoff—should do so.[14] This is an interesting comment on the still generally low state of matheamtical competence among physicists, at least in Berlin, where the incumbent in the physics chair, Magnus, remained interested chiefly in experimental work. That the editor of the *Annalen der Physik* should himself have had difficulty is indicative of a state of affairs that extended beyond Berlin. Neumann's work on electricity, as well as that of Weber, broke new ground for this more mathematical kind of study.

Neumann's 1845 treatise aimed at the analysis of currents induced in a filamentary circuit, whether by the motion of a circuit, by the motion of a magnet, or by changes in the strength of a magnet. This case, or rather set of cases, was termed " linear" induction by Neumann.[15] He began with the derivation of an equation which specified the e.m.f. induced in an element of a circuit which is moving in a magnetic field due to another stationary circuit or to a fixed magnet. Using this case as a starting point, he explained how the e.m.f. (interpreted as work per unit charge) could be found by employing the principle of virtual work. This enabled him to treat other cases of linear induction, including those involving rotational motion. More important for later developments, it enabled him to interpret the e.m.f. as resulting from a difference in potential.[16]

Neumann's analysis rests on two basic facts: the proportionality of the motionally induced e.m.f. to the velocity of the motion, and Lenz's law. Neumann's formulation of a sharpened version of the latter enabled him to calculate (with some additional assumptions) the constant of proportionality. He also employed Ohm's law, and the experimentally-established fact that the induced e.m.f. is independent of the material of the conductor. Lenz had stated his rule in in 1834[17]; Neumann interpreted it as follows:

> Lenz's law may be so expressed: the component of action of the inducing [agent] on the induced [circuit] in the direction of motion of the [induced] conductor is always negative."[18]

Here Neumann was clearly thinking of the velocity as positive.

Neumann's treatment of induced currents was also based directly on the assumption that Ohm's law is valid. In particular, the induced currents which Neumann studied initially were steady currents, to which Ohm's law applied. For this reason it was necessary for Neumann to assume at the outset that the electromagnetic force varies at a rate which is slow in comparison to the velocity of propagation of electricity. Thus he excluded from consideration currents induced by (electrostatic) spark discharge.

The first case Neumann considered was that in which a filamentary circuit moves in a magnetic field due to a stationary circuit. He posited that all components of the circuit move with the same velocity v, that is, that the motion consists of a parallel translation of the circuit. The action specified by the above quotation is the electromagnetic action given by the Ampère law. If the induced current in an element of the circuit is i, then the force component in question may be written[19] as

$$F_v i ds. \tag{3.1}$$

Since the induced e.m.f. is proportional to the velocity, Ohm's law implies that the induced current is also proportional to the velocity

$$i = Lv. \tag{3.2}$$

The problem Neumann solved is that of giving a theoretically plausible value for L which was in accord with the facts. From (3.2) we find for the component of the Ampère force in the direction of v

$$iF_v ds = LvF_v ds. \tag{3.3}$$

For the entire circuit, the component of force on ids in the direction of motion is therefore

$$i \int F_v ds = Lv \int F_v ds. \tag{3.4}$$

(Neumann writes C for what I have called F_v and C' for $\int F_v ds$).

Now Lenz's law implies that under any circumstances, $Lv \int F_v ds$ and v must have opposite signs. But, Neumann argued

> since [$\int F_v ds$] changes sign when the direction of motion changes [to the opposite direction], then L must be a function [$\int F_v ds$], and in fact such a function which changes its sign when [$\int F_v ds$] does. The simplest supposition which can be made in light of this, and one which is shown as sufficient by its consequences, is ... to set [$L = -e \int F_v ds$].[20]

This assumption of simple proportionality gives us for the induced current

$$i = -ev \int F_v ds \tag{3.5}$$

and for the induced e.m.f., by Ohm's law,

$$\begin{aligned} E &= iR \\ &= -eRv \int F_v ds. \end{aligned}$$

However, we know that the induced e.m.f. does not depend on the material of the conductor, whereas R does. Hence eR must be a constant, called ϵ by Neumann, which yields

$$E = -\epsilon v \int F_v ds \tag{3.7}$$

for the e.m.f. induced.[21]

Parenthetically, we may note that Neumann's induction law makes sense in electron-theoretic terms. The Lorentz force on a charge due to a motion of the charge with velocity $\vec{v}$ in a field $\vec{B}$ is given by $\vec{F} = e\vec{v} \times \vec{B}$, so that

$$\vec{B}_{ind} = \vec{v} \times \vec{B}$$

is the induced field. The component of this field along an element ds is

$$\begin{aligned} \vec{E} \cdot d\vec{s} &= (\vec{v} \times \vec{B}) \cdot d\vec{s} \\ &= -\vec{v} \cdot (d\vec{s} \times \vec{B}) \end{aligned} \tag{3.8}$$

But $d\vec{s} \times \vec{B}$ is the Ampère force, if $\vec{B}$ is thought of as the Biot-Savart field

$$\vec{B} = \oint \frac{d\vec{s} \times \vec{r}}{r^3}$$

The comparison of (3.8) with Neumann's

$$E = \epsilon v \int F_v ds \tag{3.7}$$

shows obvious similarities.

b) Neumann's potential law

Despite this rather comforting similarity, we may wonder what Neumann has really shown. No experimental results are given to support this law, and the justification for (3.5) is otherwise very shaky. In fact, to clarify the meaning of the formula he had given, Neumann provided a foundation for his approach in mechanics. This afforded him a pathway to generalize the relation he had found to other cases, enabling him to compare his theoretical treatment to results obtained by Weber.

Neumann's definitions of the so-called "differential current" and "integral current" were essential tools in this treatment. The terms are misleading, since neither is actually a current, but rather what we would now interpret as a charge. Using his equation, our (3.7), Neumann expressed the induced current (which he terms current strength, or current intensity) as

$$i = -\epsilon\kappa \int vF_v ds \tag{3.9}$$

where $\kappa = (1/R)$ and we have used Ohm's law.[22] (Here we leave v under the integral sign to permit treatment of cases where v varies from one element to another). This current is measured, as Neumann pointed out, but its action on a magnetic needle (e.g. in a galvanometer). However, here Neumann uses the term action (*Wirkung*) in an unusual sense. In the case of a steady induced current, a constant deflection is obtained which measures the current and hence the strength of the constant inducing "action" . If the induced current is variable, however, we find that the action is constant only for a short interval. In an interval δt, the "action" is measured by

$$-\epsilon\kappa\delta t \int vF_v ds. \tag{3.10}$$

Neumann considered the expression $-\epsilon\kappa \int vF_v ds$ as giving a measure of strength of action *per unit time*. The expression (3.10), to modern eyes, measures [current] × [time] or charge transferred in time δt, if we think of current as the amount of a single kind of free charge flowing across a cross-section per unit time. Neumann, however, called this quantity the differential *current*, $i\delta t$.[23] The integral current is obtained by treating δt as a differential and integrating the differential current:

$$\int_{t_0}^{t_1} i dt = -\epsilon\kappa \int_{t_0}^{t_1} dt \int v F_v ds. \tag{3.11}$$

But both (3.10) and (3.11) may be expressed in terms of a displacement, w such that

$$v = \frac{dw}{dt} = \frac{\delta w}{\delta t}.$$

As a differential current, we therefore obtain

$$-\epsilon\kappa \int \delta w F_v ds. \tag{3.12}$$

Here F_v is now the component of the force on ds in the direction of δw, and $-\epsilon \int \delta w F_v ds$ represents the work done by the Ampère force on the entire circuit during the displacement δw. Neumann calls this the "virtual moment" of the force which the inducing current exercises with respect to ds when the latter is conceived as bearing a current ϵ.[24] The corresponding work associated with the entire motion from w_0 to w_1 is

$$-\epsilon \int_{w_0}^{w_1} dw \oint_s F_v ds.$$

This work done should manifest itself as a change in the *vis viva* of the moving circuit, as Neumann argued. He stated the fact as follows:

> the electromotive force of the integral current is the loss in *vis viva* (*lebendiger Kraft*) which the inducing agent causes in the moving conductor on the path from w_0 to w_1.[25]

Having brought *vis viva* and the notion of virtual work into his scheme, Neumann then used the *vis viva* principle as the starting point for determining a potential, or force function, for the electromotive force. This led Neumann to the most fundamental innovation of his treatment, namely, using the time-variation of the potential to derive the e.m.f.

Neumann's colleague Jacobi had stated the principle of *vis viva* in the form

> half the *vis viva* of a system equals the force function augmented by a constant[26]

or

$$\sum \frac{1}{2} m_i v_i^2 = U + h$$

where U is such that the gradient of U gives the force.[27] By then-standard theoretical mechanics, if such a force-function exists it should be given by $\delta U = -\delta W$, where W is the work. In Neumann's analysis, the mechanical work done on a circuit by the Ampère force during a displacement $\delta w = (\delta x, \delta y, \delta z)$ is

$$\sum (F_x \delta x + F_y \delta y + F_z \delta z),$$

the sum extending over all the elements of the circuit.[28] However, the Ampère force components may be written

$$F_x = \frac{1}{2} ii' \int (C dy' - B dz'),$$

etc., where $ds' = (dx', dy', ds')$ is a circuit element of the inducing circuit. By calculating the total work done during the motion δw, Neumann showed that it was equal to the first variation of a function V, where

$$V = \frac{1}{2} ii' \iint \frac{1}{r} (dx dx' + dy dy' + dz dz') \tag{3.13}$$

or, using θ to denote the angle between the two elements,

$$V = \frac{1}{2} ii' \iint \frac{\cos \theta \, ds ds'}{r}. \tag{3.13a}$$

In this expression, i is the induced current, while i' is the current in the inducing circuit. Since the negative variation of V gives work, it followed from well-known notions of the calculus of variations that V is a force function—that is, a potential—for the Ampère force.

Neumann then identified the simple relationship between this potential and the induced current. Since from (3.12) the differential current is

$$i \delta t = -\epsilon \kappa \int \delta w (F_v ds)$$

and since we know that the work is the negative variation of V, then we may write

$$i \delta t = \epsilon \kappa \delta V,$$

where V is the potential of an inducing circuit bearing a current i' on an induced circuit which is assumed to carry a virtual current of ϵ. Consequently

$$\int i dt = \epsilon\kappa(V_1 - V_0)$$

where the subscripts refer to the positions at the end and beginning of the motion respectively. Differentiating with respect to time yields

$$i = \epsilon\kappa \frac{dV}{dt} \tag{3.14}$$

if we fix V_0 and treat V_1 as a function of time.[29] Since $\epsilon\kappa = 1/R$, Ohm's law implies that dV/dt is the induced e.m.f.

Neumann suggested a possible underlying physical model which gave a tentative interpretation to such a potential. Neumann identified the equipotential surfaces – he uses Gauss' term, *Gleichgewichtsoberflächen* – with surfaces of constant pressure in a fluid of uniform density ϵ.[30] The pressure in question is due to the Ampére force. In the event that a potential exists, Neumann said

> the electromotive force of the integral currents is defined as the difference in pressure on the two equilibrium surfaces which pass through the endpoints of the motion of the conductor.[31]

Besides its mathematical usefulness, the definition of potential had the effect of relegating questions concerning the underlying physical phenomena to the problem of correctly interpreting the induction constant ϵ. This was stated by Neumann in a letter to C. Jacobi:

> If my treatise has any merit; it lies in the fact that it has placed the concept of this quantity [ϵ] as the real physical problem of all induction phenomena. However, it [ϵ] is still shrouded in deep mystery, and is related to the unknown connection in which all [microscopic] bodies stand to one another; when the tension [*Spannung*] between them is disturbed an electric current comes into being, provided the other conditions are fulfilled. In order to permit this mystery to be grasped I gave in [section 3 of my treatise] various definitions of the induced electric current by *vis viva*, pressure etc.[32]

Although Neumann sought a method of grasping this phenomenon, he was unable to arrive at any definite conclusions about the underlying physical processes.

Neumann's potential provided a conceptually straightforward method for calculating the Ampère force and the induced e.m.f. between two circuits, even though the computations involved are not always easy.[33] However, Neumann also demonstrated its capacity for unifying the diverse cases of induced currents. In the remainder of his 1845 treatise, he showed that his potential covered cases of induction involving either translatory or rotatory motion of two circuits; and that only relative velocity was important, so that our decision to treat one circuit or the other as fixed is arbitrary.[34] He also studied currents induced by magnets, which he investigated both from Ampère's and Poisson's point of view.

c) Concluding Discussion of Neumann's Theory

Neumann's 1845 paper and the research leading up to it had several important consequences. His lectures on electrodynamics had found an interested and talented listener in G. Kirchhoff, who undertook much important work in electromagnetic theory.[35] Indeed, several of Kirchhoff's best-known achievements, which I will discuss below, were the direct result of his efforts to extend and perfect Neumann's theoretical labours. A further consequence of Neumann's work was its use by Helmholtz in his 1847 energy conservation memoir. After the widespread acceptance of energy conservation as a basic physical principle, this gave Neumann's work a unique importance.

By establishing a phenomenological theory of induced currents which allowed quantitative prediction, Neumann provided a possible test of physical hypotheses regarding the ultimate nature of electricity and electric phenomena. No theory which gave results seriously at odds with Neumann's could be admitted without modification. This proved to be a critical test for the theory due to Wilhelm Weber, which appeared shortly after Neumann's paper. The account of induction phenomena given by Weber appeared to disagree with Neumann's theory in several respects. The reconciliation of the two points of view occupied Weber, Neumann, and other physicists, in the period between 1845 and 1860. The success of this programme did much to establish the Weber-Neumann theory as the most complete account of electromagnetic phenomena, at least for the German physicists who participated in this research. That this did not proceed more rapidly can be attributed to the very small audience capable of reading Neumann's work with understanding.[36] The role of Neumann's potential in this theory considerably enhanced its importance, and stimulated the interest of mathematicians in Gauss' work.

4. Dirichlet and the extension of the Gaussian program

a) Dirichlet's lectures and their influence

Naturally, the influence of Gauss' work depended on the actual results he obtained and on their demonstrated importance. However, such influence also depended on the fact that the results were made accessible to a fairly extended audience, and that individuals within that audience had both the motivation and the talent to apply and extend Gauss' results to new situations that they were attempting to study. During the mid-1840's, the transmission and extension of Gauss' results was undertaken by P. G. Lejeune-Dirichlet in Berlin. Although Dirichlet's lectures were not published until 1876, several of his Berlin students went on to contribute to the theory and applications of potential theory in important ways; among these students were B. Riemann, R. Clausius, and probably G. R. Kirchhoff.[37]

Dirichlet's lectures provided background and details of the results of Gauss, as well as showing clearly by a number of examples (especially in electrostatics) how Gauss' theorems could be applied to physical questions. Let us note parenthetically that it was the custom for students in lecture courses to share the task of preparing complete transcriptions of the lectures, so that the absence of a published version did not prevent the more general circulation of Dirichlet's lectures.

By thus preparing the ground, Dirichlet assured the persistence of a group of men competent in the techniques Gauss had originated. Indeed, Kirchhoff, Riemann and Clausius all went on to lecture on potential theory (at Göttingen, Breslau, Heidelberg, Zürich, and Bonn), and Clausius wrote a popular textbook on the subject, the first edition of which appeared in 1859.

b) The reception of Green's Essay in Germany

A further legacy of Dirichlet was his role in the reception of the techniques invented by George Green. Green wrote his *Essay on the application of mathematical analysis to the theories of electricity and magnetism* in 1828. This work, which had anticipated a number of Gauss' 1839 results and introduced several other important new techniques in potential theory, went largely unnoticed until rediscovered by William Thomson, later Lord Kelvin, in 1845. Thomson was influential in assuring its general circulation; he sent a copy with Cayley to Berlin in 1846, where it was published in instalments in Crelle's journal beginning in 1850. Even by the date of first publication, however, Green's *Essay* had become well-known, especially in Berlin, where they were soon incorporated into Dirichlet's lectures. Equipped with the tools invented by Gauss and Green, Dirichlet's students Riemann, Kirchhoff, and Clausius applied them in their research, establishing the theory of potential as one of the msot important areas in the mathematics and mathematical physics of their day.

A few words on why Green's essay was so important are in order. Green was the first to demonstrate certain general identities, of which the two most important are

$$\int u \nabla^2 v dV + \int \nabla u \cdot \nabla v dV = \int u \frac{\partial v}{\partial n} dS \tag{4.1}$$

and

$$\int (u \nabla^2 v - v \nabla^2 u) dV = \int \left(u \frac{\partial v}{\partial n} - v \frac{\partial u}{\partial n} \right) dS \tag{4.2}$$

where u and v are sufficiently differentiable functions, and the integrals on the right are surface integrals. Green further had the key idea of employing these identities as a tool in solving boundary-value problems. Here what we now call the *Green's function* is the essential tool. Let us be reminded about what these are, by an example; a Green's function for a region D with a singularity at z_0 is a function of the form

$$g(z; z_0) = -\log|z - z_0| + g_1(z)$$

where g_1 is harmonic, i.e. satisfies Laplace's equation $\nabla^2 g_1 = 0$ and $g_1 = \log|z - z_0|$ on the boundary ∂D, so that g is 0 on the boundary. Green argued that such functions exist as *potentials*, namely as the electrostatic potential of a point charge at z_0 in the presence of a grounded conductor around the boundary.

To see what use Green and later writers made of such things, consider what happens in Green's second identity in the case where we are talking about a region in the plane, so that the integral on the left is a double integral and the one on the right is a line integral. Suppose further – and this is a critical assumption of the application at hand – that $\nabla^2 u = 0$ and $v = g(z; z_0)$. Then a routine calculation allows us to solve for u at the interior point z_0:

$$u(z_0) = -\frac{1}{2\pi} \int_{\partial D} u \frac{\partial g}{dn} ds \tag{4.3}$$

provided that we know $\partial g / \partial n$ and u on the boundary of D. That is, we can solve a boundary value problem for u. Let me emphasize that *finding* the Green's function might not be so easy, and the u must satisfy Laplace's equation.

Classical potentials for the gravitational force and the electrostatic force do, of course, satisfy Laplace's equation for regions with zero mass or charge density. Hence this method solves important potential-theoretic problems, at least it does so provided a Green's function for the region can be found.

While the foregoing has given some examples of some developments in pure mathematics associated with the publication of Green's paper, perhaps I have not sufficiently emphasized why the subject attracted such widespread attention. After all, many seminal papers - both of Gauss and of other writers – led to further important research, so that the *Allgemeine Lehrsätze* is scarcely unique in this respect. The reason for the burgeoning of activity which turned potential theory into one of the most active research specialties in analysis was its role in theoretical physics. In the hands of Kirchhoff, the notion of potential acquired a precise physical interpretation which turned it into a measurable quantity. This notion of the potential as a physical thing was then employed and extended by Helmholtz, Clausius, and others in the elaboration of Helmholtz's 1847 work on the conservation of force. Potential, defined by Green and Gauss as a mathematical entity, became potential energy in the course of a rather complex series of developments between 1850 and roughly 1860.[38]

5. Physically-motivated research in potential theory: Kirchoff, Helmholtz and Clausius

a) Kirchhoff's recasting of Ohm's law

Kirchhoff's use of the notion of potential came up in connection with his studies of electric conduction, which had already occupied him during his student years in Königsberg. In the mid-1840s, the theory of conduction was dominated by Ohm's law, formulated by G. S. Ohm two decades earlier, which expressed a relation between the electromotive force of a battery and the current the battery would produce in a filamentary circuit.[39] This relation may be stated in its familiar form, $V = IR$, where I is the current in the circuit, R is the resistance, and V is the electromotive force.

While Ohm had stated his law in largely these terms, the conceptual apparatus he employed was problematic for electricians of the following generation. Two major difficulties were involved. For one thing, Ohm proposed that the electromotive force which causes a current may be identified as in "electroscopic force." The electroscopic force of a conductor in a circuit was defined by Ohm as that quantity which is measured at the end of the open circuit with an electroscope or torsion electrometer. In Ohm's view, if a difference in electroscopic force (i.e. a tension or *Spannung*) exists between adjacent portions of a conductor, this "force" will tend to alter until the difference is zero and equilibrium is established. Ohm characterized this process as a motion of electroscopic force, or an exchange of force between the elements of a body. As long as the difference persists, there is a tension (*Spannung*) between the parts of the conductor. Since it is measured in electrostatic units, in Ohm's view it was a charge density.

Another difference between Ohm and his contemporaries was his assumption that electricity (actually, "electroscopic force") was distributed throughout the volume of a conductor even in equilibrium states. This was in contradiction to Poisson's electrostatics, at the time quite widely accepted. Poisson had argued that conductors contain an essentially infinite supply of electric fluids. Hence, in equilibrium, the force must be zero inside the conductor, or else the electricities would separate and move. This precludes the existence of free electricity inside the body of a conductor; instead, any free charge must migrate to the surface.

In his 1849 paper, "Über eine Ableitung der Ohm'sche Gesetze, welche sich an die Theorie der Elektrostatik anschliesst" Kirchhoff took up the problem of putting Ohm's law on foundations that were in agreement with then-prevailing notions in electrostatics. The critical problem for Kirchhoff was to retain the desirable features of Ohm's formulation (notably Ohm's law) while ridding it of this contradiction. His essential tool in this effort was the notion of potential.

Kirchhoff's immediate inspiration was a set of experiments described by Rudolf Kohlrausch which involved circuits containing condensors. Kohlrausch's paper, *Die elektromotorische Kraft ist der elektroscopische Spannung an den Polen der geöffneten Kette proportional*, sought to place this fact, which had hitherto been assumed, on firm experimental ground.[40] To present a mathematical analysis of such circuits Kirchhoff required an analysis of conduction phenomena which was capable of dealing with static and current electricity in a unified fashion.

Kirchhoff's derivation illustrates the importance of the potential theory he had learned from Neumann and Dirichlet; it also gave a new direction to the physical interpretation of potential. Kirchhoff pointed out that electricity can be in static equilibrium in a conductor only if the total potential of the free electricity evaluated at any point of the conductor is constant, which in turn implies that any free electricity (or charge) is distributed on the surface of the conductor. This is a non-trivial result, requiring Gauss's theorems, although Kirchhoff does not give the calculation. In the case of a bimetallic couple, the total potential must be constant for each metal, but the two potentials must be different; otherwise, Kirchhoff observed, no free electricity would be present.

> As for the difference of the potential in the two metals, this could only depend on the material of the conductors and their form; I made the assumption that it

> is independent of the latter, and that it [the potential difference] is the quantity which is called the tension (*Spannung*) of the two bodies.[41]

Kirchhoff therefore in effect redefined the tension or *Spannung* resulting from the contact of two different conductors as the difference in electrostatic potential between the two. Thus potential takes on the role of Ohm's electroscopic force, which, it will be recalled, Ohm identified with charge density.

This reinterpretation was an important step in providing a differential equation for the motion of electricity in conductors, a task on which Kirchhoff himself and Wilhelm Weber were to make substantial progress before the end of the decade. It also rendered potential an entity which, in the context of electricity theory at least, had a clear-cut experimental meaning.

b) Helmholtz, Energy Conservation and Induction

The key figure in the reinterpretation of potential as energy was Hermann Hemholtz. As a mathematician and physicist, Helmholtz was largely self-taught, reading eighteenth-century works by Euler, d'Alembert and Lagrange in his spare time while a medical student at the Freidrich-Wilhelms-Institut in Berlin (1838-1842).[42] His interest in physical questions seems to have been stimulated by his physiological studies under Johannes Müller. The question of existence of the life-force was seriously discussed in Müller's circle, and many of Müller's students were inclined to seek the answer in a study of the physical and chemical processes of the organism.[43] Helmholtz was also familiar with the work of Ernst Heinrich Weber (Wilhelm's brother) who sought to explain physiological processes by purely physical means.[44]

Helmholtz's work on energy conservation was conceived and executed during his years in Potsdam as a military doctor. This position afforded him much free time, as well as ready access to his colleagues in Berlin and their laboratories. Indeed, he spent five months in Berlin in close contact with Emil du Bois-Reymond and Ernst Brücke in 1845-46, using the laboratories of Wilhelm Magnus for research on organic decay and attending meetings of the newly-founded Berlin *Physikalische Gesellschaft*.[45]

It was to Magnus that Helmholtz submitted his memoir, *"Ueber die Erhaltung der Kraft"* in the summer of 1846, with a request that Magnus recommend it to Poggendorff for the *Annalen der Physik*. Magnus did so, but it was rejected by Poggendorff, who noted that its length and the absence of any experimental portion made it unsuitable for the *Annalen*.[46]

Helmholtz subsequently arranged for it to be published separately, and it appeared late in 1847.

The generalisation of the *vis viva* principle which Helmholtz obtained is well-known, and the history of the discovery of energy conservation has been widely discussed.[47] Helmholtz's version of the *vis viva* principle took as fundamental the denial of a certain kind of perpetual motion. Defining the *vis viva* (or *lebendige Kraft*) of a body as $(1/2)mv^2$, he asserted that the *vis viva* acquired by a body when it moves under the influence of a force must be equal to the work necessary to restore it to its original position. He then stated the *vis viva* principle as follows:

> If an arbitrary number of mobile mass points moves only under the influence of the forces which they mutually exert on one another, or which are directed toward fixed centres, then the sum of their *vires vivae* is the same at all points in time when they occupy the same position relative to one another and to any fixed centres, whatever their paths and velocities have been in the meantime.[48]

This dependence of *vis viva* on position alone implies in turn that the velocities and the forces present also depend solely on position. Helmholtz deduced from this that in such a case all the forces must be central. Helmholtz then noted a corollary: if we choose to employ a model of physical processes based on interparticulate forces, then these should be central forces. Otherwise, the *vis viva* principle will not be satisfied, implying the possibility of perpetual motion.

Helmholtz generalized the above analysis by considering the change in *vis viva* resulting from the application of a central force φ. If a body of mass m moving with initial velocity q is acted on by a central force of magnitude φ, then

$$\frac{1}{2}mQ^2 - \frac{1}{2}mq^2 = -\int_r^R \varphi dr \qquad (5.1)$$

where Q is the final velocity, and r and R are the initial and final distances between the point from which the force acts and the body. The left side represents the change in *vis viva*, or *lebendiger Kraft*. Helmholtz defined the integral on the right-hand side as the sum of the *Spannkräfte* (tension forces) regardless of the origin of the force. Helmholtz's principle of the conservation of force (*Kraft*) is the statement that, in all cases of motion of free material points under the influence of central forces whose intensity depends on distance alone, the sum of "living force" (*vis viva*, or *lebendige Kraft*) and "tension force" (*Spannkraft*) is a constant.[49]

In fact, the assumptions of centrality and dependence on position alone are not necessary for equation (4.8) to hold. Consider

$$\vec{F} = m\frac{d\dot{\vec{r}}}{dt}.$$

Calculation of the path integral

$$\int_{\vec{r}_1}^{r_2} \vec{F} \cdot d\vec{r} = m \int_{r_1}^{r_2} \dot{\vec{r}} \cdot d\dot{\vec{r}}$$
$$= \frac{1}{2}m\dot{\vec{r}}_2^2 - \frac{1}{2}m\dot{\vec{r}}_1^2$$

shows that (5.1) holds, even if $\vec{F}$ is (for example) velocity-dependent.

Helmholtz required that for a system of particles (5.1) should hold. However, he assumes in effect that the conservation equation be expressible as a sum over pairs of particles, which in turn imposes the requirement that the forces be central and derivable from a potential. This further implies that they must depend on position alone.[50]

Much of Helmholtz's memoir is devoted to showing how this principle is satisfied in various physical contexts: electricity is one of the most important examples. For the static case, the Coulomb force between two electric masses e and e' is

$$\varphi = \frac{ee'}{r^2}$$

So that the change in *vis viva* is given by

$$-\int_R^r \varphi dr = \frac{ee'}{R} - \frac{ee'}{r} \tag{5.2}$$

If we let R go to infinity, the change is $-ee'/r$, which Helmholtz noted is the electric analogue of the magnetic potential defined by Gauss. Hence, Helmholtz noted,

> the increase in *vis viva* during any motion may be set equal to the difference of potential between the beginning and end of the motion.[51]

In the case of electric currents due to batteries, Helmholtz noted that, in considering the conservation of energy, one must take into account the Joule heat, chemical action, and electrolytic polarisation. Restricting his attention to metallic conduction, Helmholtz cited the results of Joule and

Lenz to the effect that the heat generated in a metallic conductor in time dt is

$$d\Theta = I^2 R dt$$

or, using Ohm's law ($A = IR$),

$$d\Theta = AI dt \tag{5.3}$$

Note that A here remains "electromotive force" for Helmholtz, not potential difference as Kirchhoff later showed. The units are heat units, in fact kilocalories.

Now in the event that we have a magnet or closed electric circuit near the first circuit, it possesses a magnetic potential V with respect to that circuit. In this case, Helmholtz said:

> If a magnet moves under the influence of a current, then the *vis viva* it gains must result from the *Spannkräften* expended by the current. In the time interval dt, these are ... $AJdt$ in heat units or $aAJdt$, in mechanical [units] where a is the mechanical equivalent of heat. The *vis viva* engendered in the circuit is aJ^2Wdt, and that gained by the magnet is $J(dV/dt)dt$, where V is the potential of the magnet with respect to the conductor when the latter carries unit current.[52]

Thus

$$(a)AJdt = (a)J^2Wdt + J\,\frac{dV}{dt}dt.$$

Here A is the electromotive force which engenders the current J, and $AJdt$ represents the work done by A to maintain the current J; J^2Wdt is the Joule heat and

$$J\,\frac{dV}{dt}dt$$

is the *vis viva* qcquired by the magnet in the interval dt.

Helmholtz then solved for J to obtain

$$J = \frac{A - \dfrac{1}{a}\dfrac{dV}{dt}}{W}.$$

If no e.m.f. is applied in the circuit, so that $A = 0$, and the magnet is moved from a great distance, we find that there is an induced current during the motion, such that

$$\int J dt = -\frac{1}{aW}\int\left(\frac{dV}{dt}\right)dt \;=\; -\frac{1}{aW}(V_1 - V_0), \tag{5.4}$$

where V_0 and V_1 represent the potential of the magnet at the beginning and end of the motion. Helmholtz noted the resemblance to Neumann's 1845 law, with Neumann's ϵ now interpreted as the reciprocal of the mechanical equivalent of heat.[53]

Helmholtz, in the course of the above argument, overlooked the mechanical energy necessarily involved in moving the magnet. Despite this limitation, the attempted derivation of a special case of Neumann's law from energy considerations suggested the possibility of creating quite a different physical basis for the explanation of certain electrical phenomena than that of Weber.

However, the Helmholtz descrption was incomplete in many details. Furthermore, in order for such an interpretation of induced currents to be adopted by a larger community, energy conservation and energy itself had to prove themselves as physically acceptable. To mathematicians like Jacobi and Dirichlet, the energy conservation principle was readily comprehended as a generalisation of the *vis viva* principle, one which appeared promising for the purpose of bringing a wide range of forces besides gravitation into the realm of theoretical mechanics.[54] To the older generation of experimentalists (especially Poggendorff and Magnus) the conservation law appeared to be a solely mathematical construct, devoid of any clearly demonstrated physical meaning beyond its use in describing analogically and partially several different kinds of physical situations. Indeed, it was not until the 1860s that energy conservation was widely accepted as an integral part of physical theory in Germany.[55] Much of the detailed study of energy in the 1850s was due to Helmholtz himself and to Rudolf Clausius, and much of their work touched on electricity theory.

The 1847 memoir had only touched on the theory of electricity. Helmholtz's first detailed work on electrical questions was undertaken during his time as extraordinary professor of physiology at Königsberg. While his research during this period (1849-1855) concentrated on physiology, often physical problems arose in the course of his efforts to measure or describe such phenomena as the transmission of nerve impulses, or human vision. Helmholtz described his electrical research in two papers dated 1851 and 1853. Here, he investigated problems associated with induced currents and with conduction in three dimensions.[56] In both papers, Helmholtz employed results due to Neumann, among them Neumann's expression for mutual induction. Helmholtz took this as a starting point for the calculation of currents due to self-induction also obtaining an exponential function which expressed the decay of the current induced by breaking a circuit.[57]

Kirchhoff's work was also influential on Helmholtz. This was already evident in the 1851 paper, in which Kirchhoff's laws for conduction in branched circuits are used.[58] By the summer of 1852, Helmholtz was employing Kirchhoff's identification of potential and *Spannung*, rendering the rather obscure notion of "free tension" which had appeared in his 1847 paper more readily interpretable as a potential. For the case of electrodynamics, this potential was understood by Helmholtz as Neumann's potential function. This clarification led to a discussion between Helmholtz and Rudolf Clausius on the correctness of certain points in Helmholtz's 1847 work.

c) Clausius and Helmholtz on potential as mechanical work

Following and refining Helmholtz's interpretation, R. Clausius (1822–1888) seized on the role of potential as a unifying concept in theoretical physics. He expressed this clearly in his 1859 textbook, *Die Potentialfunction and das Potential.* Potential theory, he said,

> increases constantly in importance today the more it succeeds in explaining physical phenomena by the action of elementary forces and in referring them thereby to simple mechanical principles.[59]

Clausius' textbook was specifically designed to make available to physicists a brief, reasonably complete presentation of the main concepts, techniques, and areas of application of potential theory. As those responsible for elevating the subject to its present importance, he mentioned F. Neumann, Kirchhoff, Helmholtz and Thomson. Despite this importance, as he noted, the only available presentations of the subject were those given in the treatises of Green and Gauss, both of which neglect the concept of mechanical work. Yet the fact that the potential could be used (in the case of forces dependent on position alone) to express mechanical work done on or by the system of forces was, for Clausius, its key property. He expressed this as early as 1853 in a dispute with Helmholtz over the correct expression for the self-potential of a body:

> the concept of *potential* . . . has its great significance in science only by virtue of the fact that for a special but very common case [that of central inverse-square forces] it expresses the mechanical work.[60]

This significance extended beyond mechanics, because the potential could be defined for any conservative force as a measure of work, regardless of whether the interaction in question was between bodies possessing inertial mass.[61] Since the common Weberian conception of electric forces involved the interaction of massless particles, such potentials provided the critical tool for integrating these substances into the framework of theoretical mechanics. For example, the second half of Clausius' textbook deals with the use of potentials in conjunction with the principle of virtual work.

Clausius was not alone in his attempt to employ potentials in applying mechanics to physical research. Another was Bernhard Riemann, who sought in his lectures to achieve the same end. Riemann went further, attempting to generalize the Newtonian potentials to a potential for the velocity-dependent forces of action-at-a-distance electrodynamics.

6. Physically-Motivated Research in Potential Theory: the Case of Propagated Potentials.

a) Riemann's 1858 Report on Electrodynamics

In his 1858 *Beitrag zur Elektrodynamik*, Riemann introduced the notion that the force of electromagnetic interaction between circuits was propagated with finite velocity, instead of acting instantaneously. Riemann knew that Gauss had worked on a similar idea, and was somewhat relieved to learn (probably from Weber) that his own theory had turned out to be different from that of Gauss.[62] An immediate stimulus to the investigation may have been his reading of Brewster's life of Newton in August - September 1857.[63] Here he was much struck by the statement, from Newton's third letter to Bentley, saying that action-at-a-distance was to Newton "so great an absurdity, that I believe no man who has ... a competent faculty of thinking can ever fall into it".[64]

Riemann's aim was to treat the potential as a propagated static potential, and for this reason he sought to generalize the electrostatic potential of Poisson. He assumed that the potential satisfied the equation

$$\frac{\partial^2 V}{\partial t^2} - \alpha^2 \left(\frac{\partial^2 V}{\partial x^2} + \frac{\partial^2 V}{\partial y^2} + \frac{\partial^2 V}{\partial z^2} \right) + \alpha^2 4\pi\rho = 0,$$

or

$$\nabla^2 V - \frac{1}{\alpha^2}\frac{\partial^2 V}{\partial t^2} = -4\pi\rho, \tag{6.1}$$

recognizable as the non-homogeneous wave equation for the scalar potential in free space. Without giving any details of how he arrived at the result, Riemann then noted that the potential received at time t by a charge e at (x, y, z) due to a potential emitted with velocity α at time $t' = t = r/\alpha$ from a charge e' at (x', y', z') will be

$$V = -\frac{e[e']}{[r]} \tag{6.2}$$

where $[r]$ is what we now know at the "retarded" distance and $[e']$ is a retarded charge. (See Figure 6.1).

Figure 6.1

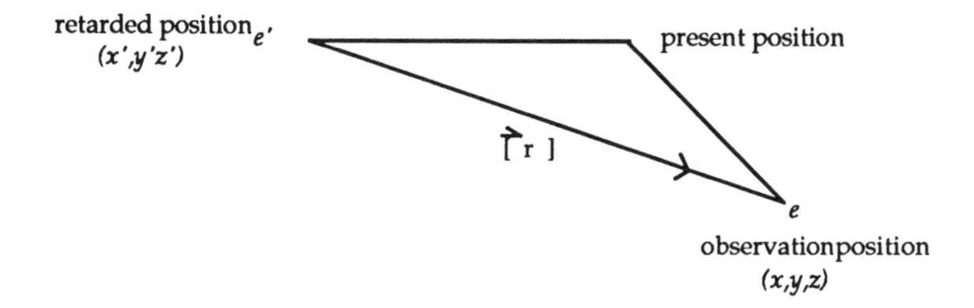

This result is non-trivial, and it is somewhat surprising that Riemann did not give the details.[65] Riemann then aimed to show that, when the total potential of two conductors is calculated, the result is Neumann's expression for the potential between two circuits. To this end, Riemann wrote Neumann's potential as

$$V = \frac{2}{c^2} \int\int \frac{uu' + vv' + ww'}{r} dSdS' \tag{6.3}$$

where (u, v, w) and (u', v', w') are current densities, c is Weber's constant, and dS and dS' are volume elements of the conductor. Riemann then transformed the above expression by writing the current densities as the product of charge densities and velocities so that

$$(u, v, w)dS = e\frac{d\vec{r}}{dt} \text{ and } (u', v', w')dS' = e'\frac{d'\vec{r}}{dt}$$

where the first derivative is presumed to result from motion of e alone, the second from motion of e' alone. (Here e and e' are the product of the densities and the volume elements).

Thus Riemann's expression for the Neumann potential becomes

$$V = \frac{1}{c^2} \sum\sum \frac{ee'}{r} \frac{dd'(r^2)}{dt^2} \tag{6.4}$$

This in turn he transformed by adding an exact differential, obtaining

$$V = \frac{1}{c^2} \sum\sum ee'r^2 \frac{dd'(\frac{1}{r}a)}{dt^2} \tag{6.5}$$

This is the expression Riemann sought to derive, using his retarded potential (6.2) and integrating over extended conductors.[66]

Riemann's procedure was flawed, however, as can be seen from the following sketch. Assume e' is at (x', y', z') at a time $t' = t - r/\alpha$ and that e is at (x, y, z) at time t. Riemann then noted that the potential of e at (present) time t due to e' should be the sum of all contributions up to that time. Choosing an initial position for e as occurring at $t = 0$, Riemann calculated the total potential by integrating (6.2) from 0 to t, obtaining

$$V = -\int_0^2 \frac{e[e']}{[r]} d\tau \tag{6.6}$$

where $[r]$ is now the retarded distance at time τ.

Riemann then claimed that the integrand in (6.6) may be expressed as

$$\frac{e[e']}{[r]} = -\int_0^{\frac{r}{\alpha}} e[e'] \frac{\partial}{\partial \sigma}\left(\frac{1}{[r]}\right) d\sigma \tag{6.7}$$

by differentiating $e[e']/[r]$ with respect to the transmission time σ between e' and e when e is at time τ. This required an argument to the effect that the integral vanishes at $\sigma = 0$. Substituting (6.7) in (6.6), Riemann obtained

$$V = \int_0^t d\tau \int_0^{\frac{r}{\alpha}} e[e'] \frac{\partial}{\partial \sigma}\left(\frac{1}{[r]}\right) d\sigma \,. \tag{6.8}$$

Riemann then interchanged the order of integration in (6.8), and, after much manipulation, obtained (6.5). However such a reversal is not permissible, since the upper limit of the inner integral in (6.8) depends on τ. This invalidates the entire procedure.

Riemann withdrew this paper after submitting it for publication, and it appeared only in 1867 after Riemann's death. Riemann's reasons for withdrawing the paper are not clear, but he may well have noticed the integration error. As well, there are fundamental problems with Riemann's approach. By considering the retarded distance directly as a function of present and retarded time, Riemann apparently failed to realize that the volume integral of the retarded charge density does *not* represent the total charge if the charge is conceived as in motion. Hence the r in the integral should be modified by a term which depends on the velocity of the charge.[67] This was first realized by Liénard and Wiechert around 1900. Consequently the law proposed by Riemann cannot be correct without additional hypotheses.

Difficulties with Riemann's paper were pointed out by Clausius soon after its 1867 publication.[68] Clausius located the source of the difficulty in the reversal of the order of integration, and his judgement was widely accepted and repeated, for example by Riemann's editor, H. Weber.[69] Weber noted, however, that he had decided to include the work in the collected papers because it might contain "germs for further fruitful thought".[70]

b) Carl Neuman on Propagated Potential

While Riemann's work was without influence when it was written, on its publication it spurred at least one other writer to investigate the question of propagated potentials. In 1865, Carl Neumann became interested in the question of the validity of the principle of *vis viva* for velocity-dependent forces. The usual formulation of the *vis viva* principle was that cited above due to Jacobi, namely that

> Half the *vis viva* of a system equals the force function augmented by a constant[71] or $\frac{1}{2}\sum M_i V_i^2 = U + h$

(where U is such that $\nabla U = \vec{F}$ that is, U is a potential for $\vec{F}$). However, for velocity-dependent forces no clear method was available to define a potential (or "force function" as Jacobi had called them) whose partial derivatives would correspond to the components of the force.

Weber had addressed this problem himself in 1848.[72] This attempt was apparently unknown to Neumann, until the mid-1860s;[73] his interest in the question was provoked by reading Gustav Theodor Fechner's 1860 *Psychophysik*.[74] There Neumann learned that Weber had told Fechner in conversation

> that he [Weber] had found the law [of *vis viva*] to be in force in all cases which he had examined, even beyond the bounds of those [i.e. conservative] forces, even if its complete validity still requires strict proof for these [velocity- dependent] forces.[75]

Neumann, motivated by these words, attempted to define a potential which would yield the force in question. By his own report, his initial attempts led him to an expression which yeilded the force components by finding its first *variation* in the directions of the coordinate axes, a procedure which he contrasted with that using partial differentiation.

Neumann's method involved positing a Lagrangian for the system. To grasp his procedure, it is useful to consider a version of Neumann's method without using the notion of propagated potentials which he was later to introduce. Let us assume the validity of Hamilton's principle, in the form $\delta \int L dt = 0$, where L is a generalized Lagrangian

$$L = T - U - V.$$

Here $U + V$ is the potential for the Weber force,

$$U + V = \frac{ee'}{r}\left(1 - \frac{1}{c^2}\left(\frac{dr}{dt}\right)^2\right),$$

where U refers to the static component, V to the dynamic component. This implies that the Euler-Lagrange equations hold, so that

$$\frac{d}{dt}\frac{\partial}{\partial \dot{x}}(T - U - V) = \frac{\partial}{\partial x}(T - U - V) \tag{6.9}$$

with corresponding expressions for y and z. The force components are of the form

$$F_x = \frac{d}{dt}\frac{\partial}{\partial \dot{x}}(U - V) - \frac{\partial}{\partial x}(U - V)$$

and a staightforward calculation shows that F_r is the Weber force.

In such a case, a conservation equation may readily be derived. First, note that

$$\dot{x}\frac{d}{dt}\frac{\partial f}{\partial \dot{x}} = \frac{d}{dt}\dot{x}\frac{d}{dt}\frac{\partial f}{\partial \dot{x}} - \ddot{x}\frac{\partial f}{\partial \dot{x}}$$

and, from (6.9),

$$\frac{d}{dt}\frac{\partial}{\partial \dot{x}}(T - U - V) = \frac{\partial}{\partial x}(T - U - V).$$

Hence

$$\dot{x}\frac{\partial}{\partial x}(T - U - V) = \frac{d}{dt}\dot{x}\frac{\partial}{\partial \dot{x}}(T - U - V) - \ddot{x}\frac{\partial}{\partial \dot{x}}(T - U - V).$$

Summing over the velocity components gives

$$\frac{d}{dt}\left(\dot{x}\frac{\partial}{\partial \dot{x}} + \cdots\right)(T - U - V) = \left(\dot{x}\frac{\partial}{\partial \dot{x}} + \cdots + \ddot{x}\frac{\partial}{\partial \dot{x}} + \cdots\right)(T - U - V). \tag{6.10}$$

Now both T and V are homogenous of second degree in dx/dt, etc., while U is independent of the velocities. Hence the left side of (6.10) becomes

$$2\frac{d}{dt}(T - U - V)$$

while the right side is just $d/dt(T - U - V)$. Hence

$$\frac{d}{dt}(T - U - V) = 0,$$

so that $T - U - V$ is constant.[76]

While Neumann did not give this example, it indicates the character of his method. In the form just described, Neumann's procedure was used later by Wilhelm Weber in his debate with Helmholtz.

(i) Neumann's 1868 Principles of Electrodynamics

Neumann extended his efforts of 1865 in an important way in 1868 after reading Riemann's posthumously-published memoir on electrodynamics.[77] Neumann grafted this approach using propagated potentials onto his 1865 research to create the theory which is described in his 1868 paper, "Principles of Electrodynamics". Riemann's approach offered Neumann two attractive features: first, the analogy with light propagation which had been mentioned explicitly by Riemann; and second, the possiblity of using the usual Newtonian potential function, $1/r$, as the starting point for his argument rather than incorporating velocity-dependence at the outset.

The main aims of Neumann's paper were to justify the new potentials and their propagation, and subsequently to derive the basic mathematical laws of electricity theory (in Neumann's view those of Weber, Ampère, and F. Neumann) as well as to prove a version of the *vis viva* principle valid for electrodynamics. After his initial discussion of the notion that, like light, the potentials may propagate with a large finite velocity, Neumann distinguished two kinds of potential, that emitted and that received. In his model, it is the latter – the received potential – which acts.

Neumann stressed the formal character of his mathematical approach. His treatment made minimal use of underlying models and in fact – while continuing to consider electricity as particulate – sought to provide an analysis which did not depend on the dualistic Fechner-Weber hypothesis, but could be applied equally well to a unitary model of electricity. Neumann expressed his hope that the formalism would be useful regardless of the possibility of the reduction of his theory to some other.

One of the beauties of Neumann's approach is that he was able to consider the emitted potential as the usual electrostatic potential. That is, if the distance between two electric particles is r, the emitted potential is

$$\frac{mm_1}{r}$$

where m and m_1 are the "electric masses" or charges of the particles. Note incidentally that these particles were considered to be without mass, imponderable, as in the older Weberian theory. Neumann expressed the potential emitted at time $t - \Delta t$ as

$$\Phi(r - \Delta r) = \frac{mm_1}{(r - \Delta r)}.$$

This potential is received at time T, when the *present* separation of the particles is r, and Neumann assumed that it acts at that time t. Let us denote the received potential by U. Assuming that U is propagated with finite velocity c, Neumann sets $\Delta t = r/c$. Since r is a function of time, $r = f(t)$,

$$r - \Delta r = f(t - \Delta t), \text{ and } r = f'(t) \text{ etc.}$$

Therefore, expanding in a Taylor series, we may write

$$\begin{aligned} r - \Delta r &= r - \Delta t r + \frac{(\Delta t)^2}{2} r + \cdots \\ &= r - \frac{r}{c} r + \frac{r^2}{2c^2} r, \end{aligned}$$

where r, dx/dt, and d^2x/dt^2 are all evaluated at time t. (Here we can neglect higher powers since c is large, so a/c^3 can be assumed to be very small).

Hence the received potential at time t is just

$$U(r, r) = \Phi(r - \Delta r) = \frac{mm_1}{r - \frac{r}{c} r + \frac{r^2}{2c^2} r}.$$

By expanding Φ in a Taylor series about the present distance r, Neumann obtained

$$U = \Phi(r) - \frac{r}{c} r\Phi'(r) + \frac{r^2}{2c^2} r\Phi'(r) + \frac{r^2}{2c^2}(r)^2\Phi''(r),$$

where he has used $\Delta t = r/c$ so that

$$\left(\frac{dr}{dt}\right) \Delta t = \Delta r.$$

This step allowed Neumann to express U as the sum of two functions, one of which is an exact differential with respect to t:

$$U = V_1 + V_2.$$

(Some manipulations are required, which I omit here.) Neumann found

$$V_1 = mm_1 \left[\Phi + \frac{r^2\Phi''}{2c^2}\left(\frac{dr}{dt}\right)^2 - \frac{(r^2\Phi')'}{2c^2}\left(\frac{dr}{dt}\right)^2\right]$$

and

$$V_2 = mm_1 \frac{d}{dt}\left[\frac{r^2\Phi'}{2c^2}\frac{dr}{dt} - \frac{\int r\Phi' dr}{c}\right].$$

These two functions are called by Neumann the effective potential (V_1) and the ineffective potential (V_2) for reasons we shall see shortly. By substituting $\Phi(r) = 1/r$, we get

$$V_1 = \frac{mm_1}{r}\left[1 + \frac{1}{c^2}\left(\frac{dr}{dt}\right)\right]$$

and

$$V_2 = mm_1\left[\frac{\log r}{c} - \frac{1}{2c^2}\frac{dr}{dt}\right].$$

These, then, are the expressions Neumann arrives at using the notion of a propagated Newtonian potential.

To show what these have to do with Weberian electrodynamics, Neumann asserted the general validity of Hamilton's principle:

$$\delta\int_{t_1}^{t_2}(T - U)dt = 0$$

where T is the usual kinetic energy, and U is the received potential. The quantity being integrated, $T-U$ is taken to be fixed at the limits of integration t_1 and t_2. This assumption serves to establish both the identification of the forces and the exact expression for those forces.

On performing the calculations implied by Hamilton's principle, the variation of V_2 vanishes, whence the name ineffective potential. Since

$$\delta\int(T - (V_1 + V_2)dt = 0$$

we may write

$$\delta\int Tdt = \delta\int(V_1 + V_2)dt.$$

But V_2, the "ineffective" potential, is an exact differential with respect to time, say $V_2 = dg/dt$, so that

$$\delta\int V_2 dt = \delta\int dg = \delta(g(t_1) - g)t_0)).$$

But the function is constant at the limits of integration, yielding

$$\delta\int V_2 dt = 0.$$

Thus we obtain that

$$\delta \int (T - V_1)dt = 0$$

and standard calculus of variations shows us that Lagrange's equations hold, namely

$$\frac{d}{dt}\left(\frac{\partial V_1}{\partial \dot{x}_i}\right) - \frac{\partial V_1}{\partial x_i} = F_i,$$

where F_i are the generalized force components. The operator on the left in the above equation is exactly that which Neumann terms the variation of V_1 in the direction x_i.

To establish a conservation principle, Neumann identified two separate components of the effective potential (V_1): the first Neumann called the static potential

$$V_{11} = \frac{mm_1}{r};$$

the second the motor potential

$$V_{12} = mm_1 \left(\frac{d\Psi}{dt}\right)^2 \quad \text{where } \Psi = \frac{2}{c} r^{1/2}.$$

Neumann proved by a straightforward calculation similar to that outlined above that

$$T - V_{11} - V_{12} = \text{constant}$$

thereby obtaining a conservation law for velocity-dependent interactions. Thus Neumann was able to obtain both a generalized potential for the Weberian force and a conservation law.[78]

Neumann's procedure, however, has several features which are puzzling on closer scrutiny, and some of these were pointed out by Clausius in an article which appeared in the *Annalen der Physik* late in 1868.[79] Before we look at Clausius' criticism, it should be mentioned that other reactions to Neumann's work were more favorable. One writer, Neumann's colleague Scheibner, sought to clarify some points in Neumann's calculation, and another, Holzmüller, attempted to complete the formulation of Hamilton-Jacobi theory for velocity-dependent forces.[80] Most interesting was Weber's reponse, which Neumann received in a letter before publishing his memoir:

> the reason why [the value of the potential] is valid for later points in time may lie in a propagation about which we could speak only on the hypothesis of a higher mechanics (as e.g. the propagation of waves in air depends on fluid mechanics).[81]

(ii) Clausius' Criticism of Neumann's Approach

Clausius' paper, titled "On the new approach to electrodynamic phenomena suggested by Gauss" , appeared in the *Annalen der Physik* in 1868, and was intended as a general critique of various approaches to electrodynamics using propagation of force or potential.[82] He examined (and found wanting) the approaches of Riemann and Riemann's Italian colleague Enrico Betti, but his most searching criticism was directed at Neumann's 1868 paper. For Clausius, there was no resemblance between Neumann's motion of the potential and the propagation of light.

If light is the model, said Clausius, then the received potential should be based on distance actually travelled by the emitted potential, i.e. the distance between m_1 at $t - \delta t$ and m at t. (See Figure 6.2).

In this view, the potential received should not depend on the position of the receiver m at the time of emission by m_1. Furthermore, Neumann had used the distance at time t as his basis for calculating δt, setting the latter equal to r/c in order to obtain his expression for the received potential at time t. Once again Clausius suggested that the time actually elapsed since emission should be used, so that

$$\delta t = \frac{r(t - \delta t, t)}{c},$$

that is, δt is now the time interval between emission and reception.

Figure 6.2: Time and distance relations for retarded potentials according to Neumann and Clausius

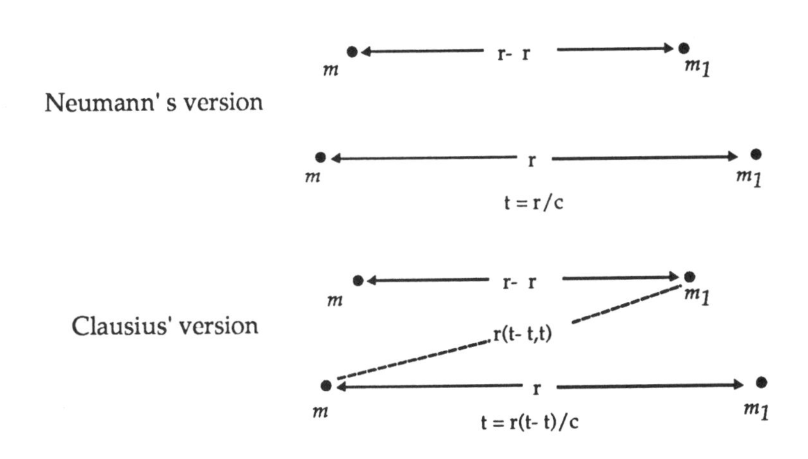

Using these assumptions, however, Neumann's approach falls apart. In particular, using his new values for distance and time, Clausius was unable to designate part of the potential as ineffective since there is no exact differential of time contained in the resulting expression. Hence, Clausius concluded that Neumann's approach is invalid if we consider that the propagation of potentials should resemble that of light, or be physically like any ordinary motion.[83]

Neumann responded to Clausius' article in a memoir dated January 19, 1869 and published in Neumann's new journal, the *Mathematische Annalen*. In this paper Neumann stated that he could not agree with Clausius' judgement, and recapitulated the main features of his treatment. In the footnotes to the paper, however, and in a discussion headed "Potential and Light" , Neumann emphasized that he had regarded the 1868 paper as only a provisional treatment, citing as evidence the fact that he had published it only in a pamphlet intended for limited distribution. Furthermore, he asserted that the similarity of his potential to light is obviously very limited.[84]

Though Neumann implied that his 1868 treatment had recognized these limitations, his repeated emphasis on similarities to light in the 1868 work seems to belie this. (For example, he went to considerable length to show that the electrodynamic potential obtained agrees in form with that employed by him in his dissertation on the Faraday effect).[85] Thus Clausius seems to have been justified in his interpretation of the weak character of Neumann's rebuttal as an indication of Neumann's acceptance of his criticisms.[86]

7. A RETURN TO MATHEMATICAL FOUNDATIONS: Hölder and Neumann.

In the research discussed so far, physical conditions made it unnecessary to further examine the fundamental assumptions of Gauss concerning the nature of the solids and their bounding surfaces, and concerning the properties of the density function. That is, the densities were assumed piecewise differentiable and the surfaces were of an appropriate rather murkily specified smoothness. The latter were usually termed 'general' surfaces. In the wake of the critical examination of the foundations of analysis associated with Weierstrass and his school, these assumptions very naturally came under scrutiny, the more so because of the equivocal status of the Dirichlet principle. Questions of this kind were taken up by many writers in the following decades, I shall illustrate the character of this research by brief reference to work by Carl Neumann and Otto

Hölder (1859-1937). Their work was basic for what followed in both the nature of the questions posed and in the methods employed.

In his 1877 *Untersuchungen über das logarithmische und Newton'sche Potential,* Neumann treated a selection of topics he had studied earlier, shifting from an emphasis on physical results to a focus on mathematical foundations and unifying concepts.[87] While applications to electromagnetic theory remained very much on Neumann's agenda, Neumann here sought to examine critically the rather loose notion of a 'general surface' which was in widespread use, not only in potential theory. Taking Green's theorems as an example, Neumann noted that

> we are in a position to prove the validity of this theorem with full rigour if the surface consists of explicitly given portions of surface of first or second order.[88]

By a surface of first or second order, Neumann means one with a rectangular equation of first or second degree. Hence in Neumann's view it was an open question as to whether the chief results of potential theory could be applied even to higher order rational surfaces. Neumann aims, in the 1877 work, at finding proofs for central results which did not depend on these conditions, and found a key hypothesis in the assumption that the bounding solid should be convex.

Hölder, on the other hand, investigated properties of the density which were sufficient to ensure that the potential at points of the mass will satisfy the Poisson equation. In his 1882 doctoral dissertation, he noted that continuity of the density is not sufficient to guarantee existence of the second derivatives of the potential. With suitable restrictions on the region, Hölder was able to show that a potential U will satisfy Poisson's equation at a point P of the mass provided that there exist positive constants, $c, A,$ and α such that

$$|U(Q) - U(P)| \leq Ar^{\alpha},$$

where $r = |PQ|$ and Q is any point such that $r \leq c$.[89]

These brief indications of the nature of pure-mathematical research in potential theory suggest that in the late 1870s and early 1880s such research was building directly on the foundations laid by Gauss in 1840. The continued importance of the work of the 1840s is further underlined by the appearance in 1876 of a published version of Dirichlet's lectures on inverse square forces, a book which was also important for the subsequent direction of mathematical research in the area.

8. Concluding Remarks.

It may seem curious that about 35 years elapsed before such theoretical studies began a critical reassessment of the hypotheses of Gauss' work. I would suggest a variety of reasons for this. Most obviously, such work is in the tradition of the Weierstrassian critique of the foundations of analysis, a critique which began to have influence only in the mid-1860s. However, instead of attempting to explain what did not happen, let us outline what *was* taking place. As we have seen the preoccupations of research in the period between 1840 and about 1875 centered very heavily on its applications notably in electromagnetic theory. These applications led to non trivial discoveries; several important breakthroughs, such as the Green's function method of solving boundary value problems, inspired pure mathematical research in themselves. Mathematical physicists were struggling simultaneously with the methods of potential theory and with the correct physical interpretation of the notion of potential. Specific efforts were made to generalize the Newtonian potential in the context of specific physical theories. Such physically-oriented research was not widely pursued after 1880, a phenomenon which also calls for comment. Evidently, this means that mathematical physicists no longer saw research in potential theory as useful, or at least, that something else was occupying their attention.

I would suggest a combination of factors. First of all, by 1880 certain basic aspects of potential theory had not altered much for some time. The integral theorems had reached a level of generality which sufficed for most physical applications. For example Green's identities had been generalized to multiply-connected domains by William Thomson.[90] Solution methods had been developed to a high level for the standard geometrical configurations which have greatest experimental application. For example, handbooks on spherical harmonics and cylindrical functions had been produced by Heine and Hankel respectively, and textbook presentations of these methods were readily available. Furthermore, the attempts to solve the problems posed by action-at-a-distance electrodynamics via the use of propagated potentials had led only to confusion, as had attempts to define velocity-dependent potentials for electrodynamic forces. In fact, this entire direction of research was considerably undermined by the appearance in 1873 of Maxwell's *Treatise on Electricity and Magnetism*, since Maxwell's approach solved several of the important problems that propagated potentials had addressed. In light of these considerations it is natural that pure mathematical research into potential theory should become more visible; and such research, which was informed by the consolidated physical interpretation of potential theoretic results but not guided by it, took potential theory increasingly far from the hands of mathematical physicists.

NOTES

1 For details on these and other developments in German electrodynamics, see Archibald (1987).

2 Gauss (1839b) This paper appeared in 1840 in the journal, jointly edited by Gauss with Wilhelm Weber, *Resultate aus den Beobachtungen des Göttinger magnetischen Vereins*. It appeared in English translation in 1842 in Taylor and Francis' journal, and in French in Liouville's Journal in the same year.

3 Gauss (1839b) page 200.

4 See, for example, Morrell and Thackray (1981), and O'Hara (1983) and (1984).

5 Gauss (1839a), page 306.

6 Gauss (1839b), page 203.

7 ibid., page 203.

8 Gauss (1839b), page 208. Gauss' proof differs strikingly from later proofs by English writers, who used essentially physical arguments for the theorems of vector analysis. See Archibald (forthcoming 1989) for a discussion of such arguments in the context of Green's theorems. Gauss, by contrast, employs arguments involving limits.

9 Gauss (1839b), page 206.

10 Königsberger (1904), p. 253.

11 Königsberger (1904), p. 355.

12 ibid.

13 ibid. Poggendorff's report to the Academy was presented on 20 October 1845.

14 ibid., pp. 355–356.

15 Neumann (1845), p. 3.

16 Helmholtz, for example, was thoroughly convinced of the correctness of Neumann's approach until 1875.

17 Lenz (1834), p. 485.

18 Neumann (1845), p. 16.

19 Here I employ notation due to Sommerfeld and Reiff (1902) in the hope that it is somewhat clearer than Neumann's.

20 Neumann (1845), p. 16.

21 Neumann wrote this as $Eds = -\epsilon v C ds$, where Eds is the e.m.f. induced in ds and C is the component of the Ampère force which I have termed F_v.

22 Neumann (1845), pp. 20–21.

23 My interpretation here is supported by Reiff and Sommerfeld (1902), p. 27. In fact, I have substituted δt for Neumann's dt, δw for dw where appropriate, since it is clear from Neuman's use of the principle of virtual work that a variation is what he intends.

24 Neumann (1845), p. 21.

25 Neumann (1845), p. 22.

26 Jacobi (1866), p. 19.

27 Note that for both Neumann and Jacobi, *vis viva* is $\sum mv^2$, and not $1/2 \sum mv^2$.

28 Neumann (1845), pp. 22–23.

29 ibid., pp. 70–71.

30 Gauss (1839b), p. 307. Neumann (1845), p.6.

31 Neumann (1845), p. 23.

32 Königsberger (1904), pp. 361–362. Letter from Neumann to Jacobi of 5 February 1846. Neumann in fact refers to this as section 2, at odds with the published version.

33 See Sommerfeld (1977), p. 97.

34 Neumann (1845), sections 4, 5.

35 Volkmann (1896), p. 57.

36 Outside Germany, Neumann's work found an interested audience in France, where it was translated *in toto* by the well known crystallographer, Bravais, doubtless already acquainted with Neumann's earlier work. Bravais made a noble attempt to render Neumann's peculiar notation—already commented on by Jacobi—more nearly standard.

37 The published version of Dirichlet's lectures, edited by F. Grube, was based on lectures of 1856, and any remarks I make concerning them are based on this transcript. It may plausibly be argued, however,

that the lectures will have altered little in this period. Dirichlet had lectured on potential theory at Berlin since at least 1845, and the main change in the decade following this was doubtless the integration of Green's results into his lectures. There is a detailed discussion of this question by Butzer in AIHS 37(1987).

38 In fact, the term potential energy was not employed until 1863, and received widespread currency only in the late 1860s.

39 Ohm(1826) and (1827). For substantially greater detal on the following discussion, see my paper "Tension and Potential from Ohm to Kirchhoff", Centaurus, to appear.

40 Kirchhoff (1849), p. 49. This was made possible by advances in instrumentation, namely, a condensor of Kohlrausch's own design, together with an electrometer constructed by Dellmann. (See Kohlrausch (1848), p. 220.)

41 ibid., p. 50.

42 Königsberger (1902), pp. 51–52.

43 Among Helmholtz's fellow students at this time were Emil du Bois-Reymond, Ernst Brücke and Rudolf Virchow. See on this, the article by J. von Kries, *"Helmholtz als Physiolog"* in *die Naturwissenschaften,* v. 9, (1921), especially pp. 673–676.

44 Königsberger (1902), p. 50. Königsberger asserts this without qualification.

45 Königsberger (1092), pp. 58, 62.

46 ibid., p. 70. Letter from Poggendorff to Magnus of August 1, 1847.

47 Kuhn (1959), p. 66; Elkana (1974).

48 Helmholtz (1847), p. 9.

49 ibid., p. 14.

50 These results are now standard. See, for example, Goldstein (1950), pp. 7–10.

51 ibid., pp.29–30.

52 ibid., p. 47.

53 ibid., pp. 47–48.

54 Königsberger (1902), p.80.

55 For example, the priority questions was not raised until 1868 at the Innsbruck *Naturforscherversammlung*, See Königsberger (1902), pp. 90–91.

56 Helmholtz (1851) and (1853).

57 According to Königsberger (1902) v. 1, p. 145, this expression was used by Neumann to study the second—order induction effects in Arago's disc. These results were not published by Neumann at the time.

58 Helmholtz (1851), p. 435.

59 Clausius (1859), p. iii.

60 Clausius (1853), p. 569. His emphasis.

61 Clausius (1859), p. 10.

62 Berlin, SPK, Nachlass Riemann Acc. 17. Letter to Ida.

63 Göttingen SUB Cod Ms Riemann 26 f36. Library borrowing slip.

64 Dedekind (1892), p. 55; Newton (ed. Thayer, (1953) p. 54).

65 Riemann (1868), pp. 290–292. For a modern derivation, see Panosfsky and Phillips (1962), pp. 242–244.

66 To grasp the transition from (4.17a) to (4.18), let us use d/dt and d'/dt in Riemann's sense, so that $d\vec{r}/dt = \vec{v}$, and $d'r/dt = \vec{v}'$, where $\vec{v}$ is the velocity of e, and $\vec{v}'$ that of e'. Hence

$$\frac{d'(r^2)}{dt} = 2\vec{r} \cdot \frac{d'\vec{r}}{dt} = 2\vec{r} \cdot \vec{v}'$$

and

$$\frac{d}{dt}\left(\frac{d'(r^2)}{dt}\right) := \frac{d}{dt}(2\vec{r} \cdot \vec{v}') = 2\vec{r} \cdot \frac{d\vec{v}'}{dt} + 2\vec{v}' \cdot \frac{d\vec{r}}{dt} \ .$$

But $d\vec{v}'/dt$, the change in $\vec{v}'$ due to the motion of $\vec{v}$, is zero, hence

$$\frac{dd'}{dtdt}(r^2) = 2\vec{v}' \cdot \vec{v} \ .$$

the integrand in (4.18) becomes

$$\frac{2}{c^2} e\vec{v} \cdot e'\vec{v}' \ ,$$

which gives the integrand in (6.3) if we translate charges and velocities into currents and volumes.

67 Panofsky and Phillips (1962), p. 342. Kaiser (1981), p. 155.

68 Clausius (1868), p. 615.

69 Riemann (1892), p. 293.

70 ibid.

71 Jacobi (1866), p. 19.

72 Weber (1848), p. 229.

73 C. Neumann (1868), p. 434.

74 *ibid.*, p. 400.

75 Fechner (1860), v.1, p. 34.

76 See the discussion in Reiff and Sommerfeld (1902), pp. 47–49.

77 C. Neumann, *op cit.*, p. 402.

78 *ibid.*, pp. 417–423.

79 Clausius (1868), p. 606.

80 Scheibner (1868), pp. 37, 47; Holzmüller (1870), pp. 69–91.

81 C. Neumann, *op cit.*, p. 433.

82 Clausius (1868), pp. 606–607.

83 See R. Clausius (1879). In his recent work on thermodynamics Clausius had devoted much effort to providing a physically plausible mechanical interpretation of the energy and entropy of systems , and the peculiar character of Neumann's potential thus held little attraction for him.

84 C. Neumann (1869), pp. 317–324.

85 C. Neumann (1858); also outlined in his later papers for example in Neumann (1878), pp. 113–114.

86 Clausius (1180), p. 622.

87 Neumann (1877) pp. v-vi lists twelve papers, five of which have problems in electrostatics on stationary heat conduction as their principal object of study.

88 ibid., p. vii.

89 Hölder (1182). See also Kellogg (1929). The resemblance of this condition to Lipschitz's condition for the existence of solutions of a differential equation suggests a direct influence, though I have not yet established this.

90 See Archibald "Connectivity and Smoke-Rings", *Mathematics Magazine*, to appear.

BIBLIOGRAPHY

Archibald, W.T. (1987). " 'Eine sinnreiche Hypothese'. Aspects of action-at-a-distance electromagnetic theory 1820-1880." Ph.D. Dissertation, University of Toronto.

Archibald, W.T., (forthcoming). "Connectivity and Smoke Rings: the history of Green's second identity in its first fifty years." *Mathematics Magazine.*

Archibald, W.T. (forthcoming). "Tension and Potential from Ohm to Kirchhoff." *Centaurus.*

Clausius, Rudolf (1853). "Ueber einige Stellen des Schrift von Helmholtz über die Erhaltung der Kraft." *Ann. d. Phys.* 89, pp. 568–579.

Clausius, Rudolf (1880). "Ueber die Anwendung des electrodynamischen Potentials zur Bestimmung des ponderomotorischen und electromotorischen Kräfte." *Ann.d.Phys.* ser.3, 11: 604–633.

Clausius, Rudolf, (1859). *Die Potentialfunction und das Potential.* Leipzig, Barth.

Clausius, Rudolf (1868). "Ueber die von Gauss angeregte neue Auffassung der elektrodynamischen Erscheinungen." *Ann. d. Phys.* 135: 606–621.

Dedekind, Richard (1892). "Bernhard Riemann's Lebenslauf." In B. Riemann, *Gesammelte Mathematische Werke.* 2nd edition. Leipzig, Teubner.

Elkana, Yehuda (1974). *The discovery of energy conservation.* Cambridge, Mass., Harvard.

Gauss, C. F. (1839a). "Selbstanzeige, Allgemeine Lehrsätze usw." In *Werke,* 5, pp. 305–308.

Gauss, C. F. (1839b). "Allgemeine Lehrsätze in Bezeihung auf die in verkehrten Verhältnisse des Quadrats des Entfernung wirkenden Anziehungs- und Abstossungs-kräfte." In *Werke,* 5, pp. 197–242.

Goldstein, H. (1950). *Classical Mechanics.* New York, Addison Wesley.

Green, George (1828). "Essay on the Application of Mathematical Analysis to the Theories of Electricity and Magnetism." From N. M. Ferrers (1871). *The Mathematical Papers of George Green.* Reprint, 1970. N.Y., Chelsea. Translation (1850) in *JRAM* 39:73.

Helmholtz, Hermann von (1847). *Ueber die Erhaltung der Kraft.* Leipzig, Engelmann. Reprint, 1902. Ostwald's Klassiker.

Kaiser, W. (1981). *Theorien der Elektrodynamik in 19. Jahrhundert.* Hildesheim, Gerstenberg.

Kellog, O.D. (1929). *Foundations of Potential Theory.* Berlin, Springer .

Kirchhoff, Gustav Robert (1848). "Ueber die Anwendbarkeit der Formeln für die Intensitäten der galvanischen Ströme in einem Systeme linearer Leiter auf Systemen, die zum Theil aus nicht linearem Leitern bestehen." *Ann.d.Phys* 75. Also in Kirchhoff (1882) 1: 33–49.

Kirchhoff, Gustav Robert (1849a). "Ueber eine Ableitung der Ohm'schen Gesetze, welche sich an die Theorie der Elektrostatik anschliesst." *Ann.d. Phys* 78. Also in Kirchhoff (1882) 1: 49–55.

Kirchhoff, Gustav Robert (1849b). "Bestimmung der Constanten, von welcher die Intensität inducirter elektrischer Ströme abhangt." *Ann.d. Phys.* 76. Also in Kirchhoff (1882). 1: 118–131.

Königsberger, Leo (1902-3). *Hermann von Helmholtz.* Vol. 1, 1902. Vol. II, 1902. Vol. III, 1903. Braunschweig, Vieweg.

Königsberger, Leo (1904). *Carl Gustav Jacobi, Festschrift zur Feier der Hundertsten Wiederkehr seines Geburstages.* Leipzig, Teubner.

Kohlrausch, R. and Weber, W. (1856a). "Ueber die Elektricitätsmenge, welche bei galvanischen Strömen durch den Querschnitt der Kette fliesst." *Ann. d. Phys.* 99: 10–25. Also in Weber (1893) III: 597–608.

Kohlrausch, R. and Weber, W. (1856b). " Elektrodynamische Maassbestimmungen insbesondere Zurückfuhrung der Stromintensitäts - Messungen auf mechanisches Maass." *Sächs. Abh.* (1857) 3: 221-290. Also in *Werke* III: 609–676.

Kuhn, T.S. (1959). "Energy conservation as an example of simultaneous discovery." In M. Clagett, ed. *Critical Problems in the History of Science.* Madison, U. of Wisconsin Press, pp. 321–356.

Lenz, E.K. (1834). "Ueber die Bestimmung der Richtung der durch elektrodynamische Vertheilung erregten galvanischen Ströme." *Ann.d.Phys.* 31: 438–494.

Morrell, Jack and Thackray, A. (1981). *Gentlemen in Science.* Oxford, Oxford University Press.

Neumann, Carl (1868a). "Resultate einer Untersuchung über die Principien der Elektrodynamik." *Gött. Nachr.* 16 June 1868: 223–235.

Neumann, Carl (1868b). *Die Principien der Elektrodynamik.* Tübingen, H. Laupp.

Neumann, Carl (1877). *Untersuchungen über das Logarithmische und Newton'sche Potential.* Leipzig, Teubner.

Neumann, Carl (1878). "Ueber das von Weber für die elektrischen Kräfte aufgestellte Gesetz." *Sächs. Abh.* 11: 79–198.

Neumann, Franz (1845). *Die mathematischen Gesetze der inducirten elektrischen Ströme.* Reprint, 1889. Ostwald's Klassiker der Exakten Wissenschaften, 10. Leipzig, Wilhelm Engelmann.

O'Hara, James Gabriel (1983). "Gauss and the Royal Society: The Reception of his Ideas on Magnetism in Britain (1832-1842)." *Notes and Records of the Royal Society of London.* 38: 17–78.

O'Hara, James Gabriel (1984). "Gauss' Method for Measuring the Terestrial Magnetic Force in Absolute Measure: Its Invention and Introduction in Geomagnetic Research." *Centaurus.* 27: 121–147.

Ohm, Georg Simon (1826). "Bestimmung des Gesetzes, nach welchem Metalle die Contact-Elektricität leiten, nebst einem Entwurfe zu einer Theorie des Voltaischen Apparates und des Schweiggerschen Multiplicators." In *Journal der Physik* 46: 137–166.

Ohm, Georg Simon (1827). *Die Galvanische Kette, mathematisch bearbeitet.* Berlin, Riemann. Reprinted (1887) in Leipzig, Toeplitz und Deuticke. Also in Ohm (1892), 61–186.

Olesko, Kathryn Mary (1980). *The Emergence of Theoretical Physics in Germany: Franz Neumann and the Königsberg School of Physics. 1830-1890).* Ph.D. Dissertation. Cornell University.

Panofsky, W. and Phillips, M. (1962). *Classical electricity and Magnetism.* 2nd edition. London, Addison Wesley.

Riemann, Bernhard G. F. (1861). *Schwere, Elektricität und Magnetismus. Nach den Vorlesungen von B. Riemann.* Edited by K. Hattendorff. 2te Ausgabe. Reprint 1880 (First edition, 1876). Hanover, Rümpler.

Riemann, Bernhard G. F. (1892). *Gesammelte Mathematische Werke.* Leipzig, Teubner.

Riemann, Bernhard G. F. (1867). "Ein Beitrag zur Elektrodynamik." *Ann.d. Phys.* 131: 237. See also Riemann (1892), 288.

Scheibner, W. (1868). *Die Principien der Elektrodynamik. Eine mathematische Untersuchung von Dr. Carl Neumann.* Essay Review of Neumann (1868b) in *ZMP* Literaturzeitung 13(4): 37–47.

Sommerfeld, A. (1977). *Elektrodynamik.* 4th edition. Frankfurt, Deutsch.

Sommerfeld, A. and Reiff, R. (1902). "Elektrizität und Optik. Standpunkt der Fernwirkung: die Elementärgesetze." *Enc.Math.Wiss.* (1904). 5(2): 3–62.

Volkmann, P.O.E. (1896). *Franz Neumann*. Leipzig, Teubner.

Joseph Liouville (1809–1882)
Courtesy of Springer-Verlag Archives

The Geometrization of Analytical Mechanics
A Pioneering contribution by Joseph Liouville (ca. 1850)

Jesper Lützen

INTRODUCTION

When Newton in his Principia Mathematica Philosophiae Naturalis [1687] described physics in mathematical terms, the mathematics he used was Euclidean geometry. A hundred years later, however, Lagrange freed mechanics from geometry and made it a part of analysis, which, in turn, he based on algebra. He expressed this program very forcefully in the introduction to his Mecanique Analytique:

> One will not find any figures in this work. The methods I present need neither constructions nor geometric or mechanical arguments but only algebraic operations. [Lagrange 1788 2.ed 1811 Avertissement].

The 20th century has witnessed a reintroduction of geometry into mechanics: in the general theory of relativity gravitation is explained by means of a curved space and classical mechanics as treated by Arnold [1978] has turned into so-called symplectic geometry. The geometry used in modern mechanics, however, is not Euclidean geometry but differential geometry.

In this paper I shall adress the question: when and how did this regeometrization take place. My interest in this problem has grown out of my work on a scientific biography of the French mathematician Joseph Liouville (1809-1882) [Lützen 1989] and so I shall particularly emphasize his role in the development. In fact it is my primary aim here to call attention to his pioneering role in the differential geometric conception of analytical mechanics.

Since this paper deals with the merge of two disciplines: mechanics and differential geometry, I shall begin by discussing them separately; in particular I shall trace those fundamental early nineteenth century developments that paved the way for the discovery of the links between them.

ISBN 0-12-599662-4.

HAMILTON-JACOBI MECHANICS

Let us first consider the mechanics of a conservative system of point masses. For such a system Joseph Louis Lagrange (1736-1813) introduced a function U of the coordinates of the point masses (x_i, y_i, z_i) such that the force on the i'th particle is equal to $-\vec{\nabla} U$. [Lagrange 1774 and 1788/1811]. He further observed that when U is added to the kinetic energy $T = \frac{1}{2}\sum_i m_i(\dot{x}_i^2 + \dot{y}_i^2 + \dot{z}_i^2)$ the result is a constant today called the total energy.

In particular Lagrange dealt with systems which are influenced by holonomic constraints. As a simple example of such a system consider a mathematical pendulum moving in a plane:

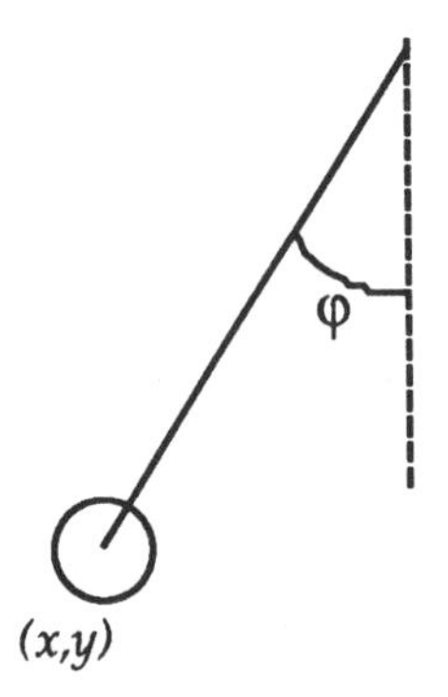

Figure 1.

Here the point mass is not allowed to move freely, and so its Cartesian coordinates (x, y) cannot take on arbitrary values. The problem this causes in the description of the motion of the pendulum, and other similar systems, was overcome by Lagrange in two ways: 1° by the use of Lagrangian multipliers and 2° by introducing generalized coordinates. Only the latter alternative is important in our story. For the pendulum the angle φ (figure 1) is a suitable generalized coordinate because it 1° completely specifies the position of the system and 2° it can vary freely. If a system of n generalized coordinates $q_1, \ldots, q_n$ satisfies these two requirements the system is said to have n degrees of freedom. The space $\mathbf{R}^n$ of $q_1, \ldots, q_n$ is now called the configuration space. Of course Newton's laws, as expressed by Euler, are differential equations in the cartesian coordinates, and so in order to treat holonomic constraints Lagrange gave the equations of motion a new form, Lagrange's equations, which also hold for generalized coordinates:

$$\frac{d}{dt}\frac{\partial L}{\partial \dot{q}_i} = \frac{\partial L}{\partial q_i}, \tag{1}$$

where $L = T - U$.

In [1834,1835] William Rowan Hamilton (1805-1865) discovered that he could give the fundamental equations of mechanics an even simpler form by supplementing the system of variables[1] $q_1, \ldots, q_n$ by n new variables $p_1, \ldots, p_n$ defined by[2]

$$p_i = \frac{\partial L}{\partial \dot{q}_i} = \frac{\partial T}{\partial \dot{q}_i}. \tag{2}$$

In terms of these $2n$ variables the equations of motion take the form

$$\dot{q}_i = \frac{\partial H}{\partial p_i}, \qquad \dot{p}_i = -\frac{\partial H}{\partial q_i}, \quad i = 1, 2, \ldots, n, \tag{3}$$

where the so-called Hamiltonian or energy function H is defined by

$$H(p_1, \ldots, p_n, q_1, \ldots, q_n) = T + U. \tag{4}$$

This formalism, which Jacobi later called canonical, is only mentioned in passing in Hamilton's papers. In Hamilton's own opinion his most important achievement in mechanics was the introduction of a characteristic function V, which satisfies two partial differential equations and has the property that once it is determined it immediately yields the solutions to Hamilton's canonical equations (3).

Carl Gustav Jacob Jacobi (1804-1851) showed [1837a] that *one* partial differential equation was sufficient. His mathematically more elegant reformulation of Hamilton's method calls for the determination of a complete solution

$$V(q_1, \ldots, q_n, a_1, \ldots, a_n), \tag{5}$$

containing n arbitrary constants $a_1, \ldots, a_n$, of the so-called Hamilton Jacobi partial differential equation

$$H\left(\frac{\partial V}{\partial q_1}, \ldots, \frac{\partial V}{\partial q_n}, q_1, \ldots, q_n\right) = h, \tag{6}$$

where h is the total energy of the system. When the characteristic function V has been determined the following $2n$ equations constitute a complete system of integrals to Hamilton's equations:

$$\frac{\partial V}{\partial a_i} = b_i, \qquad i = 1, 2, \ldots, n \tag{7}$$

$$\frac{\partial V}{\partial q_i} = p_i, \qquad i = 1, 2, \ldots, n, \tag{8}$$

where $b_i, \ldots, b_n$ are n new arbitrary constants.

Liouville's contributions

Liouville developed these ideas in at least three important directions:

1° He discovered [Liouville 1855] what is now known as Liouville's theorem (cf.[Arnold 1978 p.279-285]) to the effect that if n integrals to Hamilton's equations (3) are known and if they satisfy certain requirements (mutually vanishing Poisson brackets and independence) then the remaining n integrals can be determined by quadrature[3].

2° In 1856 he made an interesting extension of the Poisson and Lagrange brackets to systems of differential equations other than the canonical equations (3) and in this connection he began a theory of transformations, similar to Jacobi's canonical transformations. Liouville planned to dedicate this work to Poinsot but it remained unfinished in this notebooks [Ms3637(13)].

3° He made a new analysis of the principle of least action.

I have discussed the two first-mentioned achievements in my Liouville biography [Lützen 1989 Chapter XVI] so in this paper I shall concentrate on the last point. Like Liouville's theorem Liouville's analysis of the principle of least action was part of the content of a course he gave on differential equations and mechanics at the Collège de France during the spring of 1853. It was published three years later in the Comptes Rendus of the Paris Académie and in Liouville's Journal under the title: "Expressions remarquable de la quantité qui, dans le mouvement d'un système de points matériels à liaisons quelconques, est un minimum en vertu du principe de la moindre action" [Liouville 1856]. Despite its modest length (6 pages) this paper is among Liouville's most far reaching works, because it contains the first trace of a new geometrization of mechanics.

The principle of Least Action

This variational principle was formulated by Maupertuis and clarified by Euler and Lagrange (cf.[Szabo 1977, Dugas 1950]). For one free particle of mass m it states that the trajectory from one point B to another point C is the curve Γ_0 which minimizes the action integral

$$A = \int_\Gamma m\ v\ ds \tag{9}$$

among all the paths Γ beginning at B and ending at C. When Γ varies, the time it takes the particle to come from B to C may vary but the motion must satisfy the conservation of energy: $\frac{1}{2}mv^2+U=h$. In [1837b]

Jacobi pointed out that this last condition was somewhat confusing and that the implicit use of time (through v) was superfluous. Instead he suggested that one use the energy conservation to eliminate v from the action integral. For several particles he then obtained

$$A = \int_\Gamma \sqrt{2(h-U)\sum m_i(dx_i^2 + dy_i^2 + dz_i^2)} \tag{10}$$

which now has to be minimized, without further restrictions, among all paths Γ between the initial and final points B, C of the configuration space.

In [1856] Liouville adopted this expression but since he used generalized coordinates the action integral took the more general form:

$$A = \int_\Gamma \sqrt{2(h-U)\sum_{i,j} q_{ij}\ dq_i\ dq_j}. \tag{11}$$

Since $\sum_{i,j} q_{ij}\ dq_i\ dq_j$ is a positive definite quadratic form Liouville could transform the action integral into

$$A = \int_\Gamma \sqrt{2(h-U)\sum_j \left(\sum_i P_{ij} dq_i\right)^2} = \int_\Gamma \sqrt{2(h-U)\sum_j l_j^2} \tag{12}$$

where P_{ij} just as q_{ij} are functions of the q_i's and where l_j is just short hand for the form $\sum_i P_{ij}\, dq_i$. Liouville then defined what is now called the contravariant tensor P^{ij} by:

$$\sum_{i=1}^{n} P_{ij_1}\ P^{ij_2} = \delta_{j_1 j_2}. \tag{13}$$

Assuming that θ is a solution of the first order partial differential equation

$$\sum_{j=1}^{n}\left(\sum_{i=1}^{n} P^{ij}\frac{\partial\theta}{\partial q_i}\right)^2 = 2(h-U), \tag{14}$$

Liouville could finally write the action integral in the form

$$A = \int_\Gamma \sqrt{(d\theta)^2 + \sum_{j_1 > j_2} (n_{j_1} l_{j_2} - n_{j_2} l_{j_1})^2}, \tag{15}$$

where n_j is defined by $n_j = \sum_{i=1}^{n} P^{ij} \frac{\partial \theta}{\partial q_i}$.

This is the "remarcable expression" Liouville had announced in the title of his paper. As Liouville pointed out in his paper it is remarcable because it allows us to read off immediately that the action integral has a minimum if

$$\sum_{j>j_2} (n_{j_1} l_{j_2} - n_{j_2} l_{j_1})^2 \tag{16}$$

is zero which happens if each term is zero or if

$$\frac{l_1}{n_1} = \frac{l_2}{n_2} = \ldots = \frac{l_n}{n_n}. \tag{17}$$

This is a system of first order ordinary differential equations that may be used to determine $q_2, \ldots, q_n$ as functions of q_1. Here Liouville has used that $d\theta$ is an exact differential so that the integral $\int_\Gamma d\theta$ is independent of the path Γ.

It is not hard to prove that (14) is in fact the Hamilton-Jacobi equation (6) so that $\theta = V$, and that (17) can be used to determine half of the integrals of Hamilton's equations (3). Liouville did not publish the details of this demonstration but the following remark in his paper shows that he was well aware of this connection to the Hamilton-Jacobi formalism:

> On the other hand this function [θ], whose importance is known today to all geometers thanks to the works of Mr. Hamilton and Mr. Jacobi, plays in my method the greatest role [Liouville 1856].

Moreover Liouville pointed out that if θ is a complete solution of (14) containing n arbitrary constants $a_1, a_2, \ldots, a_n$ then

$$\frac{\partial \theta}{\partial a_i} = b_i \quad , \; i = 1, 2, \ldots, n \tag{18}$$

represent the n remaining integrals of the Hamilton-Jacobi equations. This corresponds to (7). Thus Liouville had derived the fundamental differential equations of mechanics from the variational principle of least action.

It was well known that such a deduction could be made; in fact Lagrange had obtained the Lagrange equations in this way. However the differential geometric nature of Liouville's argument was new. Why do I claim that Liouville's method was geometric? First of all because Liouville himself in the paper pointed to a geometric origin:

> The idea of introducing the function θ in order to express $2(U + K) \sum m ds^2$ as a sum of squares of which

> the first term is the square of an exact differential, was suggested to me when I read (in 1847) a handwritten memoir of Mr. Schläffli professor at the university of Berne, in which this geometer transformed the square of the element of length of a geodesic on an ellipsoid in the same way [Liouville 1856 p.1153].

Schläffli in turn had got the idea while studying the "beautiful works of Mr. Liouville and Mr. Chasles" [Schläffli 1847/52 p.391]. When Liouville read Schläffli's paper in 1847 he was actively applying the principles of mechanics to the study of geodesics, considering them as trajectories of free particles on the surfaces in question. (cf. e.g.[Liouville 1846, 1847 b,c]). This approach had already been suggested by Jacobi [1839]. Schläffli's paper then inspired Liouville to reverse the process, i.e. to apply geometric ideas in mechanics. In order to explain the nature of this geometric influence in more detail I shall emphasize a few trends in the preceding development of differential geometry—in particular the concept of intrinsic geometry.

INTRINSIC DIFFERENTIAL GEOMETRY

Hermann von Helmholtz (1821-1894) has nicely described intrinsic geometry of a surface as the geometry that a two dimensional surface-dweller can perceive [Helmholtz 1870/76]. One may also say that it is the study of the surface in itself without any reference to its imbedding in $\mathbb{R}^3$. Carl Friedrich Gauss (1777-1855), in whose "Disquisitiones generales circa superficies curvas" [1828] the program of intrinsic geometry was first announced, described it slightly differently:

> When a surface is regarded, not as the boundary of a solid, but as a flexible, though not extensible solid, one dimension of which is supposed to vanish, then the properties of the surface depend in part upon the form to which we can suppose it reduced, and in part are absolute and remain invariable whatever may be the form into which it is bent [Gauss 1828 §13].

Gauss stressed the importance of studying the latter type of properties, that we now call intrinsic. In particular he showed that the so-called Gaussian curvature of the surface is such an intrinsic property. Gauss defined this curvature at a point x_0 on a sufficiently smooth surface S as follows. Let Ω be a neighbourhood of x_0 on the surface S. Through each point of Ω draw the normal (vector) to S and displace each normal (vector) parallel to itself until it begins at a given point O.

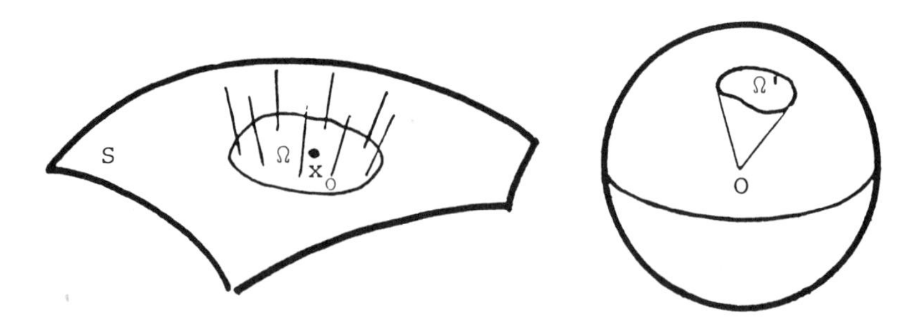

Figure 2.

All the normal (vectors) will then intersect the unit sphere around O in a point and the set of all intersection points make up a domain Ω' on the unit sphere. Gauss then defined the Gaussian curvature k in x_0 as

$$k = \lim \frac{\text{Area}\,\Omega'}{\text{Area}\,\Omega} \tag{19}$$

when Ω becomes smaller and smaller.

Gauss related this notion to Euler's earlier notion of sectional curvature. Euler had considered the curves of intersection of the surface S and planes containing the normal at x_0. When the plane rotates around the normal the intersection curve changes and so does its curvature κ at the point x_0.

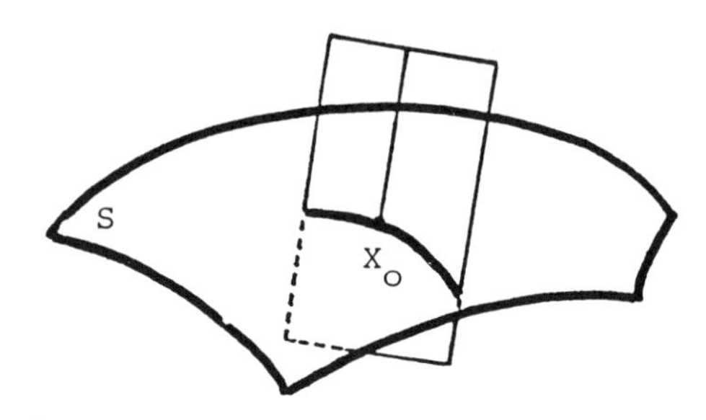

Figure 3.

Euler showed that the maximal curvature κ_1 and the minimal curvature κ_2 is obtained for two orthogonal planes, and Gauss showed that

$$k = \kappa_1 \cdot \kappa_2 \tag{20}$$

Both the definition (19) and the theorem (20) describe the Gauss curvature in terms that are highly dependent on the imbedding in $\mathbf{R}^3$ (normals cannot be perceived by surface dwellers). Yet Gauss in his "Very Important Theorem" (Theorema Egregium) stated that k is an intrinsic property or, as he put it, it is independent of bending of the surface.

Gauss' method of proof is important to us because it was to set the pattern for Liouville's later mechanical transformation. The idea was to find an intrinsic expression for k. Gauss did this by remarking that any point on the surface can be described by two intrinsic parameters (p, q) (instead of three Cartesian coordinates x, y, z). It is unclear whether Gauss saw the analogy with generalized coordinates in mechanics. In terms of these surface coordinates the line element ds^2 on the surface cannot be described by the simple Pythagorean expression

$$ds^2 = dx^2 + dy^2 + dz^2 \tag{21}$$

but by a more general quadratic form

$$ds^2 = Edp^2 + 2Fdpdq + Gdq^2, \tag{22}$$

where E, F and G are functions of p and q. Gauss' main result is then the following expression for the curvature of the surface:

$$\begin{aligned} k = \frac{1}{4(EG-F^2)^2}\bigg[& E\left(\frac{\partial E}{\partial q}\frac{\partial G}{\partial q} - 2\frac{\partial F}{\partial p}\frac{\partial G}{\partial q} + \left(\frac{\partial G}{\partial p}\right)^2\right) \\ & + F\left(\frac{\partial E}{\partial p}\frac{\partial G}{\partial q} - \frac{\partial E}{\partial q}\frac{\partial G}{\partial p} - 2\frac{\partial E}{\partial q}\frac{\partial F}{\partial q} + 4\frac{\partial F}{\partial p}\frac{\partial F}{\partial q} - 2\frac{\partial F}{\partial p}\frac{\partial G}{\partial p}\right) \\ & + G\left(\frac{\partial E}{\partial p}\frac{\partial G}{\partial p} - 2\frac{\partial E}{\partial p}\frac{\partial F}{\partial q} + \left(\frac{\partial E}{\partial q}\right)^2\right) \\ & + 2\left(EG-F^2\right)\left(\frac{\partial^2 E}{\partial q^2} - 2\frac{\partial^2 F}{\partial p \partial q} + \frac{\partial^2 G}{\partial p^2}\right)\bigg]. \end{aligned}$$

Of course the great importance of the result is not due to the precise form of this complicated expression but the very existence of such an expression which only involves E, F, G and their derivatives. For these quantities do not change when the surface is bent (without stretching) and so k has been shown to be an intrinsic quantity.

It is well known (cf.[Scholz 1980]) that Bernhard Riemann (1826-1866) in his Habilitationsvortrag [1854 publ.1867] generalized Gauss' ideas to what is now known as n-dimensional Riemannian manifolds. Riemann described such objects as "spaces" where displacements can be made in

n independent directions q_i and where the infinitely small element of distance is given by a positive quadratic form:

$$ds^2 = \sum_{i,j=1}^{n} E_{ij}\, dq_i dq_j \tag{23}$$

In fact Riemann and later Henri Poincaré (1854-1912) contemplated if physical space might be a Riemannian manifold which is not equivalent to Euclidean space $\mathbf{R}^3$ (with $ds^2 = dx^2 + dy^2 + dz^2$).

However in 1853 when Liouville made his contribution to the principle of least action Riemann's ideas were not around, and that leads us to the question: how well was the idea of intrinsic geometry understood around 1853? In Germany, Gauss' ideas had been taken up already around 1830 by Ferdinand Minding (1806-1885) and later by other mathematicians (cf.[Reich 1973]), but in France Gauss' Disquisitiones was virtually unknown until Liouville in 1847 published a new proof of the Theorema Egregium [Liouville 1847 a].

In 1850 Liouville extended Gauss' ideas further in six notes appended to the fifth edition of Monge's "Application de l'analyse à la Géométrie". In particular he proved that any surface can be equipped with coordinates α, β such that:

$$ds^2 = \lambda(\alpha, \beta)(d\alpha^2 + d\beta^2) \tag{24}$$

that is $F = 0$ and $E = G = \lambda$ in (22).

These coordinates, whose coordinate net divides the surface into little squares, are now called isothermal coordinates. They play a central role in many of Liouville's differential geometric results, such as the new proof of the Theorema Egregium.

In my opinion Liouville was among the mathematicians who best understood the idea of intrinsic geometry around 1850, only rivaled perhaps by a few Italians such as Brioschi and Chelini. A case in point is Liouville's definition of the geodesic curvature of a curve drawn on a surface (see figure 4).

Already Minding [1830] and Ossian Bonnet (1819-1892)[1848] had considered the quantity $\frac{cos\theta}{R}$ where R is the radius of curvature at a given point of the curve when it is considered as a curve in space and θ is the angle between its osculating plane and the tangent plane of the surface at the given point. They then showed with Gaussian methods that this quantity is in fact intrinsic.

Liouville [1850] on the other hand gave an intrinsic *definition* of this quantity. He drew two geodesics on the surface touching the curve in two points separated by an infinitely small arc length ds (see figure 5).

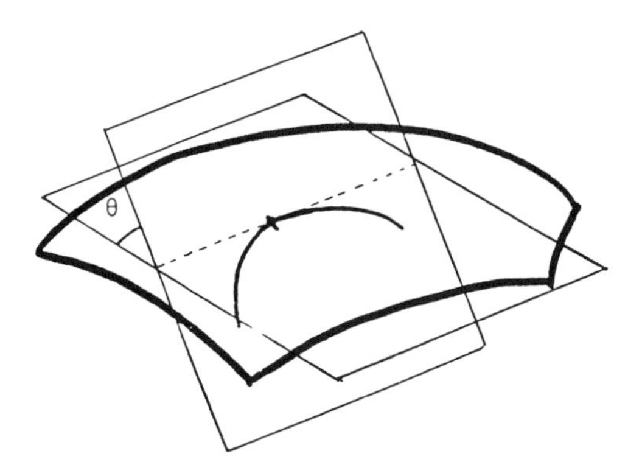

Figure 4.

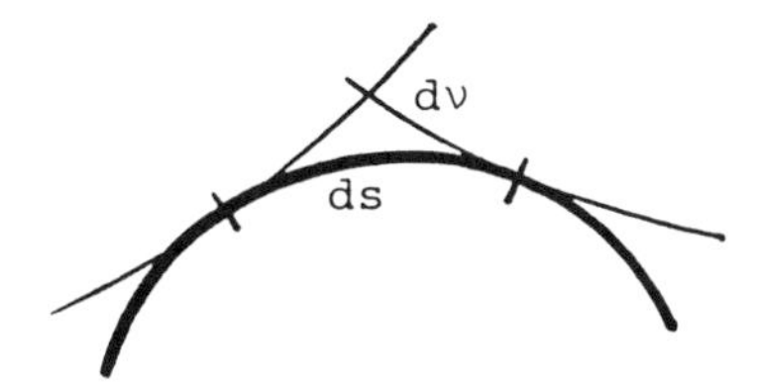

Figure 5.

If $d\nu$ denote the infinitely small angle the two geodesics make with each other, Liouville defined the "geodesic curvature" (the name is due to him) by:

$$K = \frac{d\nu}{ds} \tag{25}$$

This is in fact how a surface dweller would naturally define curvature of a curve.

By insisting that one should introduce this object with a definition that is a priori intrinsic, rather than proving a posteriori that an extrinsic definition in fact defines an intrinsic property, Liouville appears as a true protagonist of the intrinsic approach.

This impression is reinforced by the unpublished lecture notes from a course he gave in 1851, at the Collège de France (cf.Figure 6). Here Liouville emphasized:

> I have been able to arrive at these results and in particular at the formula
>
> $$ds^2 = Edu^2 + 2Fdudv + Gdv^2 \qquad (22)$$
>
> from which they result without relying on the previous formula
>
> $$ds^2 = dx^2 + dy^2 + dz^2$$
>
> and without any use of coordinates outside the surface [Liouville Ms 3640(1846-51)p.5].

Further in his brief published summary of the most important results obtained in these lectures Liouville characterized his approach as follows:

> I have treated this differential form [(22)] in itself and in all its generality, and I have tried to study it as in the theory of numbers one studies the quadratic formula
>
> $$ax^2 + 2bxy + cy^2$$
>
> to be sure with very different methods [Liouville 1852 p.478].

In the lectures Liouville also discussed three dimensional forms. For example he derived a necessary condition that a form $\sum_{i,j=1}^{3} E_{ij} dq_i dq_j$ is equivalent to the ordinary distance $ds^2 = dx^2 + dy^2 + dz^2$ in (Euclidean) space. Unfortunately Liouville did not mention whether he attached any geometric significance to the forms which are not equivalent to the Euclidean metric, so we do not know in how far he anticipated some of Riemann's more far-reaching ideas. However the fact that Liouville all his life remained uninterested in or even sceptic towards any discussion of Euclid fifth postulate and non-Euclidean geometry indicates that he did not conceive of the idea of curved higher dimensional manifolds.

Leçons au Collège de France
(1851. – 1er Semestre).

1.

1. Nous prenons pour ~~texte~~ de ces leçons l'étude des formules
différentielles entières et homogènes à plusieurs variables, comme $Mdx + Ndy$,
$Ldx^2 + 2Mdxdy + Ndy^2$, &c, qui se présentent souvent en analyse pure,
en Géométrie, en Mécanique. Nous nous occuperons surtout de la
formule quadratique à deux variables, au moyen delaquelle s'exprime le
carré de l'élément d'une ligne quelconque tracée sur une surface,
et nous dirons aussi en passant un mot des formules du même genre
à trois variables dont dépend le carré de la distance de deux points
quelconques dans l'espace. Dans une étude tout à fait régulière
et ordonnée, on devrait s'occuper d'abord des formes linéaires, puis
passer de là aux formes quadratiques, puis aux formes entières ou
fractionnaires de degré quelconque. Mais le peu de leçons que nous avons
à faire dans ce semestre nous oblige à passer de suite aux formes
quadratiques qui offrent de très nombreuses et très intéressantes applications.
Nous rappellerons seulement, relativement aux formes linéaires, la
condition connue d'intégrabilité de $Mdx + Ndy$; c'est comme on sait
l'identité $\frac{dM}{dy} = \frac{dN}{dx}$: nous aurons dans cette première leçon même à en faire
usage.

2. En désignant par $\underline{x}, \underline{y}, \underline{z}$ les coordonnées rectangulaires d'un
point quelconque d'une surface courbe, et par $\underline{ds}$ l'élément d'une
ligne quelconque tracée sur cette surface, on a

$$ds^2 = dx^2 + dy^2 + dz^2;$$

si donc on désigne par $\underline{p}$ et $\underline{q}$ les deux dérivées de la valeur de $\underline{z}$
en x, y tirée de l'équation de la surface, en sorte que

$$dz = pdx + qdy,$$

DIFFERENTIAL GEOMETRY AND LEAST ACTION

In the fourth lecture of his 1851 course Liouville treated the principle of least action. I shall translate the relevant passage from his notes (the original is shown in Figure 7) because it gives the clue to understanding his application of differential geometry to mechanics:

> *Principle of least action* $\int \sqrt{(U+C)ds^2}$; if $ds^2 = \lambda(d\alpha^2 + d\beta^2)$ this integral= $\int \sqrt{\lambda(U+C)(d\alpha^2+d\beta^2)}$, thus a reduction to the plane with force function $\lambda(U+C)$ and living force $2\lambda(U+C)$. One may reduce to another arbitrary surface. Geodesic, or shortest lines, 1° described on their surface by a point without force; 2° in the plane with λ as its force function and 2λ as its living force. One has the same relations in α and β.[Liouville Ms3640(1846-51 p.8)].

This needs some explanation. Liouville considers one particle on a surface equipped with isothermal coordinates $ds^2 = \lambda(d\alpha^2 + d\beta^2)$. His potential function has the opposite sign of our potential energy (so that $\vec{F} = \vec{\nabla} U$). By including the total energy C in the force function Liouville is allowed to let $U + C$ represent the force function as well as the kinetic energy. The living force, introduced in the quote is nothing but twice the kinetic energy. In this terminology the action integral (11) takes the form

$$A = \int (U+C)\lambda(d\alpha^2 + d\beta^2) \tag{26}$$

Now Liouville observes that one can interpret the minimization of A in two ways:

1° If we think of λ as belonging to the line element, the principle of least action states that the minimization yields the path of a point on the surface subject to the force function $U + C$.

2° If we think of λ as belonging to the force function we similarly find the trajectory of a point in the Euclidean plane ($ds^2 = d\alpha^2 + d\beta^2$) subject to the force function $(U + C)\lambda$.

In particular Liouville points out that if U is a constant we get either

1. The trajectory of a point on the surface subject to no forces, i.e. a geodesic, or

2. the trajectory of a point in the plane subject to the force function λ.

IV.

Surfaces du second degré. — Coordonnées elliptiques: $\frac{x^2}{\rho^2}+\frac{y^2}{\rho^2-b^2}+\frac{z^2}{\rho^2-c^2}=1$, trois racines ρ^2, μ^2, ν^2. Posez $\rho^2=a^2+t$, faites le produit des racines t. De là

$$x=\frac{\rho\mu\nu}{bc},\quad y=\frac{\sqrt{\rho^2-b^2}\sqrt{\mu^2-b^2}\sqrt{b^2-\nu^2}}{b\sqrt{c^2-b^2}},\quad z=\frac{\sqrt{\rho^2-c^2}\sqrt{c^2-\mu^2}\sqrt{c^2-\nu^2}}{c\sqrt{c^2-b^2}}.$$

Si l'on veut $\nu=b\cos\psi$, $\mu=\sqrt{b^2\cos^2\varphi+c^2\sin^2\varphi}$.

$$\left(\frac{dx}{d\mu}\right)^2+\left(\frac{dy}{d\mu}\right)^2+\left(\frac{dz}{d\mu}\right)^2=\frac{\rho^2\nu^2}{b^2c^2}+\frac{\mu^2(\rho^2-b^2)(b^2-\nu^2)}{b^2(c^2-b^2)(\mu^2-b^2)}+\frac{\mu^2(\rho^2-c^2)(c^2-\nu^2)}{c^2(c^2-b^2)(c^2-\mu^2)}.$$

Cette quantité est nulle pour $\mu^2=\rho^2$ ou ν^2. Il est aisé d'en conclure qu'elle est égale à $\frac{(\rho^2-\mu^2)(\mu^2-\nu^2)}{(\mu^2-b^2)(c^2-\mu^2)}$ et que

$$ds^2=(\mu^2-\nu^2)\left[\frac{(\rho^2-\mu^2)\,d\mu^2}{(\mu^2-b^2)(c^2-\mu^2)}+\frac{(\rho^2-\nu^2)\,d\nu^2}{(b^2-\nu^2)(c^2-\nu^2)}\right]=\lambda(d\alpha^2+d\beta^2)$$

avec

$$d\alpha=\frac{d\mu\sqrt{\rho^2-\mu^2}}{\sqrt{(\mu^2-b^2)(c^2-\mu^2)}},\quad d\beta=\frac{d\nu\sqrt{\rho^2-\nu^2}}{\sqrt{(b^2-\nu^2)(c^2-\nu^2)}},\quad \lambda=\mu^2-\nu^2=f(\alpha)-F(\beta).$$

<u>Principe de la moindre action</u>: $\int\sqrt{(U+C)\,ds^2}$; si $ds^2=\lambda(d\alpha^2+d\beta^2)$, cette intégrale $=\int\sqrt{\lambda(U+C)(d\alpha^2+d\beta^2)}$, d'où une réduction au plan avec $\lambda(U+C)$ fonction des forces et $2\lambda(U+C)$ force vive. — On pourrait ramener à une autre surface quelconque. Lignes géodésiques ou les plus courtes, 1° décrites sur leur surface par un point sans force; 2° sur le plan avec λ pour fonction des forces et 2λ pour force vive, on a les mêmes relations en α, β. — Cas de $\lambda=f(\alpha)-F(\beta)$, $\frac{d^2\alpha}{dt^2}=f'(\alpha)$, $\frac{d^2\beta}{dt^2}=-F'(\beta)$,

$$\left(\frac{d\alpha}{dt}\right)^2+\left(\frac{d\beta}{dt}\right)^2=2(f(\alpha)-F(\beta)),\quad \left(\frac{d\alpha}{dt}\right)^2=2(f(\alpha)-a),\quad \left(\frac{d\beta}{dt}\right)^2=2(a-F(\beta))$$

$$\left(\frac{d\alpha}{d\beta}\right)^2=\frac{f(\alpha)-a}{a-F(\beta)}=\tan^2 i,\quad f(\alpha)\cos^2 i+F(\beta)\sin^2 i=a.$$

$$\sin i=\frac{\sqrt{f(\alpha)-a}}{\sqrt{\lambda}},\quad \cos i=\frac{\sqrt{a-F(\beta)}}{\sqrt{\lambda}},\quad ds=ds_1\sin i+ds_2\cos i$$

$$ds=d\alpha\sqrt{f(\alpha)-a}+d\beta\sqrt{a-F(\beta)},\quad \delta\int ds=0 \text{ le long de la ligne.}$$

— Surfaces de révolution. — Plan, sphère, ellipsoïde. — Intégrale d'Euler pour les fonctions elliptiques. — Lignes de

Thus Liouville discovered that he could transform the problem of motion of a particle by simultaneously changing the force function and the surface – in particular the potential may become a constant so that the particle describes a geodesic. In fact, it is just one small step to observe that minimizing (26) also gives the geodesics on a surface with

$$ds^2 = (U + C)\lambda(d\alpha^2 + d\beta^2) \tag{27}$$

Of course this last step raises the question of the existence of such a surface, and that may be why Liouville did not mention this last interpretation.

Yet I believe that something like this was in his mind when he conceived of the "remarcable expression" of the action integral we discussed above. A possible reconstruction of his process of discovery runs as follows:

1° First Liouville combined Schläffli's technical discussion of geodesics on ellipsoids with the reduction discussed above. This gave him a "remarcable expression" for the action integral of a point on a surface.

2° Secondly he generalized this procedure to systems with an arbitrary number of degrees of freedom.

Unfortunately I have not found any further notes from Liouville's hand to clarify these steps.

To conclude; Liouville discovered the connection between the two problems:

Find a geodesic or

$$\text{minimize} \int_B^C ds = \int_B^C \sqrt{ds^2} \tag{28}$$

and find a trajectory or

$$\text{minimize} \int_B^C \sqrt{2(h - U)\sum_{ij} g_{ij}\, dq_i dq_j} \tag{29}$$

He stopped short of formulating the general rule: the trajectory mentioned above is simply a geodesic in a Riemannian manifold with the metric:

$$ds^2 = 2(h - U)\sum_{ij} g_{ij}\, dq_i dq_j \tag{30}$$

So although Liouville did not conceptualize the idea of a higher dimensional Riemannian manifold his familiarity with the intrinsic geometry on surfaces showed him how he could operate geometrically within such a manifold and how he could use geometric methods to yield important results in mechanics.

Liouville's successors made these ideas more precise.

LIOUVILLE'S SUCCESSORS

To my knowledge Rudolf Lipschitz (1832-1903) was the first who used Riemann's ideas to interpret the action integral in a way similar to Liouville. However he does not seem to have been aware of Liouville's work when he wrote his paper [Lipschitz 1871]. In their famous "Treatise on natural philosophy" [1879], however, Thomson and Tait explicitly referred to Liouville, but they only treated a rather uninteresting special case. So the first who made full use of Liouville's transformation was Gaston Darboux (1842-1917). In his "Leçons sur la theorie générale des surfaces" [1888] he expressed the aim of his geometrization of mechanics very clearly:

> In particular I have emphasized the connections which present themselves here between the methods used by Gauss in the study of geodesics and those that Jacobi have later applied to the problems of analytical mechanics. In this way I have been able to show the great interest of Jacobi's beautiful discoveries when these are considered from a geometrical point of view [Darboux 1888,préface].

It is interesting to note that at the same time as the mathematicians in this way got rid of forces by changing the geometry, physicists were also building up theories that avoided forces acting at a distance. This move was felt most forcefully in electro-magnetic theory where Faraday's and Maxwell's field theories gradually superseded Ampère's ideas as developed by Wilhelm Weber. Inspired by this new trend and in particular by his teacher Helmholtz, Heinrich Hertz (1857-1894) conceived of a new structure of mechanics which did not use forces acting at a distance as a basic à priori notion. In the preface to his book "Die Prinzipien der Mechanik" [1894] he acknowledged the similarity between his force-less mechanics and that developed by Lipschitz and Darboux but he claimed that he had not been aware of the work of the mathematicians until his own investigations were far advanced.

The geometrization of the equations of mechanics continued to play the role of an important application of the new absolute differential calculus (tensor calculus) that was developed around 1900 by Gregorio Ricci-Curbastro (1853-1925) and Tullio Levi-Civita (1873-1941). Inspired by Lipschitz these two authors discussed this reformulation of mechanics in their joint paper [1901] from which Einstein learned his tensor calculus (cf.[Dugas 1950 book 5 chapter 2]). Therefore it was by no means a complete novelty nor indeed a miracle (as it is sometimes stated) when Einstein was able to use tensor analysis to reduce gravitation to a property of the geometry of the space. Of course the way in which he did it was completely new: the relativistic invariance and the space-time manifold did not have any parallels in the earlier works on tensor calculus, nor did the physical reality he attached to his theory.

Concluding Remarks

The aim of this paper has been to show that Einstein's idea of explaining forces as a property of space had roots in the classical mechanics of the nineteenth century. In particular I have tried to call attention to Liouville's pioneering role. His deep insight into the intrinsic geometry of surfaces as suggested by Gauss and his contributions to Hamilton-Jacobi mechanics allowed him to make the first application of differential geometry to analytical mechanics.

Notes

1. In fact Hamilton only used Cartesian coordinates and so it was a non-trivial exercise for his successors including Liouville to transform his ideas to generalized coordinates.

2. This definition and the last of Hamilton's equations (3) had already been introduced by Poisson [1809].

3. Liouville's theorem on the volume in phase-space grew out of Liouville's earlier works in celestial mechanics; in [Liouville 1838], however, the theorem is stated as a purely mathematical theorem concerning certain determinants of solutions to differential equations (cf.[Lützen 1989]chapter XVI).

References

Arnold, V.I. [1978], *Mathematical Methods of Classical Mechanics*, New York 1978. Transl. from Russian: Mathematicheskie metody klassicheskoi mekkaniki. Nauka Moscow 1974.

Bonnet, O. [1848], *Mémoire sur la théorie générale des surfaces*, Journ. Ec. Poly. 19 (32 cahier)(1848), 1-146.

Darboux, G. [1888], *Leçons sur la théorie générale des surfaces*, 2.ème partie, Paris 1888 (2.ed.1915).

Dugas, R. [1950], *Histoire de la mécanique,* Neuchatel 1950.

Gauss, C.F. [1828], *Disquisitiones generales circa superficies curvas,* Comment.soc.reg.sci.Gottengensis recentiores 6(1828); math Classe, 99-146, Werke 4, 217-258 in [Monge 1850] 505-546.

Hamilton, W.R. [1834,1835], *On a general method in dynamics,,* Phil. Trans. (1834) 247-308; and (1835) 95-144; Math.Papers 103-161, 162-211.

Helmholtz, H. von [1870/76], *Über den Ursprung und die Bedeutung der Geometrischen Axiome. Vortrag gehalten im Jahre 1870,* Populare wissenschaftliche Vorträge vol.3 Braunschweig 1876; Vorträge und Reden vol.2 Braunschweig 1884.

Hertz, H. [1894], *Die Prinzipien der Mechanik in neuem Zusammenhange dargestellt,* Leipzig 1894.

Jacobi, C.G.J. [1837 a], *Über die Reduction der Ingegration der partiellen Differential gleichungen erster Ordnung zwischen irgend einer Zahl variabeln auf die Integration eines einzigen Systems gewöhnlicher Differentialgleichungen,* Journ. Reine, Angew.Math. 17(1837) 97-162; Werke IV, 57-127; French transl., Journ. Math.Pures et Appl. 3(1838), 60-96 and 161-201.

Jacobi, C.G.J. [1837 b], *Note sur l'intégration des équations différentielles de la dynamique,* Comp. Rend. Acad.Sci. Paris 5(1837), 61-67; Werke IV, 129-136.

Jacobi, C.G.J. [1839], *Note von der geodätischen Linie auf einem Ellipsoid und den verschiedenen Anwendungen einer merkwürdigen analytischen Substitution,* Journ. Reine, Angew. Math. 19 (1839) 309-313; Werke II 59-63; French transl. Journ. Math. Pures et Appl. 6 (1841) 267-272.

Lagrange, J.L. [1774], *Sur l'équation séculaire de la lune,* Mém.Acad.Sci. Paris. Savants étrangers 7 (1773); Oeuvres 6, 331-399.

Lagrange, J.L. [1788/1811], *Mécanique Analytique,* Paris 1788 2.ed. Paris 1811-1815.

Liouville, J. [1838], *Note sur la théorie de la variation des constantes arbitraires,* Journ. Math. Pures et Appl. 3(1838), 342-349.

Liouville, J. [1846], *Sur quelques cas particuliers où les équations du mouvement d'un point matériel peuvent s'intégrer. Premier mémoire,* Journ. Math. Pures et Appl. 11(1846), 345-378.

Liouville, J. [1847 a], *Sur un théorème de M. Gauss concernant le produit des deux rayons de courbure principaux en chaque point d'une surface,* Journ. Math. Pures et Appl. 12(1847) 291-304; Comp. Rend. Acad.Sci. Paris 25(1847) 707.

Liouville, J. [1847 b], *Sur quelques cas particuliers ou les équations du mouvement d'un point matériel peuvent s'intégrer. Second mémoire,* Journ. Math. Pures et Appl. 12(1847) 410-444.

Liouville, J. [1847 c], *Mémoire sur l'intégration des équations différentielles du mouvement d'un nombre quelconque de points matériels,* Journ. Connaissance des Temps. pour 1850 (1847) 1-40; Math. Pures et Appl. 14(1849) 257-299.

Liouville, J. [1850], *Sur les courbes à double courbure,* Note 1 in [Monge 1850].

Liouville, J. [1852], *Note sur la théorie des formules différentielles,* Journ. Math. Pures et Appl. 17 (1852), 478-480.

Liouville, J. [1855], *Note sur l'intégration des équations différentielles de la Dynamique, présentée au Bureau des Longitudes le 29 juin 1853,* Journ. Math. Pures et Appl. 20(1855), 137-138.

Liouville, J. [1856], *Expression remarquable de la quantité qui, dans le mouvement d'un système de points matériels à liaisons quelconques, est un minimum en vertu du principe de la moindre action,* Journ. Math. Pures et Appl. (2) 1(1856)297-304; Comp. Rend.Acad.Sci. Paris 42(1856) 1146-1154.

Liouville, J. Ms 3615-3640, *The Liouville Nachlass in the Biobliothéque de l'Institut de France consisting of 340 notebooks and a box of loose sheets.*

Lipschitz, R. [1871], *Untersuchung eines Problems der Variationsrechnung in welchen das Problem der Mechanik enthalten ist,* Reine. Angew. Math. 74(1871), 116-149.

Lützen, J. [1989], *Joseph Liouville 1809-1882. Master of Pure and Applied Mathematics,* to appear in the series, Studies in the History of Mathematics and Physical Sciences, Springer Verlag.

Minding, F. [1830], *Bemerkung über die Abwicklung krummer Linien von Flächen,* Journ. Reine, Angew. Math 6(1830), 159-161.

Newton, I. [1687], *Philosophiae naturalis principia mathematica,* London 1687; English ed. F.Cajori 3.ed California 1946.

Poisson, S.D. [1809], *Mémoire sur la variation des constantes arbitraires dans les questions de la mécanique,* Journ. Ec. Polyt. 8(1809), 266-344.

Reich, K. [1973], *Die Geschichte der Differentialgeometrie von Gauss bis Riemann (1828-1868),* Arch. Hist Ex.Sci. 11(1973), 273-382.

Ricci, G and Levi-Civita, T. [1901], *Méthodes du calcul diffirentiel absolu et leurs applications,* Math.Ann. 54(1901), 125-201.

Riemann, B. [1854/67], *Über die Hypothesen welche der Geometrie zu Grunde liegen,* Habilitationsvortrag 1854; Abh. Königl. Gess. der Wiss. Göttingen Math. Cl. 13(1867), 1-20; Werke 271-287.

Schläffli, L. [1847/52], *Über das Minimum des Integrals $\int \sqrt{dx_1^2 + dx_2^2 + \cdots + dx_n^2}$ wenn die Variablen $x_1, x_2, \ldots, x_n$ durch eine Gleichung zweiten Grades gegenseitig von einander abhängig sind,* Journ. Reine. Angew. Math 32(1852) 23-36; Comp. Rend.Acad.Sci. Paris 25(1847) p391.

Scholz, E. [1980], *Geschichte des Manigfaltigkeitsbegriffs von Riemann bis Poincaré*, Boston 1980.
Szabo, J. [1977], *Geschichte der mechanischen Prinzipien*, Basel 1977.
Thomson, W and Tait, P.G. [1879], *Treatise on natural philosophy*, 2 ed vol.1 Cambridge 1879.

Jan Arnoldus Schouten (1883–1971)

Tullio Levi-Civita (1873–1941)

Dirk Jan Struik (1894–) during his student days

Schouten, Levi-Civita, And the Emergence of Tensor Calculus

Dirk J. Struik

1

The flowering of the tensor calculus and its application to differential geometry, mechanics, and physics was primarily due to the impact of Einstein's general theory of relativity. Outlined in 1913, the "authorized version" of this theory appeared in a now classical paper in the *Annalen der Physik* of 1916 entitled "Die Grundlagen der allgemeinen Relativitatstheorie." It was Einstein's Zürich colleague Marcel Grossmann who had pointed out to him that the mathematical tools for his theory could be found in a paper entitled *Méthodes de calcul differentiel absolu,* published in the *Mathematischen Annalen* of 1901, the authors being the Paduan mathematicians Gregorio Ricci-Cubastro and Tullio Levi-Civita. Einstein took from this paper the mathematical inspiration he needed, baptizing this new tool the "tensor calculus." Like Kepler, who found the mathematics required for his planetary theory in Apollonius's "Konika," Einstein had come across a treasure trove that proved to be invaluable for his general theory of relativity. With it he placed his physics of space-time and gravitation into the framework of a Riemannian manifold of four dimensions, V_4, that is, a topological manifold with a system of coordinates x^k and a metrical relation expressed by a quadratic differential form

$$ds^2 = g_{\lambda\mu} dx^\lambda dx^\mu, \ \ \lambda, \mu = 1, 2, \ldots, n,$$

where the $g_{\lambda\mu}$ are functions of x^k (in Einstein's theory n = 4).[1] Such manifolds of n dimensions were introduced by Riemann in his *Habilitationsvortrag* of 1854 and had been further developed by E. B. Christoffel and others.

Let us call a topological manifold with a system of coordinates x^k an X_n. Coordinate transformations can be indicated by $x^{k'} = x^{k'}(x^k)$, and since this leads to linear transformations for the differentials

$$dx^{k'} = (\partial_k x^{k'}) dx^{k'} \quad \left(\partial_k x^{k'} = \frac{\partial x^{k'}}{\partial x^k} = P_k^{k'}\right)$$

ISBN 0-12-599662-4.

the X_n defines in each neighborhood around a point P an affine space A_n. By introducing a metric ds^2 into the X_n, this X_n becomes a Riemannian manifold V_n and the local A_n becomes a Euclidean R_n when ds^2 can be reduced to the sum of n squares of differentials.

In such an R_n one can, in the usual way, define vectors and their generalizations as linear vector functions, bivectors, tensors, etc. Passing to the V_n one can then define such entities at all points, hence as a vector or tensor field. Now by overlooking for a moment the ds^2 and considering the affine A_n at every point, one can discriminate between covariant and contravariant entities. The Ricci tensor calculus expresses this as follows for a *contravariant vector field* v^k:

$$v^{k'} = v^k P_k^{k'} \; (P_k^{k'} = \partial_k x^{k'} = \frac{\partial x^{k'}}{\partial x^k})$$

and for a *covariant vector field*:

$$w_{\lambda'} = w_\lambda \Phi_{\lambda'}^{\lambda} \; (\Phi_{\lambda'}^{\lambda} = \partial_{\lambda'} x^\lambda = \frac{\partial x^{\lambda'}}{\partial x^\lambda})$$

For more general mixed tensors like

$$P_{\mu'}^{\lambda'} = P_\mu^\lambda P_\lambda^{\lambda'} \Phi_{\mu'}^{\mu}$$

Ricci spoke of *systems* of order $1, 2, \ldots, n$. We shall call these monovalent, bivalent, trivalent systems, etc.

After the introduction of a ds^2 the X_n becomes a V_n, and covariance and contravariance will now be related. They can then be identified, as in ordinary vector analysis, where the terms covariant and contravariant need not appear:

$$v_\mu = v^\lambda g_{\lambda\mu}, \; v^\lambda = g^{\lambda\mu} v_\mu, \; g_{\lambda\mu} g^{\mu\nu} \equiv A_\nu^\lambda,$$

where $g^{\lambda\mu}$ is the inverse of $g_{\lambda\mu}$ and A_ν^λ, the unit tensor ($A_1^1 = 1$, $A_2^1 = 0$, etc.).

Tensors, like vectors in R_n, are independent of the particular coordinate system for which they are originally defined, and thus determine geometrical entities: vectors as arrows, bivalent symmetric tensors as quadrics, bivalent alternating tensors $p_{\lambda\mu} = -p_{\mu\lambda}$ as linear complexes, etc. They can also represent entities in mechanics and physics, which is what makes tensors so efficacious in Einstein's theory.

This geometrical interpretation appealed strongly to persons with an engineering mind like Schouten; it also was quite attractive to mathematicians of the Monge-Darboux school, accustomed as they were to thinking of analytical expressions (i.e. differential equations) in geometrical form. To this school can be counted Elie Cartan, who contributed so much to differential geometry and tensor calculus.

A difficulty arose with regard to the differentiation of tensor fields. This poses no problem in ordinary calculus, since in a vector field $\overline{v}(x_i)$ in R_n: one can pass from a vector $\overline{v}$ to a "neighboring" vector $\overline{v}'$ by moving *parallel* to itself and measuring $\overline{v}' - \overline{v} = d\overline{v}$. From here one obtains $\nabla\overline{v}$, ∇ being the Hamiltonian "nabla" symbol for differentiation. In a Riemannian V_n, on the other hand, neither $\partial_k v^k$ nor $dv^k = (\partial_k v^k) dx^\mu$ are tensors (unless the P and Φ are constant). Ricci overcame this difficulty by utilizing an expression due to Christoffel, namely:

$$\delta v^k = dv^k + \Gamma^k_{\mu\lambda} v^\lambda dx^\mu$$

which he took to be the *covariant differential*. Here the $\Gamma^k_{\mu\lambda}$ are the *Christoffel symbols* depending on certain first derivatives of the $g_{\lambda\mu}$. This δv^k has tensor properties as does the *covariant derivative*

$$\nabla_\mu v^k = \partial_\mu v^k + \Gamma^k_{\mu\lambda} v^\lambda.$$

Similarly,

$$\delta w_\lambda = dw_\lambda - \Gamma^k_{\mu\lambda} w_k dx^\mu, \quad \nabla_\mu w_\lambda = \partial_\mu w_\lambda - \Gamma^k_{\mu\lambda} w_k.$$

The ∇_μ behaves like a covariant vector. Other authors denote $\nabla_\mu v^k$ by $v^k_{,\mu}$ etc. Ricci himself omitted the comma and let the context explain that differentiation had taken place. The Γ are symmetric in μ and λ, hence $\Gamma_{\mu\lambda} = \Gamma_{\lambda\mu}$. From here it is now possible to derive all kinds of tensors by covariant differentiation, e.g.

$$\nabla_\mu h^k_\lambda = \partial_\mu h^k_\lambda + \Gamma^k_{\mu\sigma} h^\sigma_\lambda - \Gamma^\sigma_{\lambda\mu} h^k_\sigma.$$

Gauss, whose paper of 1826 on the differential geometry of surfaces decisively influenced Riemann, had stressed the *intrinsic* properties of surfaces V_2, i.e. such properties as the metric ds^2 and geodesic lines that are invariant under bending without stretching or tearing; also he had shown that the *Gaussian curvature* K was a bending invariant at each point of a surface. For a V_n, $n > 2$, this K is replaced by the *Riemann-Christoffel tensor*, $K_{\mu\lambda\nu k}$, which depends on the first and second derivatives of the $G_{\lambda\mu}$ (or equivalently on the Γ and their first derivatives). For this tensor the following equations hold:

$$K_{\mu\lambda\nu k} = -K_{\lambda\mu\nu k} = -K_{\mu\lambda k\nu} = K_{\nu k\mu\lambda}$$
$$K_{\mu\lambda\nu k} + K_{\lambda\nu\mu k} + K_{\nu\mu\lambda k} = 0.$$

This gives $\frac{1}{12}n^2(n^2-1)$ components, hence 1 for $n = 2$ (proportional to K), 6 for $n = 3$, and 20 for $n = 4$, the case of general relativity. In Einstein's theory the contracted curvature tensor

$$K_{\mu\lambda} = K_{\lambda\mu} = K_{k\lambda\mu\nu}g^{\mu\lambda}$$

played a crucial role. It has 10 components for $n = 4$. For general n, Riccci and Levi-Civita had already used the tensor calculus, as outlined above, in 1901 for applications to physics, mechanics and differential geometry.

2

In 1917 a new discovery widened the range of interest and applicability of the tensor calculus considerably. This was the introduction by Levi-Civita of the notion of *parallelism* now named after him. Its significance was that it gave a geometrical interpretation to the covariant differential.

Geometrical interpretations of analytical expressions have more than once proven extremely productive in the history of mathematics. We have only to think of Descartes' link between "the geometry of the Ancients" and "the modern algebra" (hence coordinate geometry), or Minkowski's four-dimensional interpretation of space-time in special relativity—or even his *Geometrie der Zahlen*.

It is at this stage that I would like to mention some of my own experiences. In 1917 I became the assistant of J. A. Schouten at the Technical University in Delft, and soon thereafter became his friend and co-worker. Schouten was an electrical engineer who had switched to mathematics. Like so many engineers and physicists in the early 1900's, he had fallen in love with vector analysis and its generalizations; not, however, in the tensor form of Ricci, but rather in the form of geometrical entities: linear vector functions, dyadics, rotors, affinors (which are also tensors, but in the physicist's sense of W. Voigt, who had introduced the term in its modern sense). This was the tradition of *direct* or *geometrical analysis*, as first introduced by Gibbs and Heaviside. The number of such generalizations with all kinds of notations had since greatly multiplied, and Schouten, in his dissertation of 1914, had applied the principles of Felix Klein's Erlangen Program to the classification of these different systems as invariants of certain transformation groups. Ordinary vector analysis, for instance, belonged to the invariant theory of the rotational group in three-dimensional Euclidean space R_3, Grassmann's *Ausdehnungslehre* to that of the affine (and then to the orthoginal) group in n-dimensional space A_n (R_n). In 1917 Schouten was busy extending these principles to Einstein's theory of general relativity, which meant constructing a direct analysis for a Riemannian V_n. He called the entities he used *affinors*: vectors are affinors of order 1, whereas symmetrical affinors of order 2 were called tensors (after Einstein adopted the term tensor, Schouten dropped the condition of symmetry).

Schouten now met with the same difficulty that Ricci had encountered, namely, how to differentiate on a "curved" manifold. In a Euclidean or affine space this is easy, since parallelism exists allowing one (as in ordinary vector analysis) to move a vector $\overline{v}$ at a point P parallel to itself to compare it with a vector $\overline{v}'$ at a "neighboring" point point P' and find $d\overline{v} = \overline{v}' - \overline{v}$, which will also be a vector. Obviously, this technique is not available in an arbitrary V_n. Schouten overcame this difficulty, however, by introducing what he called a "geodesically moving system of coordinates," (*ein geodetisch mitbewegtes Bezugsystem*). By utilizing such a system he could define the differential of an affinor, and this would again be an affinor. This approach, in fact, produced a differential that turned out to be identical to the covariant differential of Ricci.

Schouten published this work in 1918. Although he translated some of his formulas into the language of the Ricci-Einstein tensor calculus, his theory was so overloaded with symbols that it proved next to impossible to follow. Direct analysis is fine for vectors, when only two multiplications $\cdot$ and $\times$ are involved, but for higher systems one gets lost in the the maze of the dots, hooks, and crosses necessary to perform the various multiplications.

Despite all this, Schouten had succeeded in giving a geometrical interpretation of the covariant derivative, an important accomplishment. Circumstances conspired against him, however, preventing him from being credited with a major mathematical achievement. It was nearing the climax of the war, so that communications with Italy were most difficult. Thus Schouten was totally unaware of Levi-Civita's work. I still remember how Schouten came running into my office one day waving a reprint he had just received of Levi-Civita's paper. "He has it too!" he cried out. And indeed, what Schouten had established using his geodesically moving systems, Levi-Civita had also found using another approach, which he published in 1917, a year before Schouten's paper had appeared.

True, there were considerable differences between the two articles. For example, Levi-Civita had only considered a surface V_2 imbedded in an ordinary space R_3 and had obtained his *parallelism* by moving a vector $\overline{v}$ in a tangent plane to V_2 at a point P parallel to itself to a "neighboring" point P' of the surface. Thus $\overline{v}$ is no longer in the tangent plane, but one can compare the projection of onto the tangent plane at P' with the vector $\overline{v}'$ in this tangent plane at P'. This means that $d\overline{v}$ can be defined as a vector, and it turns out to be identical with the covariant differential of Ricci. Schouten and Levi-Civita had thus obtained the same result, but there were differences in the way each of them introduced parallelism. Schouten's method was entirely intrinsic, whereas that of Levi-Civita utilized a surface imbedded in space. He also had derived his result only for the case $n = 2$, although it was clear that it was intrinsic and valid for all values of n. The main difference, however, insofar as influence was concerned, was that Levi-Civita's text was elegant and employed his absolute differential calculus (the tensor calculus with which mathematicians all over the world were becoming familiar) whereas Schouten's work was difficult to read due to its unfamiliar notation. And, of course, Levi-Civita also had priority of publication, so that the discovery has since become known as the "parallelism of Levi-Civita."

This new contribution to tensor calculus was soon appreciated, and one of the first to pick it up was Hermann Weyl. Weyl was principally concerned with the removal of two *Schönheitsfehler*—one in differential geometry, the other in relativity theory, although in his eyes both were related. In Einstein's theory, gravitation was integrated into one space-time V_4, but the electromagnetic phenomena, as expressed by Maxwell's equations, were superimposed upon this structure (Einstein used the term *hineinpassen*). It was therefore not a *unified* theory. Levi-Civita's tensor calculus, on the other hand, stressed the infinitesimal displacement of a vector from a point P to a "neighboring" point P' (hence what Weyl and others called the *Uebertragung*). However, this displacement was made dependent upon a metric, since the Γ depended on the $g_{\lambda\mu}$. But since displacement is a far more fundamental concept than is a metric, Weyl wished to generalize Levi-Civita's notion of displacement and hence the form of the Γ. By doing so he could attain his other goal, namely the construction of a V_4 with gravitation and electromagnetism imbedded in its structure. Still, he did not quite do away with all dependence on the tensor $g_{\lambda\mu}$, and allowed a factor λ to enter as a multiplier. This factor he then could relate to the electromagnetic potential by using the concept of parallelism.

Weyl presented these ideas in the widely read book *Raum-Zeit-Materie*, first published in 1918. Weyl's work influenced Arthur Eddington, the British astronomer who was instrumental in catapulting Einstein's name into the headlines of 1919 by leading the famous solar eclipse expedition that proved light rays were deflected in a gravitational field. In a paper of 1921, Eddington published a generalization of Weyl's theory by making the Γ, the coefficients of displacement, entirely independent of a metric. In Eddington's theory, covariant differentiation produces a vector

$$\delta v^k = dv^k + \Gamma^k_{\mu\lambda} v_k dx^k$$

$$\delta w_\lambda = dw_\lambda + \Gamma^k_{\mu\lambda} w_k dx^\mu$$

if the $\Gamma^k_{\mu\lambda}$ are subjected to appropriate transformations of the x^k into $x^{k'}$. This approach gives more freedom for physico-astronomical interpretations. Eddington kept the Christoffel symmetry $\Gamma_{\mu\lambda} = \Gamma_{\lambda\mu}$, this being a necessary condition for what Weyl called *affine* geometry, but he also briefly considered the possibility that the Γ have no Christoffel symmetry—perhaps with the idea that these might be used to formulate an even more grandiose unified field theory incorporating quantum phenomena. Eddington gave an exposition of these ideas in his book, *The Mathematical Theory of Relativity* (1923), which was probably read as widely as Weyl's.

Schouten, in a purely mathematical investigation, then dispensed with these symmetry assumptions and even considered different Γ for contravariant and covariant vectors. In a paper of 1922, he gave a field classification of the different types of displacements that can serve as a foundation for differential geometry, including the already known Euclidean, non-Euclidean, and Riemannian geometries, as well as those of Weyl and Eddington. Schouten, like Weyl and Eddington, also published a book summarizing his results: *Der Ricci-Kalkul* (1924)[2]. In contrast to the books by Weyl and Eddington, however, this volume is purely mathematical. To these titles, another should be added: Levi-Civita's *Lezioni di calcolo differenziale assoluto* (1925). After the year 1922, the further development of the tensor calculus was deeply influenced by the work of Elie Cartan, as well as new ideas brought forth by Oswald Veblen and Luther P. Eisenhart. But this is another story.

NOTES

1. It should be $ds^2 = \sum_{\lambda,\mu} g_{\lambda\mu} dx^\lambda dx^\mu$, but at Einstein's suggestion, expressed in his paper of 1916, the $\sum$ sign is omitted when ambiguity is absent. Summation is on the same lower and upper index, in this case λ and μ. We call this the *Einstein convention*. The notation we use in this paper is the one Schouten and I adopted in our work, but it differs little from that used by others.
2. The books by Weyl and Levi-Civita have been translated into English. Schouten's English *Ricci-Calculus* (1954) is an entirely new version of the 1924 book.

Applied Mathematics
In the Early 19th-Century France

Siméon Denis Poisson (1781–1840)
Courtesy of Springer-Verlag Archives

Modes and Manners of Applied Mathematics: The Case of Mechanics

I. Grattan-Guinness

1. The modes of application

1.1 Pure and applied. The phrase 'pure and applied mathematics' is so standard in mathematical circles that it does not provoke much thought. Yet it deserves to be queried: in particular, 'applied' is so broad a term that it is almost useless. There are "internal" applications of a mathematical theory, as when the calculus is applied to number theory (or vice versa); but what is more usually intended by 'applied' is "external" applications to problems in the physical world. In the 18th and 19th centuries the adjective 'mixed' was often used, as well as 'applied', in such contexts, to refer to the mixing of mathematical and physical theories and concepts. The word seems superior in its connotations, and it is a pity that the other has supervened; and it is from that time that I draw in this paper a number of case studies from which the *varieties* of application can be exhibited and exemplified.

The particular area of applications is mechanics, the period in question is the first decades of the 19th century, and the place is France, which was then the leading country in mathematics. In the next section I shall summarise the most pertinent features of the developments there; however, since the French are not the principal concern, the survey will be brief. The third section of the paper is devoted to showing that a variety of different approaches was adopted in mechanics by outlining the development of a new one. The final section contains four examples from mechanics within which not only variety but even disagreement of purpose will be evident. Further details, in the context of these case studies, may be studied in my papers (1981a,1981b), and especially in the book (1990) (chs. 1 and 19 for overall appraisals). Before that, however, a little philosophical background is required

1.2 Some philosophical questions. One of them is the question of whether new results were being sought or whether ones already established were being set in a systematic framework. This corresponds to (but is *not* to be identified with) the modern practise of developing new results or of axiomatising old ones.

ISBN 0-12-599662-4.

Another issue is the question of style, where I prefer to use the German word *Denkweise* as more exact: modes of developing theory within some particular epistemological framework. Axiomatisation itself is one such; but freer systems display several kinds of *Denkweisen*. Two of them are of major importance here: the geometric approach, where diagrams (or at least thought-pictures) are readily formulated, where the calculus is used in a literally differential form due to Euler, and where functions are thought of as curves and surfaces; and the algebraic, advocated by Lagrange, in which diagrams are eschewed as much as possible, generality is sought in basic formulae expressed in an algebraic manner, in which (for example) functions are regarded as expressions.

A third issue relates centrally to applications themselves: whether general behaviour and properties of the physical phenomena are being investigated; or whether formulae amenable to numerical calculation are sought for testing against experimental data. Pitfalls await both kinds of analysis. The general theory will probably already involve simplifications of its own (typically, ignoring thermal and electrical effects); and it could end up with what I call 'notional' solutions, in which results concerned with the phenomena at hand are found but in which the problem-cases are artificial and/or the solutions are incoherent in some way or intractable even for the purposes of theoretical assessment. Meanwhile, the number-friendly result may entail further simplifications of the general theory, with the attendant danger that *over*-simplifications may occur, thus rendering useless the whole analysis.

2. The French communities, 1800-1840

2.1 Professional institutions. This piece of historical space-time is of quite especial importance; for it changed the intellectual and professional faces of mathematics and related sciences in substantial ways which are still evident today. On the educational side, after the Revolution a range of new or restructured institutions were brought in (see my (1988) for a detailed summary). At the top of the tree of prestige was the *Ecole Polytechnique*, founded in 1794 as a preparatory engineering school for the suite of specialist *écoles d'application* in which more specialised engineering studies were then undertaken by the students. Then in the late 1800s was founded the *Université Impériale*, a misleading name for an Empire-wide structure responsible for education at the school level, with a few *Facultés des Sciences* here and there (especially in Paris, of course) to provide the highest level of instruction; but relative to the other *écoles* it was a second-rate affair: even its elite institution, the *Ecole Normale*, did not come into eminence in science until mid century.

After graduation comes the profession (see Crosland (1967) for a survey of the institutions of the time). The establishments named above provided employment (as did also the *Collège de France,* among others: although it is not significant for this story). In addition, there were various *Corps* of civil engineering, especially *Ponts et Chaussées* and *Mines;* special bodies such as the *Bureau des Longitudes* for astronomy and navigation and the *Dépôt Générale de la Guerre* for cartography and geodesy; and the army and navy. Among scientific societies the *Académie des Sciences* (known as the first *classe* of the *Institut* from 1795 to 1816) was the chic establishment, the principal target of professional ambition; but various other organisations operated within science and engineering. In particular, the *Société Philomatique de Paris* usually managed to publish promptly in its monthly *Bulletin* summary accounts of recent work; thus it is one of the major serials for French science at that time.

2.2 THE SCIENTIFIC ACHIEVEMENTS. Within this structure a galaxy of (newly) professionalised *savants* learnt and taught, and researched and pursued their careers—usually in Paris, and with a disposition for rivalry that only the French manifest. Over thirty major figures can be specified across four generations, from those who were already prominent before the Revolution through those who emerged with the new institutions and around the time of the Restoration in the mid 1810s to the new generation who began their activity in the late 1820s. In addition, another score of *savants* are of some note, and in most other communities would have been quite prominent. The whole ensemble provides an instance of that fascinating historical occasion when a constellation of major quality is copresent at one time and place. Table 1 gives their names and dates; the division into two groups, and the meaning of the asterisks, are explained in the next sub-section.

The achievements in research may be summarised as follows. The main accomplishment was that in "pure" mathematics the calculus of the day, with its attendant theories of functions and series, broadened out into mathematical analysis, in which all these topics were subsumed under this umbrella discipline founded upon a theory of limits, and exhibiting a new standard of rigour, especially in its more careful attention paid to the necessary and/or sufficient conditions required to establish the truth of theorems. (Cauchy is the principal founder of this movement.) At the same time, and of more relevance here, mechanics broadened out into mathematical physics, with new mathematicisations being introduced into heat theory, physical optics, and electricity and magnetism (and especially their interconnections in electromagnetism). Mechanics itself also developed very substantially: indeed, examples of both the old and the new results will feature in the next sections. This galaxy of talent

TABLE 1
Principal Mathematicians in the Period 1800-1830
Divided by Research Interests

Calculus/Mechanics/Engineering	Mathematical Analysis/Physics
C.S.J. Bossut (1730-1814)	A.M. Ampère (1775-1836)
L. Carnot (1753-1823)*	J.P.M. Binet (1786-1856)
B.P.E. Clapeyron (1797-1864)	J.B. Biot (1774-1862)
C.P.M. Combes (1801-1872)	A.L. Cauchy (1789-1857)*
G.G. Coriolis (1792-1843)*	J.M.C. Duhamel (1797-1872)
J.B.J. Delambre (1749-1822)	J.B.J. Fourier (1768-1830)
F.P.C. Dupin (1784-1873)	A.J. Fresnel (1788-1827)
L.B. Francoeur (1773-1849)	S. Germain (1776-1831)*
P.S. Girard (1765-1836)*	J.L. Lagrange (1736-1813)*
J.N.P. Hachette (1769-1834)	G. Lamé (1795-1870)
G. Monge (1746-1818)	P.S. Laplace (1749-1827)*
A.J. Morin (1795-1880)	A.M. Legendre (1785-1833)
C.L.M.H. Navier (1785-1836)*	E.L. Malus (1775-1812)
T. Olivier (1793-1853)	L. Poinsot (1777-1859)
J.V. Poncelet (1788-1867)*	S.D. Poisson (1781-1840)*
G. Riche de Prony (1755-1839)*	P.G. le D. de Pontécoulant (1795-1874)
L. Puissant (1769-1843)	C. Sturm (1803-1855)

worked within a career pattern called *cumul*, in which a *savant* would hold several posts at once: indeed, in order to make a decent living he would have to do so, but then competition for these posts arose regularly, and often excited fresh occasion for competition. The point is relevant for our concerns, for the intellectual differences outlined in the previous section were often accompanied by different patterns of career and even personal dislike.

2.3 TWO COMMUNITIES. Indeed, this last feature manifests itself collectively as well as individually. For in its intellectual research the community of major figures was divided concerns into two halves of nearly equal size, and the difference constitutes the first principal demarcation between the varieties of application which are our main concern in this paper. The group indicated in the left hand column of Table 1 worked almost exclusively in mechanics, and contributed but little to the new areas; but they were strongly driven by the needs of engineering, and were often employed on one of the *Corps* mentioned above. They were particularly likely to seek numerical solutions to compare with experimental data. The members of the other group tended more towards the general and theoretical applications, both within mechanics and the new areas which some of them created (and also the mathematical analysis mentioned in the previous sub-section): they were usually the greater mathematicians as such, they held a somewhat greater proportion of the teaching posts, and (an interesting statistic) they were often elected to the *Académie* at an earlier age than their engineering colleagues. Numerical calculation was less likely to feature in their studies, although it was not absent. The members of each group with whom we are concerned in this paper are marked with an asterisk in Table 1.

In addition, two figures should be mentioned whose active contributions had entirely or virtually finished by 1800, but who relate closely to the concerns of the first group: J.C. Borda (1733-1799) and C.A. Coulomb (1736-1806) (on them see respectively Mascart (1919) and Gillmor (1971)). Interestingly, Coulomb does not fit into the division of the community described above, for his other major efforts were in electricity and magnetism; but of course he is an 18th-century figure.

Coulomb can serve as an example of the last general point which I wish to make in this section. It is well known that up to the end of the 18th century the sciences were fairly sharply divided into two kinds: the classical, in which mathematics was prominent and experimentation played a relatively minor role; and the Baconian, where experiments took

Figure 1.

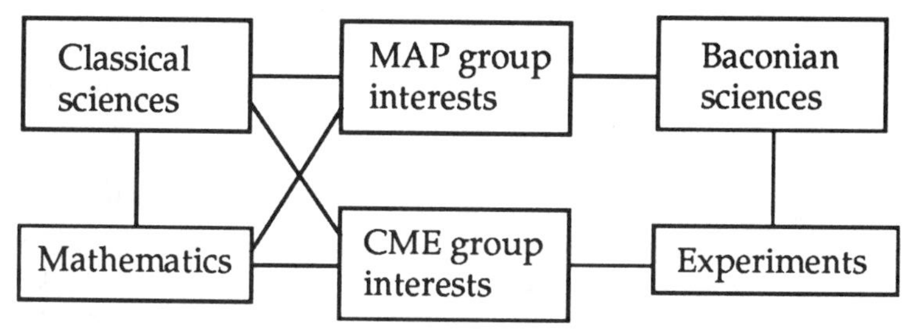

the stage and mathematics was almost entirely absent. After the inauguration of mathematical physics, of course, a new distinction is needed; and the characteristics of membership to the two groups gives us the key to it. The first group tackled only the classical sciences, but concerned themselves with mathematics and experiments; the second group handled both classical and Baconian sciences, but usually from a mathematical point of view; they were not usually talented at experiments (Fresnel is an important exception). Figure 1 represents both the old and new connections in diagrammatic form.

3. An engineering approach to mechanics

3.1 The varieties of mechanics. It is customary to divide this subject into its terrestrial and celestial kinds, but this division is too coarse for our purposes. Five categories of mechanics are involved: 1) celestial, in which the heavenly bodies are usually treated as mass-points: 2) planetary (and also satellitary), where the shape of the body has to be allowed for (and indeed may be the main object of study); 3) corporeal, where ordinary-sized objects are considered, and under which I also place the (various) foundational principles of the subject; 4) engineering and technology, including pertinent instruments; and 5) molecular, where the presumed intimate structure of matter (and also the aether) are under investigation. The examples to be considered below relate to the second, third and fifth categories. Only principal items from the primary literature are cited, and wherever possible secondary sources are cited. All these cases are dealt with in my (1990), especially chs. 5–8, 10, 15 and 16.

3.2 Equilibrium or disequilibrium? By the late 18th century three different traditions were in competition in Continental mechanics. Newton's laws of motion were frequently used, but they did not enjoy the degree of dominance that one might imagine. For d'Alembert and especially Lagrange advocated d'Alembert's principle and the principles of virtual velocities (as it was then called) and of least action, cast in the algebra of variational mechanics (see Fraser (1983) and Lindt (1904)); and most figures paid some attention to the principle of the conservation of energy (or *forces vives*, in the jargon of the time).

Two points need to be stressed. Firstly, in all these traditions, especially the second, preference was normally given to equilibriate situations, although it was not always clear whether both static and dynamic equilibrium were intended. (d'Alembert's principle was intended to reduce dynamic situations to static ones, although the legitimacy and generality of the reduction were two of the issues involved in the competition between the traditions.) Secondly, of these three traditions the one based upon energy was the least in favour as the basis for a general approach.

3.3 Energy and its loss. All this was to change in the gradual emergence of a new view of mechanics based upon considerations of energy and advocated by several members of the engineering group. The story is a very fine example of engineering influencing the development of science; it is summarised in my (1984).

The father figure was Carnot, who perceived the *limited* "generality" of current approaches to mechanics, especially the views of Lagrange; for they could not handle cases of shock and percussion, and the effects of friction, which were very important in the engineering of that time. Carnot developed a position in which statics had to be seen as a special case of dynamics, involving no motions. *Dis*-equilibrium was the primary concern; the energy approach was to be preferred, but extended to allow for loss as well as conservation, thanks to an expression for lost *forces vives* proposed by Borda. In this connection the work expressions (to use the modern term) were *not* assumed to admit a potential.

Carnot outlined his approach in the 1780s (see Gillispie (1971)), but it gained circulation after the appearance of his book (1803) on machines. Among those *savants* who took it up, the most important was Navier, who applied it extensively in the late 1810s to hydraulics and thence to mechanics in general. When *forces vives* were lost, they went into *quantité d'action*, or work in our parlance (see especially his (1819)). His approach was extended further during the 1820s in independent studies by Coriolis and Poncelet, who each produced books in 1829. The concept of work (the word ('travail') was introduced as a technical term by Coriolis, incidentally) was now taken on a par with *forces vives*, each convertible into the other. General principles of energy mechanics were presented, and the applications laid especial stress on frictions studies and all sorts of machine; in addition, ergonomics (with man or animal treated as a work-rating machine) gained some attention. However, gas and vapour mechanics were sideshows in the panorama of concerns of this time, with energy produced by consumption being envisioned *only* as convertible into work, without the inverse conversion being investigated. The insights of Carnot's son Sadi were to wait for twenty years for adoption after their publication in 1824, although Sadi also belonged (not as a mathematician, of course) to the engineering community.

The mathematical *Denkweise* attached to this new tradition was strongly geometrical in form. The *savant* was encouraged to think of bodies, fluids, or whatever moving in this or that direction, with components of velocity calculated by the usual composition rules. Even the basic concept $\int P\, ds$ of work was explicitly put forward as an area.

A final point is worth making: as often has happened in the history of mathematics, education provided a major stimulus for the research. Navier and Poncelet both developed their ideas during the 1820s at *écoles d'application* at which they were employed, while Coriolis's book had some bearing on the teaching of machines at the *Ecole Polytechnique*, although he did not take the course in machines. There were even further consequences, connected with the attention paid to work-rate and ergonomics: the education of the working classes in these ideas, and the conditions of their employment, belonged to this movement.

4. Four case-studies from mechanics

4.1 Projectile theory. A classic problem in mechanics, especially because of its importance in military engineering, was the calculation of the path of a projectile. Borda (1772) gave a nice treatment in which a tractable solution was obtained. Allowing for air resistance but ignoring the rotation of the earth, he produced two ordinary differential equations of motion and found parametric solutions for them. Poisson gave an account of the method in the first edition of his textbook in mechanics ((1811, vol. 1, 346–352). He did not cite Borda (a somewhat common feature of Poissonian scholarship); and he also allowed the simplification caused by the omission of the rotation of the earth. This decision was taken presumably in tune with a statement in the preface of his book, in which he promised 'for clarity and simplicity' appropriate for a text to be used at the *Ecole Polytechnique*, thereby allowing 'geometrical considerations and formulae of Algebra mixed in the same question' instead of following his preference for Lagrange's algebraic approaches.

This passage is noteworthy, for Poisson's other mentor, Laplace, had recently given a less simplified treatment of the closely related theme of 'the motion of a body which falls from a great height' described in a paper (1803) published by the *Société Philomatique* and somewhat elaborated in the fourth volume (of 1805) of his *Mécanique céleste*. Laplace's theory did allow for the rotation of the earth; and in doing so he showed particularly well his ability to handle fixed and moving frames of reference. Using the principle of virtual velocities to furnish three ordinary differential equations of motion, and assuming a squared- velocity law for air resistance, he came up with explicit expressions for the coordinates of the body that were a good deal more complicated than Borda's.

The so-called 'Coriolis force' was not included as such in Laplace's analysis, but its components were included at the appropriate places. This is a case where the attached name is appropriate, for in (1832) and (1835) Coriolis showed with greater generality than any predecessor the general consequences for mechanics—especially the energy mechanics outlined in section 3.3 above—of allowing for moving frames of reference and relative motion. Laplace's older analysis received important attention a few years after Coriolis when Poisson extended it by considering the rotation of the body while in flight; and out of that came Foucault's demonstration of the rotation of the earth by means of his pendulum (see Acloque (1981)).

4.2 Pendulum theory. The simple pendulum was a most serious instrument of study in the early 19th century (see Wolf (1889)), becoming indeed a very sophisticated instrument in a design made by Borda. The analyses were stimulated especially by the use of the instrument in fine-detail geodesy; thus effects were studied very carefully for their order of smallness. Laplace and the ever dutiful Poisson were particularly active. Poisson proved the near equality of the periods of the upward and downward half-oscillations (again assuming a squared-velocity law for air resistance) and the negligible extensibility of the supporting wire during the motion ((1808) and (1809): he omitted these refinements from his textbook); then in (1816a) he analysed the effect of the motion of the torsion of the wire. At that time Laplace (1817) showed that the motion of that wire about the pivoting knife-edge was indeed tiny.

These studies are typical of the theoreticians' concerns; quite complementary are those of the engineers. Borda's design of the pendulum was inspired by the geodetic survey of the 1790s from which the metre was designated (indeed, he proposed that word). In the posthumous paper (1810) he gave a analysis of Boscovich's method of 'coincidences', where the oscillations of the pendulum were compared with those of a regulating clock by timing the interval between the moments when both pendula passed the vertical together. The difficulty was that one could not expect to choose exactly the appropriate lengths of the pendula to furnish exact isochrony; thus a correction term was needed. For this purpose, Borda made the assumption, suggested by experimental evidence, that the amplitudes of the oscillations decreased in geometric progression from its values at the moments of coincidence. Then, adopting the second-order approximative expression for the period of oscillation, he produced a simple expression for the desired correction.

De Prony was another engineer *savant* concerned with pendulum design for the geodetic survey. His own design was of a 'reversible' type now associated with Kater's proposal of the late 1810s; but de Prony had priority, and was rather aggrieved by the lack of interest shown in his own design by his contemporaries; his paper (1792) remained unpublished. This type of pendulum used two pivots and three weights, of which the lighter two were adjustable along the bar in order to yield the desired time of oscillation. Three readings were taken from different positions; they could be put into the quadratic equation relating the period, the radius of gyration of the pendulum, and the location of its centre of gravity.

4.3 HYDRAULICS VERSUS HYRODYNAMICS? De Prony's calculations were characteristic of his lifelong desire to tailor mechanics to the needs of engineering. During the 1790s he directed an enormous project of calculating logarithmic and trigonometric tables, run as a team effort following the principles of Adam Smith, but calculating the entries for the tables to such numbers of decimal places as to render the project notional in the sense of section 1.2 (see Lefort (1875): who wants to calculate log sine functions to 29 places?). In his lectures as a founder professor of mathematics at the *Ecole Polytechnique,* he taught difference equations rather than differential equations, in the grounds that measurements were taken only from time to time. (His colleague professor, Lagrange, was serving up a very different fare!) Then in the early 1800s, he applied his preference for numerical methods to hydraulics, in connection with the construction of the Ourcq canal to Paris, the digging of a port at La Villette outside the walls of the city, and the construction of two supplementary canals from the Seine river to the port. Various questions needed attention, concerning the measurement of the flow of large bodies of water and the filling and emptying of locks.

Coulomb (1801) had used his torsion balance recently to study experimentally 'the coherence of fluids and the laws of their resistance', and had proposed a law of resistance involving terms proportional to both velocity and squared-velocity. de Prony (1804) used these results to obtain a similar law relating the mean velocity of a body of water to its slope. Then, to calculate some coefficients from data, he presented a geometrically conceived version of a theory of errors given by Laplace in 1799 in the second volume of his *Mécanique céleste* (and in fact stolen from Boscovich). de Prony chose this *Denkweise* because 'the geometrical constructions are much more familiar than abstract analysis to a great number of engineers'.

The director of the Ourcq canal and related projects was de Prony's friend Girard (see his 1831-1843); and he also contributed some analyses of water-flow at this time. He is a scientist distinguished by a love of analogy, even rather outlandish ones. In an essay (1803) he studied the motion of a canal as if it were a flexible chain lying on a smooth surface, taking account also Coulomb's law of of water resistance (the term of squared velocity being caused by 'asperities' always found on surfaces).

Another of Girard's and de Prony's assumptions was the 'hypothesis of parallel slices', which decreed that the flow of water could be analysed as moving in echelon in adjoining straight differential slices. The verisimilitude of this theory to the actaual motions of waterbecame quite a point of contention between members of the two communities of *savants.* The engineers, especially de Prony, held it in high esteem (see, for example. his use of it in an extensive discussion (1802) of waterflow). By contrast, even though Poisson (1807) had provided a nice mathematical analysis of such a supposed motion, he did not draw upon it when he came to study fluid flow in a general way—for hydrodynamics, more than hyraulics, belonged to his mathematical world.

An occasion for Poisson to demonstrate his prowess in hydrodynamics was provided by a prize problem proposed by the first *classe* for 1815. It dealt with the differential equations of the motion of a deep body of fluid, which had not yet been subjected to mathematical study. Young Cauchy submitted a paper, which won the prize; but it did not appear until (1827), after the author had doubled its length by the addition of seven additional appendicial notes. Poisson was a member of the examining committee, but in his typical fashion he presented his own paper on the matter during the time of the competition, and had it quickly published as (1818). The two papers are compared in Burkhardt (1912), and in Dahan-Dalmedico's essay in these volumes.

In contrast to de Prony's and Girard's use of empirical formulae such as Coulomb's, Poisson and Cauchy started out from the differential equations for hydrodynamics, using respectively Euler's and Lagrange's forms. Both men proposed integral solutions, Poisson ending up with a convolution form (apparently the first such) and Cauchy with Fourier integrals. Fourier had presented these integrals a few years earlier, in connection with heat diffusion in infinite bodies; but he had not published them by 1815. Cauchy's discovery seems to have been independent, and when he found Fourier's integral theorem in 1817 he acknowledged Fourier's priority. In neither his nor in Poisson's analysis of hydrodynamics did the integrals carry any physical significance, unlike the work integral of section 3.3.

Both men drew similar and important conclusions. They determined that waves were propagated with constant acceleration and with constant velocity under various conditions. For the former type they indulged in numerical calculations, to calculate the locations of the tops of the waves. For Poisson the latter type was 'a type of *nodes*, mobile at the surface...which appears to move on the surface' (1818, 120). Cauchy stated that he had also found such results but had not included them in the paper (1827, 78); but neither man matched the understanding of Fourier (1818) on these 'furrows' (as Fourier called them), which move with what is now call the 'group velocity' of a wave train. Thus the theoreticians' analyses could lead to profound consequences.

4.4 ELASTICITY THEORIES. Despite the intensive work of Euler, Lagrange and others in the 18th century, general theories of the motion of elastic solids and fluids were still wanting. In fact, elasticity theory became quite a favourite among the savants of both groups. I summarise here the two main episodes: for more details, see Todhunter (1886), chs. 3-5; and Burkhardt (1908), ch. 9.

The first *classe* proposed a prize for 1811 on the differential equations of the motion of an elastic lamina. Poisson had not yet been elected, and would have been expected to win; but in the end only Germain submitted, and in 1815 she eventually won with her third entry, of which she published a version as (1821). Meanwhile, Poisson had been elected anyway, and did in fact present his own study of the problem, which appeared as (1816b).

The only feature common to both papers was the topic: the procedures followed reveal quite different content and *Denkweisen.* Following some work of Euler, Germain made several attempts to express by geometrical reasoning the balance of forces, or of moments, as the membrane was deformed. Neither the derivation of the differential equation nor the calculation of the solutions was competently executed. Poisson followed a mathematicised molecularist programme which had been initiated by Laplace a decade earlier; thus by a mixture of central-force mechanics and differential geometry, but with no geometric visualisation employed as such, he arrived at an extraordinary partial differential equation which he showed did reduce to known forms for certain special cases. His analysis was somewhat notional in several ways: he did not specify the manner in which the boundary of the membrane was held, and he had the wrong power of the thickness of the "thin" membrane.

Elasticity theory received a more important impulse when Navier came into the picture. In section 3.3 we saw him make important contributions to energy mechanics in 1819: in this *annus mirabilis* he also began a suite of papers on elastic bodies (rod, surface, body) and viscous fluids (hence the (mis-)named 'Navier-Stokes equations'). Applying Hooke's law to the extension and deformation of a body in any direction, he derived in the lithograph (1820) the differential equations expressing the equilibrium and motion of an elastic surface. His approach was noticeably geometrical, but was incoherent in some ways; in particular, the units and dimensions of his (only) elastic parameter are hard to understand.

Since Navier was not a member of the *Académie*, he had to submit his lithograph there, where it was seen by Cauchy. Upon studying Navier's model, Cauchy was inspired to develop a remarkable alternative to it, partly influenced by Fresnel's study of the elastic aether in an analysis of double refraction. In Navier's scheme, resultant forces acted along the normal to a surface; but Cauchy perceived that this resultant could lie in some oblique direction. From this insight he conceived of an approach which Rankine was to name the 'stress-strain' model of linear isotropic and anisotropic elasticity theory, in which a portion of a body was isolated and the action across its imagined surface was analysed.

Cauchy wrote a prosodic account (1823) of his new theory for the *Société Philomatique*. He gave a (minimal) acknowledgement of Fresnel's prior insight, and remarked that Navier's paper had occasioned the study (which did not prevent Navier from issuing a sour accusation of plagiarism). Five years later Cauchy began to publish a detailed theory, in a long suite of papers; I conjecture that the delay was caused by doubts over the generality of his theory, which he resolved by realising that *two* parameters were needed for isotropic elasticity theory, and not the one which Navier (and all other predecessors) had used. Poisson had to join in, of course, with even longer papers presenting only the isotropic theory based on a certain modification of the Laplacian programme mentioned above. Navier came back with bad-tempered accusations again, and penetrating criticism of the assumptions made in Poisson's theory.

From various points of view the trio differ in different ways. For example, both Cauchy and Navier used a geometric approach to conceive of their theories, although Cauchy rapidly relied thereafter only on calculations; none of his papers used any diagrams. Again, like Navier, Poisson's elastic parameters are difficult to understand in terms of units and dimensions; but in addition, while he followed Cauchy in adopting two parameters, he argued that one of them was actually zero (see Poisson (1828), 397-400: Navier (1829) was to attack the reasoning severely).

Finally, Cauchy and Poisson came to study elasticity from a theoretician's standpoint, whereas Navier was motivated by the concerns of engineering. In fact, he was teaching properties of flexure of materials at the *Ecole des Ponts et Chaussées*, and had also secured the agreement of his Corps to design and construct a steel suspension bridge—which was then a fairly new kind of artefact—in Paris. Unfortunately the bridge ruptured in 1826, when it was near completion, although he always maintained that the accident was due to bad luck rather than bad design (see his (1830)). The contrast between the engineer at the river-bank and the theoretician in his study is particularly marked in this case!

5. Concluding remarks: complement or competition?

Even this small selection of examples from one branch of applied mathematics studied in one country during a short period of time shows that a wide range of concerns can be tackled and a variety of methods can be deployed: indeed, I chose a restricted piece of space-time precisely to emphasise the diversity of possible approaches and *Denkweisen*. It is clear that in most of the cases described in the last section the concerns of the two groups are complementary. Borda wanted to give good working results for ballistics and pendulum use, while Laplace and Poisson were applying general principles of dynamics to particular effects; de Prony and Girard wanted good practise in the handling of large quantities of water, while Poisson and Cauchy sought the differential equations and their solutions appropriate to a deep fluid body.

But the anti-Lagrangian mechanics initiated by Carnot and taken up by several other engineering *savants* (section 3.3) shows that the intellectual concerns of the two groups *could* overlap, and to a substantial measure. For that approach held implications for the subject as a whole, and therefore for the more general theoreticians such as Lagrange himself. The lack of response made by the theoreticians reflects more the narrowness of the conceptions than any limitations on Carnot's insights.

A specific example of clashes concerns the hypothesis of parallel slices favoured by the engineers in their study of waterflow (section 4.3). Cauchy explicitly replaced it in his teaching at the *Ecole Polytechnique* by a theory of 'curved filaments'; but in so doing he earned the ire of de Prony, as a graduation examiner of the school, in one of a long series of criticisms made between 1826 and 1829 about the inappropriateness of the professor's teaching at an educational establishment for aspiring engineers (see my 1981b, 684–690).

Finally, the elasticity theories summarised in section 4.4 show that members of the two groups of *savants* could clash with some violence. Here the controversies were partly motivated by personal factors, especially the (apparent) similarity of the results found by Navier and Cauchy; but the difference of motivations (materials versus differential equations) is very striking. We note also the later competition between Cauchy and Poisson as a (typical) example of differences of methods and results between members of the *same* group.

In all these situations there *is* competition, which has never received the attention that it deserves. Down to our time "mechanics" and engineering mechanics have often been treated in separate books and taught in different courses, with little or no notice taken in each tradition of the other line. Education is thereby impoverished, of course; for the wonderful variety in the spectrum of applications of mathematics is hidden from sight to all except those students who have the nous and the courage to feel dissatisfied and puzzled at their instruction.

Bibliography

The dating codes are those of publication. '=' link reprintings of an item.

Aclocque, P.
(1981) *Oscillations et stabilité selon Foucault...*, Paris (CNRS).

Bucciarelli, L.L. and Dworsky, N.
(1980) *Sophie Germain...*, Dordrecht (Reidel).

Burkhardt, H.F.K.L.

(1908) *Entwicklungen nach oscillirenden Functionen und Integration der Differentialgleichungen der mathematischen Physik*, Jahresbericht der Deutschen Mathematiker-Vereinigung, 10, pt.2, xii + 1804 pp.

(1912) *Untersuchungen von Cauchy und Poisson über Wasserwellen*, Sitzungsberichte, Akademie der Wissenschaften zu München, mathematisch-physikalische Klasse, 97–120.

Carnot, L.

(1803) *Principes fondementaux de l'équilibre et du mouvement*, Paris (Bachelier).

Cauchy, A.L.

(1823) *Recherches sur l'équilibre et sur le mouvement intérieur des corps solides ou fluides, élastiques ou non élastiques'*, Bulletin des sciences, par la Société Philomatique de Paris, 9–13 = Oeuvres complètes, ser.2, vol.2, 1958, Paris (Gauthier-Villars), 300–304.

(1827) *Théorie de la propagation des ondes la surface d'un fluide pesant d'une profondeur indéfinie*, Mémoires présentés par savants étrangers l'Académie Royale des Sciences, (2)1 (1827), 3-312 = Oeuvres complètes, ser.1, vol.1, 1882, Paris (Gauthier-Villars),4–318 [this printing cited].

Coriolis, G.G.

(1832) *Sur les principe des forces vives dans les mouvements relatifs des machines*, Mémoires présentés par savants étrangers l'Académie Royale des Sciences, (2)3, 573–607 = Journal de l'Ecole Polytechnique, (1)13, cah. 21, 268–302.

(1835) *Sur la manière d'établir des différens principes de mécanique...*, Journal de l'Ecole Polytechnique, (1)15, cah. 24, 93–125.

Coulomb, C.A.

(1801) *Expériences destinées déterminer la cohérence des fluides...*, Mémoires de la classe des sciences mathématiques et physiques de l'Institut de France, 3, 264–305.

Crosland, M.P.

(1967) *The Society of Arcueil ...*, London (Heinemann).

de Prony, G.C.F.M. Riche

(1792) *Recherches sur les moyens de déterminer la longueur du pendule*, ms. in Ecole Nationale des Ponts et Chaussées (Paris), ms.1357.

(1802) *Mémoire sur le jaugeage des eaux courantes*, Paris (Imprimerie Impériale). [Summary, and commentary by others, in Mémorial de l'officier du génie, 2 (1st ed.), 151–177,178–203 = 2 (2nd ed.)(1821), 48–73, 73–103.]

(1804) *Recherches physico-mathématique sur la théorie des eaux courantes*, Paris (Imprimerie Impériale).

Fourier, J.B.J.
(1818) *Note relative aux vibrations des surfaces élastiques ...*, Bulletin des sciences, par la Société Philomatique de Paris, 129–136 = Oeuvres, vol. 2, 1890, Paris (Gauthier-Villars),255–265.

Fraser, C.
(1983) *J.L. Lagrange's early contributions to the principles and methods of mechanics*, Archive for history of exact sciences, 28, 197-241.

Germain, S. (1776-1831)
1821a *Recherches sur la théorie des surfaces élastiques*, Paris (Courcier).

Gillispie, C.C.
(1971) (Ed.) *Lazare Carnot, savant*, Princeton (Princeton University Press).

Gillmor, C.S.
(1971) *Charles Augustin Coulomb ...*, Princeton (Princeton University Press).

Girard, P.S.
(1803) *Rapport l'assemblée des Ponts et Chaussées, sur le projet général du canal de l'Ourcq*, Paris (Imprimerie de la République). [Summary, with revisions in (1831-1843), vol. 1, 313–357.]
(1831-1843) *Mémoires sur le canal de l'Ourcq...*, 2 vols. and 2 vols. atlas, Paris (Carilain-Goeury). [Vol. 2 posthumous, ed. L.-J. Favier.]

Grattan-Guinness, I.
(1981a) *Mathematical physics in France, 1800-1840: knowledge, activity and historiography*, in J.W. Dauben (ed.), *Mathematical perspectives...*, New York (Academic Press), 95-138.
(1981b) *Recent researches in French mathematical physics of the early 19th century*, Annals of science, 37, 663–690.
(1984) *Work for the workers: advances in engineering mechanics and instruction in France, 1800-1830*, Annals of science, 41,1–33.
(1988) *Grandes écoles, petite Université: some puzzled remarks on higher education in mathematics in France, 1795-1840*, History of universities, 7, 197–225.
(1990) *Convolutions in French mathematics, 1800-1840. From the calculus and mechanics to mathematical analysis and mathematical physics*, three volumes, to appear, Basel(Birkhäuser) and Berlin(DDR) (Verlag der Wissenschaften).

Laplace, P.S.
(1803) *Mémoire sur le mouvement d'un corps qui tombe d'une grande hauteur*, , Bulletin des sciences, par la Société Philomatique de Paris, 3 (1801-05), pt.2,109-115 = Oeuvres complètes, vol. 14, 1912, Paris (Gauthier-Villars), 267–277. [See also Traité de mécanique céleste, vol. 4 (1805), Book 10, ch.5.]

(1817) *Sur la longueur de pendule séconde*, Connaissance des temps, (1820/17), 265–280 = Oeuvres complètes, vol. 13, 1904, Paris (Gauthier-Villars), 121–139.

Lefort, P.A.F.

(1875) *Observations relatives aux remarques...sur les grandes tables of de Prony*, Proceedings of the Royal Society of Edinburgh, 8 (1872-75), 563–581.

Lindt, R.

(1904) *Das Prinzip der virtuellen Geschwindigkeiten...*, Abhandlungen zur Geschichte der Mathematik, 18, 145-195.

Mascart, J.M.

(1919) *La vie et les travaux du Chevalier Jean-Charles Borda (1733-1799)...*, Lyon = Annales de l'Université de Lyon, n.s., sec. 2 (droit, lettres), fasc.33.

Navier, C.L.M.H.

(1819) *Editorial notes, passim* in his edition of B. Forest de Bélidor, Architecture hydraulique, Paris (Didot).

(1820) *Mémoire sur la flexion des plans élastiques*, Paris (lithograph). [Rare: copies in Ecole Nationale des Ponts et Chaussées; University of London library.]

(1829) *Lettre ...M. Arago* on elasticity theory, Annales de chimie et de physique, (2)40, 99–107.

(1830) *Notice sur le Pont des Invalides*, in *Rapport... et mémoire sur les ponts suspendus*, 2nd ed., Paris (Carilain-Goeury), 249–301.

Poisson, S.D.

(1807) *Sur le mouvement d'un fluide ... en admettant l'hypothèse du parallèlisme des tranches horizontales*, Correspondance sur l'Ecole Polytechnique, 1 (1804-08), 289–294.

(1808) *Sur les oscillations du pendule dans un milieu résistant ...*, , Journal de l'Ecole Polytechnique, (1)7, cah. 14, 143–158.

(1809) *Addition* to (1808), Journal de l'Ecole Polytechnique, (1)8, cah. 15, 345–353.

(1816a) *Sur les oscillations du pendule composé*, Connaissance des temps, (1819/16), 332–343.

(1816b) *Mémoire sur les surfaces élastiques*, Mémoires de la classe des sciences mathématiques et physiques de l'Institut de France, (1812), pt. 2, 167–192, 191–225 [sic].

(1818) *Mémoire sur la théorie des ondes*, Mémoires de l'Académie Royale des Sciences, 1 (1816/18), 71–186.

(1828) *Mémoire sur l'équilibre et le mouvement des corps élastiques*, Mémoires de l'Académie Royale des Sciences, 8 (1829), 357-570. [Offprints available in 1828.]

Prony See de Prony

Todhunter, I.

(1886) *A history of the theory of elasticity and the strength of materials*, vol. 1 (ed. K. Pearson), Cambridge = 1960, New York.

Wolf, C.

(1889) (Ed.) *Mémoires sur le pendule...*, pt. 1, Paris (Gauthier-Villars).

Augustin Louis Cauchy (1789–1857)
Courtesy of Springer-Verlag Archives

La Propagation des Ondes en Eau Profonde Et ses Développements Mathématiques (Poisson, Cauchy 1815–1825)

Amy Dahan Dalmedico

Abstract*

On 27 December 1813, the French Academie des Sciences announced a mathematical prize competition on the phenomenon of wave propagation on the surface of a liquid of indefinite depth. Questions about wave motion had long occupied scientists: Newton had dealt with such problems in the *Principia;* Laplace had published memoirs on the subject; and Lagrange had formulated a mathematical theory in the *Méchanique Analytique* and elsewhere. Poisson recognized that the general theory Lagrange had put forth led to contradictions if the depth of the liquid medium were assumed indefinite, and so, as a member of the Academie's question selection committee in 1813, Poisson was in a position to influence the ultimate choice of topic.

In July 1815, the twenty-five year old Augustin-Louis Cauchy presented a solution to the wave problem. Later that summer in August, Poisson deposited a memoir of his own on the subject at the Academie to be read only after the close of the competition. Thus, although Cauchy officially won the prize in 1816, both mathematicians went on record as having solved the problem at essentially the same time. The actual published version of their work came out much later, however, with Poisson's paper appearing in 1818 and Cauchy's following only in 1827.

Despite the fact that Poisson's solution filled a hundred pages while Cauchy's spanned well over three hundred, the papers these two men produced were really quite similar. They each reflected a phenomenological point of view relative to the wave problem, and they each utilized Fourier transforms and series expansions as their primary mathematical tools. These similarities aside, though, various mathematical difficulties led the two researchers to different characterizations of wave propagation. Cauchy's initial work resulted in his conclusion that a liquid of indefinite depth sustained one type of wave only, that is, the uniformly accelerated wave, while Poisson's mathematics pointed to the existence of this kind

* Prepared by Karen Parshall

ISBN 0-12-599662-4.

of wave as well as the wave in uniform motion. In developing some of his ideas further in 1819, Poisson not only presented the now classical form of the integral of the wave equation in two and three variables and explored techniques for transforming it but also accentuated the symbolic calculus involved and proposed a general method of integration applicable to linear equations of partial derivatives with constant coefficients.

Prior to the publication in 1827 of his own wave researches, Cauchy responded both to Poisson's findings and to his own contemporaneous work in preparing a lengthy set of accompanying notes (150 of the over 300 pages) to his manuscript of 1815. By the mid 1820's, Cauchy had come to appreciate and to deal with many of the subtleties inherent in the mathematics of wave propagation, subtleties which had led him to his incomplete conclusions of 1815. Thus, spurred by Poisson's papers, Cauchy used the notes to the 1815 research as a vehicle for carefully examining issues such as series convergence, series manipulation, and the rules governing the substitution of imaginary-valued variables. The mathematics Cauchy laid down in these notes represented important early contributions to the rigorous development of analysis in the nineteenth century.

Résumé:

L'article présente les travaux contemporains de Poisson et Cauchy sur la théorie des ondes (1815), et les principales idées mathématiques nouvelles qui s'y trouvent. C'est la première fois que Cauchy est confronté à un problème de propagation d'une perturbation ou de théorie ondulatoire (théorie à laquelle il se heurtera plus tard, au moment de ses multiples essais d'élaboration d'une théorie élastique de la lumière) et il commet des erreurs, dont certaines résident dans une manipulation incorrecte des séries et des intégrales.

Incontestablement Poisson, dont le premier mémoire décrit justement deux types d'ondes en propagation, et dont le deuxième mémoire contient la formule intégrale de l'équation des ondes, marque des points sur son jeune collègue. Cauchy reviendra sur l'ensemble de ces résultats dans des Notes ajoutées pour la publication de son mémoire (1827). Pendant près de dix ans, les deux savants vont développer et préciser leurs techniques mathématiques (transformation de Fourier, résolution de l'équation des ondes et interprétation par le principe de Huygens, développement du calcul symbolique, utilisation de la variable complexe etc.) dans une proximité de points de vue et suivant une compétition serrée.

I. Introduction

1) Le Grand Prix sur les Ondes

Notre histoire commence le 15 Novembre 1813, quand la classe de Mathématiques élit une commission chargée de choisir un sujet pour le Prix de Mathématiques, et c'est Poisson, Legendre, Laplace, Delambre et Poinsot qui sont élus. Le 27 Décembre la commission annonce qu'elle a proposé le *Problème des Ondes à la Surface d'un liquide de profondeur indéfinie*, et elle témoigne du désir que Laplace se charge de la rédaction du programme.

Le 24 Juillet 1815, Cauchy lit devant l'Académie un "Mémoire sur le Problème des Ondes", pour lequel Poinsot et Ampère sont élus commissaires. En fait Cauchy semble avoir donné connaissance à cette date des lois de propagation qu'il avait trouvées. Puis le 28 Août, Poisson dépose sur le Bureau un mémoire sur la Théorie des Ondes pour être lu après la fermeture du concours. C'est le 2 Octobre et le 18 Décembre que Poisson lira finalement son mémoire.

La commission du Prix est d'ailleurs élue le 2 Octobre et comprend Legendre, Poisson,Laplace, Biot et Poinsot. Elle décernera le Prix pour 1816, le 26 Décembre, à la pièce qui une fois découverte, se trouve être celle de Cauchy. Il y avait une autre pièce mais nous ne l'avons pas retrouvée.

On voit que Poisson qui a présidé au choix du sujet, s'est mis immédiatement au travail lui-même, dans une compétition de fait avec Cauchy. Le mémoire de Poisson sera publié en 1818, tandis que celui de Cauchy ne le sera qu'en 1827, avec de longues Notes supplémentaires. Dans l'intervalle, les deux savants publieront des articles sur les points techniques nouveaux de leurs mémoires, afin de prendre date de leurs résultats.

Plusieurs points restent peu clairs : comment se fait-il que Cauchy qui participait au Concours ait publiquement lu son mémoire et que l'anonymat n'ait pas été respecté? quel a été le niveau des échanges entre Poisson et Cauchy? enfin notons, bien que ceci ne soit pas extraordinaire, qu' il n'y a eu aucun rapport sur ce Prix.

2) L'histoire du problème physico-mathématique

Lorsqu'on agite l'eau en un endroit de sa surface, on voit se former des ondes qui se propagent circulairement autour d'un point commun, et qui sont dûes aux élévations et abaissements successifs du fluide, au dessus et au dessous de son niveau naturel. Le problème des ondes est de trouver les lois de ces oscillations. Poisson, en introduction à son mémoire, retrace les étapes les plus marquantes de sa résolution.

Newton, dans les *Principia*[1] avait comparé ces oscillations à celles de l'eau dans un syphon renversé et en avait déduit, par une analogie mal fondée avec un pendule, que la vitesse de propagation des ondes, était proportionnelle à la racine carrée de leur largeur[2]. Puis Laplace[3] a traité, à partir des équations différentielles des fluides incompressibles et pesants, et dans l'hypothèse où les vitesses et les oscillations soient assez petites pour qu'on néglige les éléments d'ordre supérieur à un, le cas d'une "déformation trochoïdale de la surface" (c'est-a-dire une ligne serpentante); mais Poisson juge qu'il est plus intéressant de traiter le cas des déformations de *petite* étendue.

Dix ans plus tard, Lagrange[4] traite le cas où la "profondeur du fluide est supposée très petite" et trouve que la propagation des ondes obéit aux mêmes lois que celles du son. Il en déduit que la vitesse de propagation est:
- proportionnelle à la racine carrée de la profondeur du fluide, quand celui-ci est dans un canal de largeur constante,
- constante au cours du temps,
- et indépendante de l'ébranlement primitif.

Puis Lagrange postule que le mouvement excité à la surface d'un fluide de profondeur quelconque, ne se transmet qu'à de très petites profondeurs, et donc son analyse antérieure serait encore valide quelle que soit la profondeur du fluide; la vitesse de propagation des ondes serait encore constante au cours du temps et *"proportionnelle à la racine carrée de la profondeur à laquelle le mouvement est insensible."*[4a]

Poisson conteste d'abord cette extension de Lagrange: pour lui il y a décroissance progressive des ondes et donc on ne peut déterminer de profondeur-frontière, au-delà de laquelle il n'y a plus de propagation (c'est un problème mathématique de limite). Puis par une "analyse dimensionnelle" des équations (au sens de Feynman et des physiciens d'aujourd'hui), Poisson conclut :

> - soit la vitesse est indépendante de l'ébranlement primitif, c'est-à-dire de la forme et du volume du corps plongé, et alors le mouvement des ondes sera semblable à celui des corps graves, à accélération constante;
>
> - soit le mouvement des ondes est uniforme et il faudra que la vitesse dépende de l'ébranlement primitif.

Dans les deux cas, dit Poisson, il y a contradiction avec la *Mécanique Analytique*. Le cas du fluide de profondeur indéfinie restait posé. Il fait l'objet du concours, et nous allons voir que cette contradiction sera un point crucial dans les deux mémoires de Cauchy et Poisson et dans les développements ultérieurs.

En dépit de quelques différences de style[5], les mémoires de Cauchy et de Poisson de l'année 1815, sur la théorie des ondes, sont relativement proches:

- même point de vue dans la démarche physico-mathématique pour aboutir aux équations différentielles du Problème; tous deux font appel à un potentiel de vitesse ou impulsion dont le vecteur-vitesse de chaque élément de fluide est le gradient, et ils arrivent aux mêmes équations à des différences secondaires de notations, près[6]

- les deux séparent les équations de surface et les équations générales (un point qui semble acquis depuis Fourier); mais Cauchy étudie séparément le système différentiel définissant les conditions initiales et accorde une attention spéciale à ce problème.

- les deux savants avancent pas à pas dans l'intégration, en faisant d'abord abstraction de l'une des dimensions horizontales du fluide[7]. Poisson étudie de plus la propagation dans le sens de la profondeur du fluide.

- les techniques mathématiques sont très voisines (transformée de Fourier à plusieurs variables développement systématique en séries de Taylor des fonctions et des intégrales). Poisson utilise plus systématiquement les analogies entre calcul différentiel et calcul des puissances, ce que Cauchy ne fera qu'un peu plus tard.

Pourtant un certain nombre de difficultés mathématiques conduisent à des divergences sur l'interprétation finale des lois de propagation. Cauchy aboutit seulement à des ondes d'accélération constante. Poisson conclut à l'existence de deux types d'ondes.

L'élucidation des points de divergence n'est pas immédiate et s'étale sur plusieurs années; elle accompagne le mouvement d'approfondissement de la rigueur en analyse (manipulation des séries, intégrales singulières, calcul des résidus), qui trouve un premier aboutissement dans les Cours de Cauchy à l'Ecole Polytechnique. Nous mentionnerons certains exemples tirés de ces travaux et les réflexions du mathématicien à leur propos, pour indiquer qu'ils ont sûrement mis Cauchy sur la voie de quelques uns de ses résultats les plus connus mais notre objet n'est pas ici de faire une histoire précise du mouvement d'approfondissement de la rigueur en analyse à cette époque.

Cauchy rectifiera plusieurs points de sa théorie dans des Notes supplémentaires ajoutées au mémoire au moment de sa publication, qui n'intervient pas avant 1827. Sans qu'il soit possible de les dater tout à fait sûrement, les Notes XIV et XV étaient rédigées en 1821, et les Notes XVI à XX l'ont été au plus tard en 1824[8]. L'ensemble est très touffu, désordonné, contradictoire même sur quelques points, Cauchy n'ayant pas voulu reformuler tous ses résultats.

Le mémoire de Poisson sera publié bien avant, en 1818, et sera bientôt suivi d'un mémoire d'analyse sur l'intégration des équations aux différences partielles (lu en 1819 et publié en 1820). Ce dernier représente incontestablement un moment de synthèse des recherches menées par Poisson au cours de cette période. Les progrès suivants appartiendront alors plutôt à Cauchy.

II. Le Mémoire de Cauchy de 1815

1) La modélisation physico-mathématique

Le mémoire de Cauchy est divisé en trois parties: établissement des équations aux dérivées partielles pour l'état initial, puis équations générales du mouvement en distinguant celles qui déterminent les conditions à la surface, enfin énoncé des lois de propagation.

Cauchy suit la démarche d'Euler dans ses mémoires d'hydrodynamique (1755-57) et applique une méthode de bilan des forces à un parallélipipède de fluide. Il considère en fait un "potentiel de vitesse" qu'il appelle *impulsion* q_0 et tel que le champ des vitesses initiales dérive du potentiel q_0/δ (δ est la densité du fluide).

Il retrouve ainsi les équations de Clairaut, la condition d'incompressibilité en variables de Lagrange, aboutit à l'équation de Laplace pour q_0 et cherche les équations qui déterminent les conditions initiales de la surface extérieure.

Pour l'étude du fluide à une époque quelconque du mouvement, il procède de la même faon. A partir de la propriété caractéristique des fluides, il aboutit aux équations d'Euler de l'hydrodynamique (puis à celles de Lagrange si l'on prend, comme variables, les valeurs initiales a, b, c, et le temps t). Il montre que s'il y a impulsion à l'état initial, il y aura impulsion à tout instant, c'est-à-dire qu'un mouvement de fluide irrotationnel à un instant donné le sera à tout instant[9]. Cauchy énonce d'ailleurs l'interprétation physique du potentiel de vitesse puisqu'il note qu'un état réel de mouvement d'un liquide, pour lequel un potentiel de vitesse existe, peut être produit instantanément à partir du repos, en appliquant un système approprié de pressions ; les équations fournissent aussi le système requis de pressions qui pourraient stopper complètement le mouvement.

Avec un potentiel de vitesse noté de façon générale q (q_0 étant le potentiel de vitesse pour l'état initial), les équations d'Euler se réduisent à l'équation dite de Cauchy-Lagrange

$$\frac{1}{\delta}\left(p - \frac{dq}{dt}\right) + \frac{u^2 + v^2 + w^2}{2} - \phi = 0,$$

où ϕ est le potentiel dont dérive la force extérieure appliquée.

Dans le cas de la gravité et d'un vecteur-vitesse infinitésimal, elles s'écrivent (système I):

$$(I)\qquad \begin{aligned} &\Delta q = \frac{\partial^2 q}{\partial x^2} + \frac{\partial^2 q}{\partial y^2} + \frac{\partial^2 q}{\partial z^2} = 0 \\ &u = -\frac{1}{\delta}\frac{\partial q}{\partial x}; v = -\frac{1}{\delta}\frac{\partial q}{\partial y}; w = -\frac{1}{\delta}\frac{\partial q}{\partial z}; p = \frac{\partial q}{\partial t} - \delta y, \end{aligned}$$

où u, v, w sont les composantes de la vitesse en un point (de coordonnées x, y, z), p la pression, δ la densité du fluide (g la constante de gravitation). Elles redonnent (sauf la dernière) en faisant $t = 0$, les équations aux derivées partielles pour l'état initial, de la première partie.

Pour étudier les équations de surface, Cauchy considère que x, y, z, et t, ne sont plus des variables indépendantes; il suppose y fonction des trois autres. En substituant cette fonction y, les quantités u, v, w, p, q deviennent respectivement U, V, W, P, Q qui sont les quantités correspondantes au bout du temps t, au point de la surface de coordonnées x et z. Le problème peut se résumer, dans le cas d'une surface initiale peu perturbée (c'est-a-dire y et ses coefficients différentiels assez petits pour qu'on néglige les termes de second ordre) en le système:

$$(II)\qquad \Delta q = 0; \quad gy - \frac{1}{\delta}\frac{\partial q}{\partial t} = 0; \quad g\frac{\partial q}{\partial y} + \frac{\partial^2 q}{\partial t^2} = 0.$$

Ce sont les mêmes équations auxquelles aboutira Poisson.

En fait Cauchy préfère englober le problème dans la formulation plus générale suivante: q étant une fonction de x, $\alpha + y$, z et t assujettie à vérifier $\Delta q = 0$ et à rendre les deux autres équations susceptibles d'être vérifiées par une seule valeur de y en x, α, z, et t, déterminer la valeur Q que cette fonction q acquiert, lorsque on y substitue la valeur de y en fonction de x, α, z, et t. Q doit alors vérifier le système différentiel:

$$(II')\qquad \begin{aligned} &\frac{\partial^4 Q}{\partial t^4} + g^2\left(\frac{\partial^2 Q}{\partial x^2} + \frac{\partial^2 Q}{\partial z^2}\right) = 0; \quad y = \frac{1}{g\delta}\frac{\partial Q}{\partial t}, \\ &U = -\frac{1}{\delta}\frac{\partial Q}{\partial x}; \quad V = \frac{1}{g\delta}\frac{\partial^2 Q}{\partial t^2}; \quad W = -\frac{1}{\delta}\frac{\partial Q}{\partial z}, \end{aligned}$$

où P est supposée nulle à la surface du fluide.

2) L'intégration des équations par transformée de Fourier

L'outil principal de cette partie est la transformation de Fourier à une ou plusieurs variables, les solutions devant satisfaire les conditions initiales. Rappelons qu'en 1815, Cauchy ne connaissait sans doute pas directement le mémoire de Fourier de 1811 resté inédit, dans lequel Fourier avait commencé à utiliser systématiquement la transformation qui allait porter son nom pour résoudre divers exemples d'équations de diffusion de la chaleur. Mais Cauchy en avait certainement parlé d'une part avec Laplace, dont à cette date il est l'élève, d'autre part sans doute avec Poisson, tous deux ayant été examinateurs du mémoire de Fourier. Si Cauchy devait reconnaître plus tard le rôle de Fourier[10], il ne reconnut jamais clairement la priorité de ce dernier.

Plusieurs développements mathématiques sont traités de façon séparée dans les Notes complémentaires au mémoire. Les 13 premières figuraient dès le début. Par exemple, dans la longue Note IX[11], il donne brutalement la solution de l'équation des ondes (à 2 dimensions spatiales x et y), $\Delta q = 0$, sous la forme:

$$q = \Sigma \int_0^\infty \cos mx e^{+ym} f(m) dm$$

où le signe Σ signifie la somme de deux termes semblables, le deuxième s'obtenant par substitution d'un sinus au cosinus. (Nous supposerons qu'il y a symétrie par rapport à x et ne garderons que le seul terme pair en x). Puisque $y = 0$ sur la surface du fluide, q se réduit à une fonction $F(x)$, telle que:

$$F(x) = \int_0^\infty \cos mx f(x) dm.$$

Alors Cauchy montre que:

$$f(m) = \frac{2}{\pi} \int_0^\infty \cos mu F(u) du;$$

donc

$$F(x) = \frac{2}{\pi} \int_0^\infty \int_0^\infty \cos mx \cos mu F(u) du dm.$$

On note qu'à cette date Cauchy possède ainsi non seulement la représentation en intégrale double de Fourier, mais la démonstration de la transformation réciproque[12]. En voici l'essentiel: Cauchy écrit que

$$\int_0^\infty \int_0^\infty \cos mx \cos mu f(u) dm du =$$
$$\lim_{\alpha \to 0} \int_0^\infty \int_0^\infty \cos mx \cos mu e^{-\alpha m} f(u) dm du.$$

Or on sait que

$$2\int_0^\infty \cos mx \cos mu e^{-\alpha m} dm = \frac{\alpha}{\alpha^2 + (u-x)^2} + \frac{\alpha}{\alpha^2 + (u+x)^2}.$$

Si on suppose α très petit, la deuxième fraction est négligeable pour u réel positif. Il reste:

$$\int_0^\infty \int_0^\infty \cos mx \cos mu e^{\alpha m} f(u) dm du = \frac{1}{2} \int_0^\infty f(u) \frac{\alpha}{\alpha^2 + (u-x)^2} du.$$

Cette dernière intégrale a été appelée *singulière* par Cauchy en 1814, et il montre par la méthode du *Mémoire sur les Intégrales Définies* qu'elle vaut $(\pi/2) \cdot f(x)$. La formule de transformation réciproque est ainsi établie.

Si donc on se donne $F(x)$ qui définit l'état initial, $f(m)$ est déterminée et peut être utilisée pour trouver q, par la formule

$$q = \Sigma \int_0^\infty \cos mx e^{+ym} f(m) dm.$$

Cauchy trouve le même type de solutions avec intégrales doubles de Fourier pour l'équation en dimensions trois. Poisson dérive d'ailleurs l'intégrale de Fourier de la même façon que Cauchy.

Quand ce ne sont pas les conditions initiales mais les conditions aux limites qui sont données, la résolution des équations aux différences partielles se ramène souvent au problème suivant[13]: F et h étant des fonctions données d'une seule variable et la fonction f étant assujettie à:

$$\Sigma \int_0^\infty \cos am f(m) dm = F(a)$$

il faut déterminer la quantité

$$\Sigma \int_0^\infty \cos am h(m) f(m) dm$$

exprimée seulement en fonction de h et de F (le même problème à deux variables est traité). La solution complète du problème est donnée par transformée de Fourier; elle vaut:

$$\Sigma \int_0^\infty \cos amh(m)f(m)dm = \frac{1}{\pi}\int_{-\infty}^\infty \int_0^\infty \cos \mu(\varpi - a)h(\mu)F(\varpi)d\mu d\varpi.$$

Ainsi Cauchy obtient les valeurs des inconnues q, Q, u, v, w, p, x, y, z sous une forme intégrale. Il donne la solution du système précédemment introduit (II'). Il obtient:

$$y = \frac{G}{\pi}\int_0^\infty \cos\sqrt{\mu g}t \cos \mu x d\mu; \quad Q = \frac{G\sqrt{g}\delta}{\pi}\int_0^\infty \sin\sqrt{\mu g}t \cos \mu x \frac{d\mu}{\sqrt{\mu}}$$
$$q = \frac{G\sqrt{g}\delta}{\pi}\int_0^\infty \sin\sqrt{\mu g}t \cos \mu x e^{\mu y}\frac{d\mu}{\sqrt{\mu}}$$

(avec si $y = 0$, $q = Q$)

Cauchy vérifie que q est homogène (l'intégrale est indépendante des unités de mesure de temps et de densité) et par suite x/gt^2 est un nombre abstrait.[14]

Il pose:

$$\frac{x}{\frac{1}{2}gt^2} = \frac{1}{k}$$

La solution du système, dans le cas simplifié de deux variables d'espace s'écrit:

$$Q = \frac{G\delta}{\pi t}\int_0^\infty \sin\sqrt{\mu}\cos\frac{\mu}{2k}\frac{d\mu}{\sqrt{\mu}}; \quad y = \frac{G}{\pi g t^2}\int_0^\infty \cos\sqrt{\mu}\cos\frac{\mu}{2k}d\mu$$
$$U = \frac{G}{\pi g t^3}\int_0^\infty \sin\sqrt{\mu}\sin\frac{\mu}{2k}\sqrt{\mu}d\mu; \quad V = \frac{-G}{\pi g t^3}\int_0^\infty \sin\sqrt{\mu}\cos\frac{\mu}{2k}\sqrt{\mu}d\mu.$$

Cauchy pose

$$\int_0^\infty \cos\sqrt{\mu}\cos\frac{\mu}{2k}d\mu = 2k \cdot K$$

ce qui revient à considérer l'intégrale *a priori* divergente (mais Cauchy n'en dit mot):

$$K = \int_0^\infty \cos\sqrt{2k\mu}\cos\mu d\mu.$$

Alors on a, par exemple, $y = \frac{2Gk}{\pi g t^2}K$. C'est à partir de cette dernière expression qu'il interprètera les solutions des équations aux différences partielles et énoncera les lois des ondes.

3) Lois générales du Mouvement des Ondes (1815):

Cauchy écrit: *"Si pour un instant déterminé, on veut fixer le nombre et la position des ondes à la surface du fluide, il faudra considérer le temps comme une constante; et chercher les valeurs de x qui rendent la valeur de y un maximum absolu, (c.a.d. $\partial(kK)/\partial x = 0$). Si k_1, k_2, ...sont racines réelles de l'équation, on trouve pour valeurs maximum $x = \frac{1}{k_1}\frac{1}{2}gt^2, x = \frac{1}{k_2}\frac{1}{2}gt^2, \ldots$."*[14a] Pour avoir la vitesse de propagation des ondes, on considère t comme variable, et x croît alors comme le carré du temps. La vitesse de la première onde vaudra $(1/k_1)gt$, la vitesse de la seconde sera $(1/k_2)gt$ etc..

Cauchy peut conclure: le mouvement des ondes n'est pas uniforme, contrairement à ce que Lagrange avait écrit dans la *Mécanique Analytique*, mais il est uniformément accéléré, et la vitesse de chaque onde croît indéfiniment.

Dans le mémoire de 1815, ce résultat est commenté ainsi:

> *Avant d'obtenir les intégrales générales des équations du mouvement, j'avais déjà été conduit par des considérations particulières à soupçonner ce résultat, et j'en avais fait part à M.Laplace. Mais je n'osais encore m'arrêter à cette idée, lorsque M. Poisson m'y confirma par cette considération que, pour satisfaire à la condition d'homogénéité, les espaces parcourus par les ondes, supposés qu'ils fussent indépendants de la forme de la surface, devaient être proportionnels à l'espace parcouru par un corps grave, c.a.d. à $1/2gt^2$.*[14b]

L'énoncé descriptif du mouvement des ondes, que donne Cauchy dans son mémoire, peut se résumer ainsi[15]:

1) -la vitesse de chaque onde est indépendante de la petite portion de fluide soulevée ou déprimée à l'origine, et de la courbure de la surface qui terminait cette portion;
   -elle n'est pas constante mais proportionnelle au temps;
   -par suite, l'espace parcouru par chaque onde, et la distance comprise entre les sommets successifs (qu'on peut prendre pour largeur de l'onde), croissent comme le carré du temps ;
2) -les hauteurs d'une onde décroissent comme les carrés des temps.

(1) et (2) impliquent que le volume du fluide renfermé par chaque onde est constant.

3) - la hauteur et la vitesse de chaque onde dépendent du volume que renferme la portion de liquide soulevée ou déprimée, et croissent proportionnellement à ce volume.

4) - il y a symétrie par rapport au centre du mouvement.

Poisson trouve des résultats semblables[16] dans la première partie de son mémoire (lue le 2 Octobre), comme il l'annonçait dans l'introduction que nous avons citée. Mais il aboutit à des résultats différents dans la deuxième partie (lue le 18 Décembre). En effet Poisson montre que les lois de propagation ne sont pas les mêmes à toutes les époques du mouvement. Dans les premiers instants, le mouvement des ondes est uniformément accéléré (suivant les lois de Cauchy) mais lorsque le temps croît, Poisson obtient des ondes propagées à vitesse constante. Celles-ci sont les plus sensibles et donc les plus importantes à considérer.[17]

Cauchy ne semble avoir compris clairement ces résultats qu'ultérieurement. Il ne s'en explique que dans les Notes XV, XVI et suivantes, rédigées en 1824 pour la publication. Une des principales raisons est d'ordre mathématique, comme nous allons le voir; l'autre est dans la reconnaissance d'un phénomène ondulatoire non encore perçu.

III. Les Résultats de Poisson (1815)

1) Les calculs d'approximation des intégrales

Poisson adopte une modélisation très proche de celle de Cauchy. Il se place dans le cas où la perturbation initiale est dûe à l'immersion d'un corps peu enfoncé[18], de façon à pouvoir considérer que les mêmes molécules sont restées à la surface du fluide. Alors, *"la courbe qui termine le segment plongé se confondra sensiblement avec sa parabole osculatrice au point le plus bas; dans ce cas on pourra prendre pour $f(x)$, qui représente l'ordonnée verticale de cette courbe, une valeur de cette forme*

$$f(x) = \frac{h(l^2 - x^2)}{l^2}$$

h étant la flèche du segment plongé, et l représentant la demi- largeur de sa base."[19] Avec cette valeur de $f(x)$ qui détermine l'état initial, Poisson aboutit alors aux résultats suivants[19a]:

$$\phi = \frac{h\sqrt{g}}{\pi l^2} \int_0^\infty \int_{\alpha=-1}^{\alpha=+1} (l^2 - \alpha^2) e^{-ay} \cos(ax - a\alpha) \frac{\sin t\sqrt{ga}}{\sqrt{a}} da d\alpha$$

$$\eta = \frac{1}{g}\left(\frac{d\phi}{dt}\right)_{y=0} = \frac{4h}{\pi l^2} \int_0^\infty \frac{(\sin al - al\cos al)}{a^2} e^{-ay} \cos ax \cos \frac{t\sqrt{ga}}{\sqrt{a}} da$$

où ϕ désigne le potentiel-vitesse et η est l'élévation verticale à la surface.[20] Poisson trouve aussi les valeurs des vitesses des molécules de surface (qui sont égales aux dérivées de l'impulsion par rapport aux variables d'espace) sous une forme analogue.

Quand x est très grand par rapport à α, on peut remplacer $x-\alpha$ par x (et réciproquement) dans les cosinus qui sont sous les signes d'intégration et procéder à des développements en séries et des intégrations. Poisson aboutit en substance à l'expression suivante de l'impulsion[21]:

$$\phi = \frac{gt}{\pi}\int_0^\infty \left[1 - \frac{gt^2}{3!}k + \frac{(gt^2)^2}{5!}k^2 - \cdots\right] e^{ky}\cos kx dk$$

et à celle de l'élévation au dessus de la surface

$$(\star) \qquad \eta = \frac{1}{\pi x}\left(\frac{gt^2}{2x} - \frac{1}{3\cdot 5}\left(\frac{gt^2}{2x}\right)^3 + \frac{1}{3\cdot 5\cdot 7\cdot 9}\left(\frac{gt^2}{2x}\right)^5 - \cdots\right)$$

C'est à partir de ces expressions[22] que Poisson trouve les lois de propagation identiques à celles de Cauchy que nous avons citées. Il est en effet clair qu'une phase particulière de la perturbation de surface (en particulier un zéro, un maximum ou un minimum de η), associée à une valeur définie de $(gt^2/2x)$, voyage sur la surface à accélération constante. D'ailleurs comme l'avait fait Cauchy, Poisson détermine la première onde, celle dont le mouvement est le plus rapide ou qui précède tous les autres. Elle vérifie $(d\eta/dx) = 0$, et il trouve pour le sommet de cette première onde:

$$x = \frac{gt^2}{2}(0,3253)$$

ce qui montre que ce point se propage avec une vitesse égale à un peu moins du tiers que celle des corps pesants[23].

Mais quand (x étant toujours grand par rapport à α) $gt^2/4x$ est du même ordre que x/α, les substitutions précédentes, c'est-a-dire de $x-\alpha$ par x dans $\cos(ax - a\alpha)$, ne sont plus possibles. En particulier la série définissant η est très lentement convergente et l'expression ne convient pas en dehors des moments initiaux de la perturbation[24]. Si l'élévation initiale est confinée au voisinage immédiat de l'origine, $f(\alpha)$ est tel que $\int_{-\infty}^{+\infty} f(\alpha)d\alpha = 1$, le potentiel de vitesse est donné par:

$$\phi = \frac{g}{\pi}\int_0^\infty \frac{\sin \sigma t}{\sigma} e^{ay}\cos kx dk$$

et ϕ à la surface est de la forme[25]:

$$\phi_0 = \frac{g}{\pi}\int_0^\infty \frac{\sin\sigma t}{\sigma} e^{ay} \cos kx dk$$
$$= \frac{1}{\pi}\int_0^\infty \sin(\frac{\sigma^2 x}{g} + \sigma t) d\sigma - \frac{1}{\pi}\int_0^\infty \sin(\frac{\sigma^2 x}{g} - \sigma t) d\sigma$$

En fait pour obtenir cette dernière décomposition, on pose:

$$\zeta = \frac{\sqrt{x}}{\sqrt{g}}(\sigma \pm \frac{gt}{2x}).$$

Il vient:

$$\int_0^\omega \sin(\frac{\sigma^2 x}{g} + \sigma t) d\sigma = \frac{\sqrt{g}}{\sqrt{x}}\int_\omega^\infty \sin(\zeta^2 - \omega^2) d\zeta,$$

et

$$\int_0^\infty \sin(\frac{\sigma^2 x}{g} - \sigma t) d\sigma = \frac{\sqrt{g}}{\sqrt{x}}\int_{-\omega}^\infty \sin(\zeta^2 - \omega^2) d\zeta, \text{ oú } \omega = \sqrt{\frac{gt^2}{4x}}.$$

Il vient:

$$\phi_0 = \frac{2\sqrt{g}}{\pi\sqrt{x}}\int_0^\omega \sin(\zeta^2 - \omega^2) d\zeta \text{ et } \eta = \frac{\sqrt{g}t}{\pi x^{3/2}}\int_0^\omega \cos(\zeta^2 - \omega^2) d\zeta,$$

soit

$$\eta = \frac{\sqrt{g}t}{\pi x^{3/2}}\left[\cos\omega^2 \int_0^\omega \cos\zeta^2 d\zeta + \sin\omega^2 \int_0^\omega \sin\zeta^2 d\zeta\right]$$

Les intégrales définies sont pratiquement celles qu'utilisera Fresnel[26] et Poisson aboutit finalement à l'expression:

$$(\star\star) \qquad \eta = \frac{4h\sqrt{kl}}{k^3\sqrt{2\pi x}}(\sin k - k\cos k)[\cos\frac{gt^2}{4x} + \sin\frac{gt^2}{4x}]$$

On peut encore écrire $\eta = S \cdot T$ où $T = [\cos\frac{gt^2}{4x} + \sin\frac{gt^2}{4x}]$. C'est cette expression ($\star\star$) qui est utile dès que t est assez grand, et que $gt^2/4$ est grand par rapport à x.

2) L'onde dentelée et la notion de "groupe de vitesse":

Poisson remarque que le facteur T entre crochets dans η est une fonction périodique, dont le maximum est égal à $\pm\sqrt{2}$ et a lieu quand $gt^2/4x$ vaut $n\pi + \pi/4$. On en déduit facilement que les amplitudes des oscillations d'égales durées sont inversement proportionnelles à $\sqrt{x}$ (c'est-a-dire à la racine carrée de la distance du point où on les mesure, à l'ébranlement primitif). La période de ces oscillations vaut

$$t_0 = \frac{\pi\sqrt{l}}{\sqrt{gk}}$$

au premier ordre près. De plus la fonction T varie aussi très rapidement par rapport à x. Poisson note que ses maximas et minimas se succèdent alternativement à de très petits intervalles, dans tout l'étendue de la surface fluide et *"la petite distance entre deux sommets consécutifs* (en relief et en creux) *peut être prise pour largeur de ces ondes".*[27] En la désignant par λ (et en négligeant son carré), il vient:

$$t_0 = \sqrt{\frac{\pi\lambda}{g}}$$

ce qui prouve que la durée des oscillations, en un point quelconque, est proportionnelle à la racine carrée de la largeur des ondes en ce point, et au même instant.[28] Enfin l'équation $\sin k - k\cos k = 0$, qui annule l'amplitude des oscillations verticales, permet de *"déterminer sur la surface des points qui n'ont aucun mouvement vertical et qu'on pourra regarder comme des espèces de noeuds, mobiles à cette surface: l'espace compris entre deux noeuds consécutifs forme un* groupe d'ondes, *que l'on peut considérer comme une seule onde,* dentelée *dans toute son étendue, laquelle paraît se mouvoir à la surface, en s'élargissant à raison de la différence de vitesse des deux noeuds qui la termine."*[29] Poisson vérifie que le mouvement de ces noeuds est uniforme et leur vitesse constante.

Poisson étudie ensuite les points de la surface qui font les plus grandes oscillations verticales. Ceux-ci vérifient l'équation:

$$\frac{dS}{dx} = 0 \text{ où } S = \frac{\eta}{T} = \frac{4h\sqrt{kl}}{k^3\sqrt{2\pi x}}(\sin k - k\cos k).$$

Les calculs conduisent à l'équation

$$(4k^2 - 9)\tan k + 9k = 0.$$

Chacune des racines de cette équation définit une onde, dont le sommet sera à une distance du centre de la perturbation initiale, égale à:

$$x = \frac{t\sqrt{gl}}{2\sqrt{k}}$$

ce qui prouve que les points d'amplitude maximum se meuvent, comme les noeuds, uniformément et avec une vitesse proportionnelle à $\sqrt{l}$. Le premier maximum précède le premier noeud, puis il y a un maximum entre le premier et le second noeud, etc. Poisson définit la vitesse d'une telle "onde dentelée" comme la vitesse apparente du point de cette onde qui répond aux plus grandes oscillations verticales[30]. Poisson calcule cette amplitude maximale: c'est une fonction de k qui croît de $k = 0$ à une certaine valeur comprise entre les deux plus petites racines de l'équation

$$(4k^2 - 9)\tan k + 9k = 0$$

puis elle décroît indéfiniment quand k augmente[31].

En résumé dans la région où la deuxième formule pour η $(\star\star)$ est valide, à tout instant, les changements en longueur et en hauteur d'une onde à une autre, sont très graduels; un nombre considérable d'ondes consécutives, de longueurs d'onde proches, peut être approximativement représenté par une courbe sinusoïdale dentelée qui voyage à vitesse constante. En fait dans ce groupe d'ondes, il y a des ondelettes qui se propagent plus vite que le paquet d'ondes lui-même, semblent naître derrière lui et le traverser, pour mourir devant lui. En effet les ondelettes sont uniformément accélérées et continuellement étirées en longueur, mais leurs amplitudes sont inversement proportionnelles à la distance à l'origine des ondes. Poisson n'a pas étudié ces ondelettes, c'est Cauchy qui le fait dans la Note XVI de son mémoire.[32]

Cet exemple des rides provoquées dans un étang par l'impact d'un caillou jeté dans l'eau est devenu l'archétype pédagogique des phénomènes ondulatoires dans un milieu dispersif (soit quand la vitesse de propagation dépend de la longueur d'onde) et Poisson est le premier à mettre en évidence et à commencer à mathématiser ce phénomène qui aura une très grande importance ultérieurement, tant dans la théorie ondulatoire et la physique classique du XIXème siècle, qu'en mécanique quantique[33].

3) Les oscillations de niveau

Il n'est sans doute pas inutile de proposer ici une visualisation des perturbations[34]

1ère Figure: les variations de η par rapport au temps, en un lieu donné. (Pour différents lieux, les intervalles entre phases assignées varient comme $\sqrt{x}$, tandis que les élévations correspondantes varient comme $1/x$)

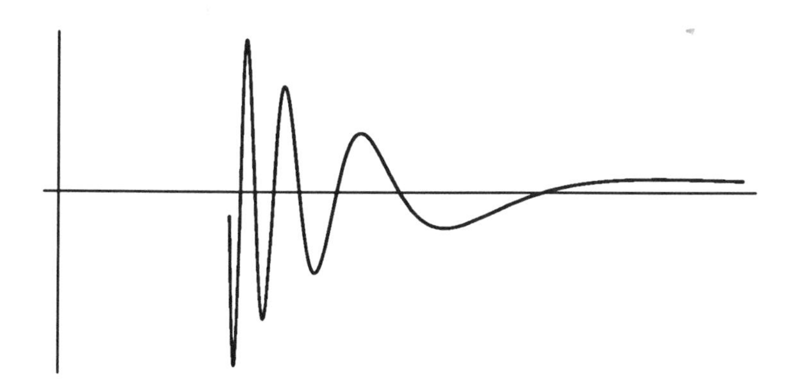

(l'unité sur l'échelle horizontale est $\sqrt{2}x/g$; celle sur l'échelle verticale est $A/\pi x$, si A est l'aire de la section initialement perturbée.)

2ème Figure: Profil de la vague en un instant donné.
(en différents temps, les distances horizontales entre points correspondants varient comme t^2, tandis que les élévations correspondantes varient comme $1/t^2$, si t est le temps écoulé depuis le début de la perturbation)

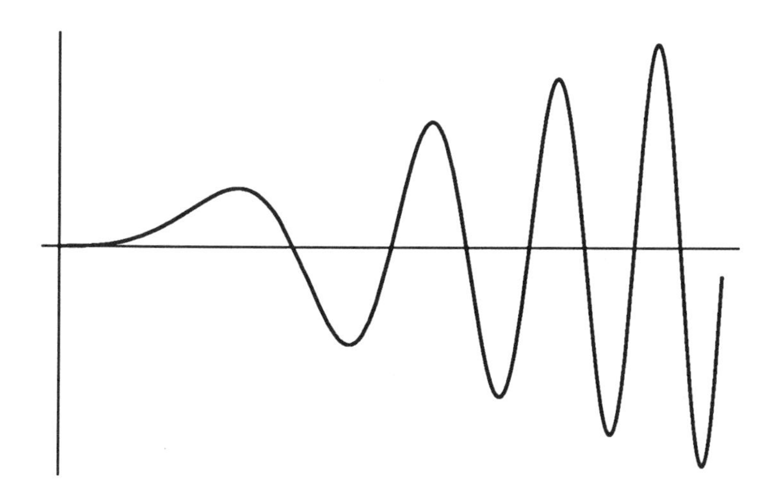

(Ici l'unité horizontale est $1/2 \cdot gt^2$, l'unité verticale est $2A/\pi gt^2$)

IV. Cauchy, l'elucidation des Divergences

C'est dans les Notes XIV à XX ajoutées pour la publication de son mémoire, que Cauchy revient sur ses résultats de 1815 et se trouve cette fois en accord avec les résultats de Poisson. La note XVI est un véritable deuxième mémoire sur les ondes. Les autres notes portent plutôt sur des points mathématiques et Cauchy indique quelques unes de ses erreurs antérieures.

1) La Note XVI

Reprenant les calculs de son mémoire initial, Cauchy rappelle la formule qui permet de déterminer les sommets des ondes et qui se trouvait déjà dans la Troisième Partie[35]:

$$y = \frac{1}{4\sqrt{2\pi}} \int_{-\alpha}^{+\alpha} \frac{\sqrt{g}t}{(x-\varpi)^{3/2}} (\sin \frac{gt^2}{4(x-\varpi)} + \cos \frac{gt^2}{4(x-\varpi)}) F(\varpi) d\varpi$$

où $F(\varpi)$ est la fonction initiale de perturbation qui ne conserve des valeurs sensibles que pour ϖ compris entre $-\alpha$ et $+\alpha$.

Quand t croît, écrit Cauchy, on ne peut plus remplacer le développement de

$$\frac{gt^2}{4(x-\varpi)} = \frac{gt^2}{4x} + \frac{gt^2\varpi}{4x^2} + \frac{gt^2\varpi^2}{4x^3} + \cdots \text{ par son seul premier terme } \frac{gt^2}{4x};$$

il faut considérer la somme des termes qui conservent une valeur sensible finie entre les limites $+\alpha$ et $-\alpha$. Et dans ce cas *"on se trouvera conduit aux formules que Poisson a données dans son second Mémoire présenté à l'Institut en décembre 1815, et à l'aide desquelles il a établi le premier l'existence d'une série d'ondes propagées avec des vitesses constantes."* Et il ajoute: *"J'étais parvenu moi-même à de semblables formules, en examinant le cas où t prend une valeur considérable. Mais croyant que l'espèce de mouvement qu'elles exprimaient était trop peu sensible pour qu'on pût en tenir compte, et me trouvant d'ailleurs pressé par le temps, je n'avais pas cherché à les discuter, et ne les avais pas transcrites sur mon Mémoire."*[36]

Ainsi la première question qui avait initialement fourvoyé Cauchy, relève du domaine de convergence des séries, et donc de la validité des formules d'approximation. Rectifiant les formules antérieures du mémoire de 1815, Cauchy a cette phrase fort surprenante: *"La nature du mouvement change avec la méthode d'approximation"*,[37] phrase qui en dit long sur sa conception du rapport de la description mathématique aux phénomènes physiques. On ne peut pas dire que Cauchy était préoccupé par l'ontologie des phénomènes quand il faisait des mathématiques![38]

Le reste de la Note XVI est consacré à des calculs similaires à ceux de Poisson, mais Cauchy les mène dans des cas plus généraux que celui où le corps immergé est représenté par sa parabole osculatrice. Il écrit: *"il n'est nullement nécessaire de prendre pour $F(x)$ une fonction qui conserve, pour toutes les valeurs de x, la même forme analytique. Il en résulte qu'on peut donner à cette fonction telle forme que l'on jugera convenable, entre les limites $x = -\alpha, x = +\alpha$ et supposer qu'elle devienne constamment nulle en dehors de ces limites"*[39]

Ainsi Cauchy retrouve les groupes d'ondes à vitesse constante, exhibés par Poisson. Il appelle *sillons* ce que Poisson avait désigné par *dents*, c'est-à-dire les ondelettes d'un même paquet d'ondes[40]. Cauchy montre que le mouvement des sillons n'est pas uniforme comme celui des groupes d'ondes mais uniformément accéléré. Ce qui veut dire que si on fixe son attention sur une onde particulière dans le groupe d'ondes, on voit qu'elle avance à travers le groupe, mourant graduellement en atteignant le front avant, tandis que la première place dans le groupe est occupée successivement par des ondes qui viennent de l'arrière. Les notions de groupe d'ondes et de vitesse de groupe qui n'avaient pas été initialement vues par Cauchy, sont cette fois plus complètement décrites que chez Poisson.

2) Les difficultés mathématiques

Rappelons que pour déterminer la vitesse et la position des ondes, Cauchy devait calculer les racines de l'équation $\partial(kK)/\partial k = 0$. Mais l'intégrale K est non-convergente (sans que ceci soit explicitement dit) et Cauchy avait cherché à l'approximer par des développements asymptotiques. Il introduit un facteur de convergence exponentiel et définit la valeur de K par:

$$\int_0^\infty \cos\sqrt{2k\mu}\cos\mu d\mu = \lim_{\alpha\to 0}\int_0^\infty \cos\sqrt{2k\mu}\cos\mu e^{-\alpha\mu} d\mu.$$

Par suite il développait en séries de Taylor $\cos\sqrt{2k\mu}$ et intègrait ce développement, à condition toujours de considérer

$$\int_0^\infty \mu^a \cos\mu d\mu = \lim_{\alpha\to 0}\int_0^\infty \mu^a \cos\mu e^{-\alpha\mu} d\mu$$

$$\text{et d'utiliser } \int_0^\infty \mu^a e^{-\alpha\mu} d\mu = \Gamma(a+1)$$

résultat établi par Legendre, et qu'il avait rappelé dans la Note III.[41] Cauchy aboutit au développement en séries suivant:

$$K = \frac{2k}{2} - \frac{(2k)^3}{4\cdot 5\cdot 6} + \frac{(2k)^5}{6\cdot 7\cdot 8\cdot 9\cdot 10} + \cdots$$

Toujours dans cette Note III, il calcule d'autres intégrales comme:

$$K_2 = \int_0^\infty e^{-\sqrt{2k\mu}} \cos \mu d\mu; \quad K_4 = \int_0^\infty \sin \sqrt{2k\mu} \sin \mu d\mu$$

toujours en introduisant le facteur de convergence exponentiel. Il trouve[42]:

$$K_4 = \int_0^\infty \sin \sqrt{2k\mu} \sin \mu d\mu = \frac{\sqrt{\pi k}}{2}(\sin \frac{k}{2} + \cos \frac{k}{2}),$$

et aboutit à la formule $K = K_4 - K_2$, qui s'écrit:

$$K = \frac{\sqrt{\pi k}}{2}(\sin \frac{k}{2} + \cos \frac{k}{2}) - \int_0^\infty e^{-\sqrt{2k\mu}} \cos \mu d\mu.$$

Pour la valeur de k devenant très grande, Cauchy propose de se servir de cette dernière expression de K, en négligeant le deuxième terme formé par l'intégrale. Ces manipulations posent plusieurs problèmes d'interversion: entre le signe Σ de la série, et le signe $\int$ de l'intégrale, et entre la série et le passage à la limite. En effet, le calcul de Cauchy peut s'écrire:

$$K = \lim_{a \to 0} \int_0^\infty e^{-a\mu} (\sum \frac{(-1)^n (\sqrt{2k\mu})^n}{n!}) \cos \mu d\mu.$$

Tous les instruments "élémentaires" à notre disposition aujourd'hui, comme le théorème d'interversion pour les séries uniformément convergentes, ou le théorème de convergence dominée de Lebesgue, échouent à justifier ces calculs. Il faut une analyse encore plus fine, évidemment tout à fait inexistente à l'époque.

Cauchy revient sur cette intégrale K dans la Note XVII[43],

$$K = \int_0^\infty \cos \sqrt{2k\mu} \cos \mu d\mu,$$

et il écrit que si l'on développe successivement en séries:

1°) le facteur $\cos \sqrt{2k\mu}$

2°) le facteur $\cos \mu$ ou bien on procède par intégrations par parties successives, on obtient respectivement deux expressions pour le développement de K:

$$K = \frac{2k}{2} - \frac{(2k)^3}{4 \cdot 5 \cdot 6} + \frac{(2k)^5}{6 \cdot 7 \cdot 8 \cdot 9 \cdot 10} + \cdots$$

et

$$K = -\frac{1}{k}(1 - \frac{3 \cdot 4 \cdot 5}{(2k)^2} + \frac{5 \cdot 6 \cdot 7 \cdot 8 \cdot 9}{(2k)^4} - \cdots),$$

et Cauchy affirme que le deuxième développement est erroné[44]. Pour prouver cela, Cauchy se contente d'exhiber une conséquence absurde.

Ainsi il vérifie que ce second développement est opposé au développement en série de $\int_0^\infty e^{-\sqrt{2k\mu}} \cos \mu d\mu$. Or K et $-\int_0^\infty e^{-\sqrt{2k\mu}} \cos \mu d\mu$ ne peuvent être égaux quand k tend vers l'infini puisque d'après le résultat établi dans la note III, on a

$$K = \frac{\sqrt{\pi k}}{2}(\sin \frac{k}{2} + \cos \frac{k}{2}) - \int_0^\infty e^{-\sqrt{2k\mu}} \cos \mu d\mu.$$

Il note: *"On s'expose à de graves erreurs lorsqu'on détermine les fonctions par le moyen de leurs développements en série, sans tenir compte de leurs restes"*[44a] Cauchy remarque qu'il est essentiel qu'après un certain nombre d'intégrations partielles, la valeur que représente le reste soit fort petite. Or ceci est évidemment vrai pour $\int_0^\infty e^{-\sqrt{2k\mu}} \cos \mu d\mu$ mais non pas pour $\int_0^\infty \cos \sqrt{2k\mu} \cos \mu d\mu$, d'où encore un appel à la vigilance dans l'usage des séries.

Soulignons que Cauchy n'était pas arrivé en 1815 à trouver la forme intégrale de Poisson pour l'équation des ondes (c.f. ci-dessous) ; dans la Note XV (sans doute de 1821), il impute son échec à des erreurs mathématiques, toujours du même type. Il indique qu'il avait initialement trouvé, dans le cas de deux dimensions, l'intégrale

$$\begin{aligned} Q = 2\Sigma \int_0^\infty & \cos mx \cos \sqrt{ghm} t \phi(m) dm \\ & + 2\Sigma \int_0^\infty \cos mx \sin \sqrt{ghm} t \psi(m) dm \end{aligned}$$

et qu'on peut réduire le premier terme à

$$\int_0^\infty \int_0^\infty \cos 2k\sqrt{\mu\nu} \sin(\mu + \nu) d\mu d\nu.$$

Reste à calculer cette dernière intégrale.

En 1821, Cauchy écrit qu'on y parvient en remplaçant k par $2k\sqrt{-1}$ dans le développement en série de

$$\int_0^\infty \int_0^\infty \sin(\mu + \nu) e^{-k\sqrt{\mu\nu}} d\mu d\nu = \frac{\pi k}{4}(1 - \frac{3}{2}\frac{k^2}{4} + \cdots) = \frac{\pi k}{4}(1 + \frac{k^2}{4})^{-3/2}$$

(mais en 1815, Cauchy n'effectuait pas ces substitutions avec la variable imaginaire...).

Cauchy écrit: "*Mais au lieu d'établir ces équations, desquelles on passe facilement à [l'intégrale de Poisson] je m'étais arrêté devant cette considération que, si l'on développe l'expression* $\cos 2k\sqrt{\mu\nu}$ *et par suite l'intégrale*

$$\int_0^\infty \int_0^\infty \cos 2k\sqrt{\mu\nu}\sin(\mu+\nu)d\mu d\nu.$$

en séries ordonnées suivant les puissances ascendantes de k, tous les termes du développement de l'intégrale se réduiront constamment à zéro. Cette circonstance est d'autant plus remarquable que le développement du cosinus produit une série toujours convergente, et elle fait voir que, dans la solution des problèmes, on ne doit user qu'avec beaucoup de circonspection des développements en séries."[45]

Il est donc intéressant de souligner que cette intégrale double a fourni probablement à Cauchy l'exemple d'une fonction (de k) dont le développement en série de puissances est identiquement nul alors que la fonction ne l'est pas. Souvenons nous ici que Cauchy ne fournit le fameux exemple analogue

$$e^{-\frac{1}{x^2}}$$

qu'en 1823, dans une foulée de notes qui paraissent dans le *Bulletin de la Société Philomatique*, et qui portent sur des questions mathématiques faisant suite à ce mémoire sur la théorie des ondes. C'est dans celle intitulée "Sur le développement des Fonctions en séries et sur l'Intégration des Equations différentielles ou aux Différences partielles"[46], que Cauchy soulève le problème de cette exponentielle, en notant qu'à une série de MacLaurin donnée (c'est-à-dire $f(0)$, $f'(0)$, $f''(0)$,... système de valeurs données) ne correspond pas toujours une seule fonction $f(x)$; en particulier on peut trouver une infinité de fonctions différentiables dont les développements en séries ordonnées suivant les puissances ascendantes de x se réduisent à zéro. Et il souligne que ceci est vrai même si la série est convergente. Cauchy conclut, là encore, à l'insuffisance des méthodes d'intégration fondées sur le développement en séries. La Note se situe bien dans une problématique de l'unicité (et non de l'analyticité), propre à l'étude des équations différentielles.

Ces difficultés mathématiques ont eu des répercussions immédiates sur l'obtention des lois de propagation des ondes. En effet rappelons qu'il s'agit des mêmes calculs avec $k = \dfrac{gt^2}{2(x-\varpi)}$, et on remplace μ par

$\mu(x - \varpi)$, l'expression qui permet de déterminer les sommets des ondes étant:

$$y = \frac{1}{4\sqrt{2\pi}} \int_{-\alpha}^{+\alpha} \frac{\sqrt{g}t}{(x-\varpi)^{3/2}} (\sin \frac{gt^2}{4(x-\varpi)} + \cos \frac{gt^2}{4(x-\varpi)}) F(\varpi) d\varpi$$

où $F(\varpi)$ est la fonction initiale de perturbation qui ne conserve des valeurs sensibles que pour ϖ compris entre $-\alpha$ et $+\alpha$.

En résumé, dans ces calculs, Cauchy a manipulé des intégrales divergentes, en introduisant des facteurs de convergence de type exponentiel, et en développant systématiquement les fonctions en séries entières. Toutes ces manipulations sont évidemment soumises à des restrictions qui ne sont pas maîtrisées. Les résultats "élémentaires" sont énoncés correctement dans les traités de l'Ecole Polytechnique, mais ici il ne s'agit pas seulement de cela. En fait l'aboutissement de ces recherches réside plutôt dans les résultats énoncés par Cauchy[47], en 1842, à propos des intégrales de Fresnel. Il s'agit d'une méthode pour le développement en séries des intégrales de Fresnel,

$$\int_v^\infty \cos \frac{\pi}{2} v^2 dv \text{ et } \int_v^\infty \sin \frac{\pi}{2} v^2 dv$$

qui conduit au résultat singulier que des séries, divergentes pour toute valeur de la variable v, convergent cependant rapidement lorsqu'on se borne à considérer leurs premiers termes, dans le cas où v a une valeur supérieure à 1 et assez grande. Cauchy fait un calcul qui rappelle beaucoup ce à quoi il fait référence assez confusément, plusieurs fois dans ces Notes au mémoire sur les ondes. Voici en substance ce calcul:

$$\int_v^\infty \cos \frac{\pi}{2} v^2 dv = \int_v^\infty \pi v \cos \frac{\pi}{2} v^2 \frac{dv}{\pi v} = \left(\frac{1}{\pi v} \sin \frac{\pi}{2} v^2 \right|_v^\infty + \int_v^\infty \sin \frac{\pi}{2} v^2 \frac{dv}{\pi v^2}$$

$$\int_v^\infty \sin \frac{\pi}{2} v^2 dv = - \left(\frac{1}{\pi^2 v^3} \cos \frac{\pi}{2} v^2 \right|_v^\infty - 3 \int_v^\infty \cos \frac{\pi}{2} v^2 \frac{dv}{\pi^2 v^4}$$

$$\int_v^\infty \cos \frac{\pi}{2} v^2 \frac{dv}{\pi^2 v^4} = - \left(\frac{1}{\pi^3 v^5} \sin \frac{\pi}{2} v^2 \right|_v^\infty + 5 \int_v^\infty \sin \frac{\pi}{2} v^2 \frac{dv}{\pi^3 v^9}$$

d'où par substitutions successives:

$$\int_v^\infty \cos\frac{\pi}{2}v^2 dv = \cos\frac{\pi}{2}v^2\left[\frac{1}{\pi^2 v^3} - \frac{1\cdot 3\cdot 5}{\pi^4 v^7} + \frac{1\cdot 3\cdot 5\cdot 7\cdot 9}{\pi^6 v^{11}} - \cdots\right] +$$
$$\sin\frac{\pi}{2}v^2\left[-\frac{1}{\pi v} + \frac{1\cdot 3\cdot 5}{\pi^3 v^5} + \frac{1\cdot 3\cdot 5\cdot 7}{\pi^5 v^9} + \cdots\right] + R$$

ou le reste R est égal à:

$$1\cdot 3\cdot 5\cdots(2n-1)\int_v^\infty \sin\frac{\pi}{2}v^2\frac{dv}{\pi^n v^{2n}}$$

où à

$$-1\cdot 3\cdot 5\cdots(2n-1)\int_v^\infty \cos\frac{\pi}{2}v^2\frac{dv}{\pi^n v^{2n}}$$

suivant que le rang du terme, auquel on s'arrête, est pair ou impair. D'où:

$$|R| < \frac{1\cdot 3\cdot 5\cdots(2n-3)}{\pi^n v^{2n-1}}$$

et les "séries de Cauchy" ne peuvent servir au calcul des intégrales de Fresnel que pour des valeurs assez grandes de v.

V. Le Mémoire de Poisson de 1819

Revenons un peu en arrière avec le mémoire de Poisson[1819/1820], *"Sur l'Intégration des équations linéaires aux différences partielles et particulièrement de l'équation générale du mouvement des fluides élastiques"*, mémoire particulièrement remarquable.

En effet non seulement Poisson y donne l'intégrale de l'équation des ondes à 2 et 3 variables d'espace sous une forme devenue classique et qui permettait l'interprétation physique du phénomène, et y développe encore les mêmes techniques de transformée de Fourier, mais il met aussi l'accent sur le calcul symbolique et propose une méthode générale d'intégration pour de nombreuses équations linéaires aux dérivées partielles. Poisson devance ici incontestablement Cauchy qui donnera de multiples travaux à sa suite, à partir de 1821.[48]

1) L'intégrale de l'équation des ondes sphériques

La solution de l'équation des ondes vibrantes (à une seule variable)

$$\frac{d^2\phi}{dt^2} = a^2\frac{d^2\phi}{dx^2}$$

avait été donnée par d'Alembert: $\phi = f_1(x - at) + f_2(x + at)$.

Dans le cas de la propagation dans l'espace, la solution n'était connue que pour $\phi = \phi(t, r)$, c'est-à-dire pour l'équation des ondes sphériques isotropiques

$$\frac{d^2\phi}{dt^2} = a(\frac{d^2\phi}{dr^2} + \frac{2}{r}\frac{d\phi}{dr}),$$

et la solution avait été trouvée par Lagrange:

$$\phi = \frac{1}{r} \cdot [f_1(r - at) + f_2(r + at)].$$

Poisson s'occupe de l'équation des petits mouvement des fluides à 4 variables indépendantes: *"l'intégrale à laquelle je suis parvenu dans ce mémoire, est d'une forme très simple: elle ne contient que des intégrales définies doubles; et les deux fonctions arbitraires s'y déterminent immédiatement d'après l'état initial du fluide ; Le procédé qui m'y a conduit est très simple: il est fondé sur un théorème relatif aux intégrales définies et sur les* analogies connues des puissances et des différences, *que j'ai employées dans tout mémoire pour trouver d'une manière plus rapide, les sommes des séries par lesquelles j'ai d'abord exprimées les intégrales des équations que j'ai considérées... Quand on substitue les coordonnées polaires des molécules fluides [...] elle montre clairement que quelquesoit l'ébranlement primitif, le mouvement se propage avec* la même vitesse dans tous les sens, *quoique les vitesse propres des molécules ne soient pas les mêmes suivant toutes les directions."*

Suivons la méthode de Poisson, pour mettre en évidence le fonctionnement du calcul symbolique des opérateurs de dérivation. Soit:

$$\frac{d^2\phi}{dt^2} = a^2\Delta\phi = a^2(\frac{d^2\phi}{dx^2} + \frac{d^2\phi}{dy^2} + \frac{d^2\phi}{dz^2}).$$

Il introduit la notation:

$$\Delta q = \delta q$$
$$\Delta(\delta q) = \delta^2 q$$
$$\delta^n q = \delta\delta^{n-1} q \text{ etc..}$$

L'intégrale complète de l'équation des ondes s'écrit:

$$\phi = U + \frac{a^2t^2}{2!}\delta U + \frac{a^4t^4}{4!}\delta^2 U + \frac{a^6t^6}{6!}\delta^3 U + \cdots + tV + \frac{a^2t^3}{3!}\delta V + \frac{a^4t^5}{5!}\delta^2 V + \cdots$$

U, V, sont des fonctions arbitraires de x, y, z. Si T est la deuxième somme infinie, on remarque que la première s'en déduit en différenciant par rapport à t et en remplaçant V par U. Poisson utilise les analogies connues des puissances et différences:

$$\delta^n V = (g^2 + h^2 + k^2)^n V$$

pourvu qu'on regarde les puissances de g, h, k comme des signes d'opérations qui indiquent les différentielles relatives à x, y, z, c'est-à-dire:

$$Ag^i h^{i'} k^{i''} V = A\frac{d^{i+i'+i''}V}{dx^i dy^{i'} dz^{i''}}.$$

D'où

$$T = (1 + \frac{a^2t^2}{2!}\frac{p^2}{3} + \frac{a^4t^4}{4!}\frac{p^4}{5} + \cdots)Vt, \text{ où } p^2 = g^2 + h^2 + k^2.$$

Si on pose: $\alpha = g\cos u + h \sin u \sin v + k \sin u \cos v$, on peut établir que

$$\int\int \alpha^{2n+1} \sin u du dv = 0 \text{ et } \int\int \alpha^{2n} \sin u du dv = \frac{4\pi p^{2n}}{2n+1}.$$

Poisson en déduit alors:

$$T = \frac{1}{4\pi}\int\int (1 + at\alpha + \frac{a^2t^2\alpha^2}{2!} + \cdots)Vt \sin u du dv = \frac{1}{4\pi}\int\int e^{at\alpha}Vt \sin u du.$$

En vertu des mêmes analogies entre calcul différentiel et calcul des puissances, on a:

$$e^{gx'} \cdot e^{hy'} \cdot e^{kz'} V = f(x + x', y + y', z + z,).$$

Finalement il vient:

$$\phi = \int_0^{2\pi}\int_0^{\pi} f(x + at\cos u, y + at \sin u \sin v, z + at \sin u \cos v)t \sin u du dv$$
$$+ \frac{d}{dt}\int_0^{2\pi}\int_0^{\pi} F(x + at\cos u, y + at \sin u \sin v, z + at \sin u \cos v)t \sin u du dv.$$

Les fonctions f et F sont liées aux conditions initiales par[49]:

$$\phi(x,y,z,0) = 4\pi F(x,y,z); \quad \phi_{,t}(x,y,z,0) = 4\pi f(x,y,z).$$

2) L'interprétation physique: le principe de Huygens

Nous allons faire une digression pour rappeler l'explication physique de ce résultat, en précisant qu'elle ne se trouve pas dans le mémoire de Poisson. Elle émerge lentement, d'abord au cours des divers mémoires successifs de Poisson et Cauchy, et surtout de Fresnel, dans lesquels elle est utilisée, mais elle n'est véritablement maîtrisée qu' au début du XXème siècle[50].

Avec un changement de coordonnées sphériques, on voit qu'on intègre sur une sphère S_{at} de rayon at, centrée au point P de coordonnées x, y, z. La formule intégrale de Poisson s'écrit de faon condensée:

$$4\pi U(P,t) = \int_{S_{at}} t u_{,t}(M,0) d\sigma + \frac{d}{dt} t \int_{S_{at}} u(M,0) d\sigma$$

où M est le point courant de cette sphère S_{at}. Soit une perturbation initiale dûe à un corps V telle que F et f soient définies sur V et nulles en dehors de V [51]. Physiquement une onde part de V et se répand dans l'espace. La formule de Poisson dit ce qui se passe en n'importe quel point $P(x,y,z)$ extérieur à V:

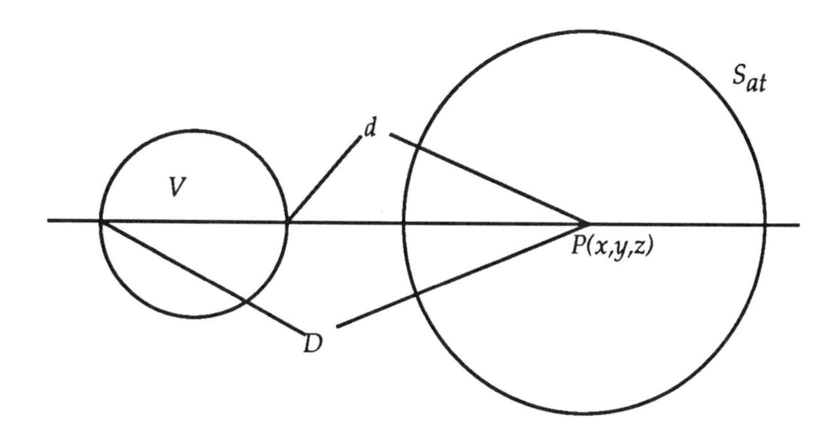

Soient d et D les minimum et maximum des distances de P aux points de V.

- quand $t < d/a$, les intégrales sont nulles car le domaine d'intégration (c'est-à-dire S_{at}) ne coupe pas le corps V. Ceci veut dire que l'onde partant de V n'a pas encore atteint le point P;
- à $t = d/a$, la sphère S_{at} touche le corps V.
- de $t = d/a$ à $t = D/a$, S_{at} coupe le corps V et donc $\phi(P,t) \neq 0$
- pour $t > D/a$, S_{at} englobe V ce qui veut dire que la perturbation est passée par P, donc $\phi(P,t) = 0$ à nouveau.
- l'instant $t = D/a$ correspond au passage de la queue de l'onde à travers P.

En un instant t donné, la tête de l'onde prend la forme d'une surface qui sépare les points non atteints par la perturbation de ceux qui sont atteints. C'est l'enveloppe d'une famille de sphères centrées sur la frontière de V et de rayon at. Ceci traduit la *linéarité* de la théorie: chaque point du front d'onde peut être pris comme nouveau point de départ de la perturbation, il y a superposition des effets de ces "sources" fictives.

La queue de l'onde au temps t est une surface séparant les points affectés par la perturbation de ceux pour qui elle est déjà passée. On voit donc que la perturbation, localisée dans l'espace, donne naissance en chaque point P à un effet qui ne disparaît qu'après un temps fini. L'onde a un bord initial et un bord final.

Le phénomène entier s'appelle le *"principe de Huygens"*.

Jacques Hadamard[52] a montré que le principe de Huygens pouvait être pris dans plusieurs sens différents, qui n'avaient pas été suffisamment distingués entre eux ; pour analyser chacun d'entre eux il propose de le mettre sous forme d'un syllogisme que nous allons également utiliser:

- (A) *(Majeure) "L'action des phénomènes produits à l'instant $t = 0$, sur l'état des choses au temps ultérieur $t = t_0$, s'opère par l'entremise de chaque instant intermédiaire $t = t'$, c'est-à-dire que (en supposant $0 \leq t' \leq t_0$) pour trouver ce qui se passe au temps $t = t_0$, on peut déduire de l'état en $t = 0$ celui qui existera pour $t = t'$, et de ce dernier, l'état cherché à l'instant final $t = t_0$.*
- (B) *(Mineure) Si, à l'instant unique $t = 0-$ ou au moins fictivement, pendant un court intervalle $\epsilon \leq t \leq 0$ - on produit une perturbation lumineuse[53] localisée au voisinage immédiat de 0, son effet sera localisé, pour $t = t'$, au voisinage immédiat de la surface d'une sphère de centre 0 et de rayon ct': c'est-à-dire dans une couche sphérique très mince de centre 0 contenant la sphère précédente."*
- (C) *(Conclusion) "Pour calculer l'effet du phénomène lumineux initial produit en 0 au temps $t = 0$, on peut le remplacer par un système convenable de perturbations se produisant en $t = t'$, distribuées sur la surface de la sphère de centre 0 et de rayon ct'."*

Hadamard remarque que la majeure (A) est une loi de la pensée, qu'il désigne à plusieurs reprises comme la loi du "déterminisme scientifique", soit une loi inévitable de la raison, un truisme incontournable. Il montre que la proposition (C), quoique n'étant pas aussi évidente, est une propriété générale des équations aux dérivées partielles linéaires hyperboliques. Mais tel n'est pas le cas de (B) ; il écrit: *"elle est une propriété toute spéciale de certaines équations particulières: en fait on ne sait pas encore si l'équation des ondes sphériques(et d'autres qui n'en sont pas au fond toutes distinctes), ne sont pas les seules à posséder cette propriété."*[53a]

Considérons le cas où l'équation du mouvement des ondes ne contient que deux coordonnées cartésiennes par exemple x et y; la fonction d'onde ϕ représente alors une perturbation qui est la même dans tous les plans perpendiculaires à Oz et est appelée une onde cylindrique puisque les fronts d'onde sont des cylindres parallèles à Oz. Elle vérifie l'équation:

$$\frac{\partial^2 \phi}{\partial x^2} + \frac{\partial^2 \phi}{\partial y^2} = \frac{1}{a^2}\frac{\partial^2 \phi}{\partial t^2}.$$

La formule intégrale de Poisson, valable pour les ondes cylindriques, prend la forme[54]:

$$2\pi\phi(P,t) = \frac{d}{dt}\int \frac{\phi(M,0)}{\sqrt{a^2t^2-\rho^2}}d\sigma + \int \frac{\phi_{,t}(M,0)}{\sqrt{a^2t^2-\rho^2}}d\sigma$$

où ρ est la distance du point courant à l'axe du cylindre et les intégrales sont prises sur le disque $\rho < at$.

Si la perturbation initiale est confinée dans le cylindre $\rho < \rho_0$, la formule de Poisson montre qu'en un point P à une distance $\rho > \rho_0$ de l'axe des z, l'effet est nul jusqu'à l'instant $t = (\rho - \rho_0)/a$; mais après cet instant, la perturbation n'est jamais nulle car la sphère de centre P et de rayon at coupe toujours le cylindre $\rho = \rho_0$, t aussi grand soit il et donc les valeurs moyennes qui apparaissent dans la formule de Poisson ne sont pas nulles. Dans ce cas, la perturbation meurt lentement en amont du front d'onde lorsque t croît. Il n'y a plus de front arrière, l'onde traîne[55]. Ceci s'apparente à un phénomène de "diffusion" et le principe de Huygens ne s'applique plus.[56] Ainsi bien que les équations du mouvement des ondes à deux ou trois variables d'espace respectivement, soient deux équations aux dérivées partielles hyperboliques linéaires, elles ne se comportent pas du tout de la même façon, relativement à la propriété (B). Il est frappant de constater que tant Poisson[57], que Cauchy[58] ne perçoivent pas à ce moment cette différence importante de comportement du phénomène ondulatoire.

3) La méthode symbolique de Poisson

Revenons au mémoire de Poisson. Celui-ci applique la même méthode à d'autres équations aux dérivées partielles de la physique mathématique, déjà très classiques à son époque. Citons quelques exemples.

a) Particularisations de l'équation des ondes.

- Pour une fonction à une variable, il retrouve l'intégrale de d'Alembert.
- Pour les $\phi(t, r)$, il retrouve l'intégrale:

$$\phi = (1/r) \cdot [f(r - at) + F(r + at)]$$

- Pour l'équation des ondes dans le plan

$$\frac{d^2\phi}{dt^2} = a^2(\frac{d^2\phi}{dy^2} + \frac{d^2\phi}{dz^2}).$$

il vient l'intégrale:

$$\phi = \int_0^{2\pi} \int_0^{\pi} f(y + at \sin u \sin v, z + at \sin u \cos v) t \sin u du dv$$

- Pour l'équation des fluides incompressibles,

$$\Delta\phi = \frac{d^2\phi}{dx^2} + \frac{d^2\phi}{dy^2} + \frac{d^2\phi}{dz^2} = 0$$

il est intéressant de noter que Poisson en donne une intégrale, en la déduisant de l'équation précédente en changeant t en x, et en faisant $a = \sqrt{-1}$. Son intégrale complète sera:

$$\phi = \int_0^{2\pi} \int_0^{\pi} f(y + x \sin u \sin v\sqrt{-1}, z + x \sin u \cos v\sqrt{-1}) t \sin u du dv$$
$$+ \frac{d}{dx} \int_0^{2\pi} \int_0^{\pi} F(y + x \sin u \sin v\sqrt{-1}, z + x \sin u \cos v\sqrt{-1}) t \sin u du dv,$$

mais à cause des imaginaires, Poisson indique qu'il vaut mieux s'en tenir à la valeur de ϕ en séries (de Fourier) infinies, comme celle utilisée dans sa Théorie des Ondes[59]. Ce type de cas poussera Cauchy vers la théorie des intégrales de la variable complexe.

b) l'équation de la chaleur:

$$\frac{d\phi}{dt} = a(\frac{d^2\phi}{dx^2} + \frac{d^2\phi}{dy^2} + \frac{d^2\phi}{dz^2}).$$

On sait (Fourier, Poisson [1813]) que cette équation bien que du 2ème ordre n'admet qu'une seule fonction arbitraire dans l'intégrale complète. Avec la même signification pour les notations qu'avant, il vient:

$$\phi = U + \frac{a^2t^2}{2!}\delta U + \cdots \text{ etc. donc:}$$

$$\phi = \left[1 + at(g^2+h^2+k^2) + \frac{a^2t^2}{2!}(g^2+h^2+k^2) + \cdots)\right].$$
$$U = e^{atg^2}e^{ath^2}e^{atk^2}U.$$

Or

$$e^{atg^2} = \frac{1}{\sqrt{\pi}}\int_{-\infty}^{+\infty} e^{-\alpha^2}e^{2g\alpha\sqrt{at}}d\alpha$$

résultat qui figure dans les mémoires de 1815 de Poisson et de Cauchy.[60] Donc

$$\phi = \frac{1}{\pi\sqrt{\pi}}\int\int\int e^{-\alpha^2-\beta^2-\gamma^2}e^{2(g\alpha+h\beta+k\gamma)\sqrt{at}}U\,d\alpha d\beta d\gamma.$$

Et en vertu des mêmes analogies que précédemment,

$$e^{2(g\alpha+h\beta+k\gamma)\sqrt{at}}f(x,y,z) = f(x+2\alpha\sqrt{at}, y+2\beta\sqrt{at}, z+2\gamma\sqrt{at}).$$

Finalement, l'intégrale complète s'écrit:

$$\phi = \int_{-\infty}^{+\infty}\int_{-\infty}^{+\infty}\int_{-\infty}^{+\infty} e^{-\alpha^2-\beta^2-\gamma^2}f(x+2\alpha\sqrt{at}, y+2\beta\sqrt{at}, z+2\gamma\sqrt{at})d\alpha d\beta d\gamma$$

où $\phi(x,y,z,0) = \pi^{3/2}f(x,y,z)$. Cette expression de l'intégrale de l'équation de la chaleur, est aussi trouvée par Fourier dans la *Théorie Analytique de la Chaleur*.[61]

c) l'équation des surfaces élastiques vibrantes[62]

$$\frac{d^2z}{dt^2} + b^2(\frac{d^4z}{dx^4} + 2\frac{d^4z}{dx^2dy^2} + \frac{d^4z}{dy^4}) = 0.$$

Poisson pourrait traiter sa résolution directement mais il préfère la déduire de l'équation de la chaleur à deux variables. En effet si on différencie cette dernière par rapport à t, on retrouve l'équation des surfaces élastiques avec $a^2 = -b^2$. Il vient donc:

$$z = \int_{-\infty}^{+\infty} \int_{-\infty}^{+\infty} e^{-\alpha^2-\beta^2} f\left(x + 2\alpha\sqrt{bt\sqrt{-1}}, y + 2\beta\sqrt{bt\sqrt{-1}}\right) d\alpha d\beta$$
$$+ \int_{-\infty}^{+\infty} \int_{-\infty}^{+\infty} e^{-\alpha^2-\beta^2} F\left(x + 2\alpha\sqrt{-bt\sqrt{-1}}, y + 2\beta\sqrt{-bt\sqrt{-1}}\right) d\alpha d\beta,$$

et en chassant les imaginaires, Poisson aboutit à l'intégale:

$$z = \int_{-\infty}^{+\infty} \int_{-\infty}^{+\infty} \sin(\alpha^2 + \beta^2) f\left(x + 2\alpha\sqrt{bt}, y + 2\beta\sqrt{bt}\right) d\alpha d\beta$$
$$+ \int_{-\infty}^{+\infty} \int_{-\infty}^{+\infty} \cos(\alpha^2 + \beta^2) F\left(x + 2\alpha\sqrt{bt}, y + 2\beta\sqrt{bt}\right) d\alpha d\beta.$$

Elle avait été trouvée par Fourier dans un mémoire sur les surfaces élastiques, non publié[63].

A la fin de son mémoire Poisson formule un résumé général de sa méthode: pour une équation aux dérivées partielles linéaire à coefficients constants sans terme indépendant de la fonction inconnue ϕ ou de ses dérivées, on pourra toujours avoir une solution de la forme:

$$\phi = Ae^{pt+gx+hy+\cdots}$$

où $A, p, g, h, \ldots$ sont des constantes indéterminées. Si on substitue cette expression de ϕ dans l'équation, A reste indéterminé et une seule constante, par exemple p, s'exprime en fonction des autres $g, h, \ldots$. L'équation qui détermine p sera d'un degré égal à l'indice de la plus haute dérivée partielle par rapport à t; soient $p, p', p'', \ldots$ ses racines, on pourra les prendre successivement pour ϕ,

$$\phi = \Sigma Ae^{pt+gx+hy+\cdots} + \Sigma Ae^{p't+gx+hy+\cdots} + \cdots,$$

les Σ désignant les sommes qui s'étendent à toutes les valeurs réelles ou imaginaires, de $A, g, h, \ldots$. On a l'intégrale complète développée en somme d'exponentielles, et chaque terme satisfait à l'équation aux dérivées partielles. Comme il est permis de penser, dit Poisson, que ces quantités $g, h, \ldots$ changent par degrés infiniment petits, d'un terme à l'autre de la série, il écrit l'intégrale complète ϕ sous la forme

$$\int e^{pt+gx+hy+\cdots} f(g, h, \ldots) dgdh \cdots + \int e^{p't+gx+hy+\cdots} f_1(g, h, \ldots) dgdh \cdots + \cdots$$

qui est équivalente à la précédente (les limites des intégrales sont indéterminées, les f, f_1, ... sont arbitraires); Poisson reconnaît que, dans cette généralité, cette formulation n'est pas très intéressante mais elle peut permettre de résoudre certains problèmes aux valeurs intiales("de Cauchy"). C'est également la marche suivie par Fourier dans son mémoire sur la chaleur.

En fait la méthode, résumée ici par Poisson, était déjà partiellement à l'oeuvre chez Fourier, chez Cauchy (1815), chez Poisson lui-même pour la théorie des ondes, mais elle n'avait jamais été énoncée en tant que telle. Le mémoire de Poisson donne le véritable point de départ de *l'algébrisation des équations aux dérivées partielles linéaires, à coefficients constants* qui sera ensuite développée par Cauchy.

En effet outre les Notes au *Bulletin de la Société Philomatique*[64] de 1817 et 1818, Cauchy publie en 1821, deux Notes sur "l'Intégration des Equations linéaires aux Différences Partielles à coefficients constants." Puis en 1823, il publie un très important mémoire dans le Journal de l'Ecole Polytechnique sur ce sujet dans lequel il explique comment l'intégrale de Fourier donne une solution formelle pour la valeur initiale du "problème de Cauchy", ceci pour n'importe quelle équation aux dérivées partielles à coefficients constants, et pour n'importe quelle valeur initiale. Dans le même temps, il rédige les Notes supplémentaires à son mémoire sur la Théorie des Ondes qui ne seront publiées qu'en 1827. Cauchy, cette fois, assimile les résultats de Poisson et les développe largement. Par rapport au mémoire précédent de son collègue, on peut souligner chez Cauchy:

- une utilisation systématique de la variable imaginaire dès 1821
- une algébrisation poussée du calcul symbolique.
- une réflexion sur plusieurs points délicats (substitution de la variable imaginaire, question des intégrales singulières, unicité des développements en séries de puissances etc.) passés sous silence dans la partie originaire du mémoire de 1815.

L'importance de ces mémoires sera étudiée par ailleurs[65].

BIBLIOGRAPHIE

Belhoste B. [1989]: Augustin-Louis Cauchy, Biography Springer-Verlag à paraître.

Brisson[1808]:"Mémoire sur l'Intégration des équations différentielles partielles" Journal de l'Ecole Polytechnique T 7. 1808. p 191–261.

Cauchy [1814/1827]: " Mémoire sur les Intégrales définies" avec 2 Suppléments. O.C. (1è s.) T 1, p 329–506.

[1815/1827]: "Théorie de la propagation des Ondes à la surface d'un Fluide pesant d'une profondeur indéfinie", suivi de Notes. Mém. div. Savants Etrangers. T 1. 1827 O. C. (1è s.) T 1, p 5–318.

[1817]: "Sur une loi de réciprocité qui existe entre certaines fonctions" Bull. Soc. Phil. 1817 p 121–24. O. C. (2è s.) T 2, p 223–27.

[1818]: "Seconde Note sur les Fonctions réciproques" Bull. Soc. Phil. 1818 p 188–91. O. C. (2è s.) T 2, p 228–32

[1821]: "Mémoires sur l'Intégration des Equations linéaires aux Différences Partielles à Coefficients Constants et avec un dernier terme à la variable". Deux Notes au Bull. Soc. Phil. 1821. p 101–112 et p 149–152. O. C. (2è s.) T 2, p 253–266 et 267–275.

[1823]: "Mémoire sur l'Intégration des Equations Linéaires aux différentielles partielles et à coefficients constants" J. de l'Ecole Polytechnique. 19è Cahier, T 12 p 511. 1823. O. C. (2è s.) T 1, p 275–357.

Fourier [1822]: Théorie Analytique de la Chaleur Oeuvres. T 1. Gauthier-Villars. 1888–90. Paris.

Fresnel [1818/1826]: Mémoire sur la Diffraction. Mém. de l'Académie t 5. 1826. Oeuvres, t I.

Grattan-Guiness & Ravetz[1972]: Joseph Fourier (1768–1830) M I T Press. Cambridge (Mass.)- Londres. 1972. Cet ouvrage contient les mémoires de 1807 et de 1811 de Fourier sur la théorie de la chaleur.

Koppelman E. [1971]: "The Calculus of Operations and the Rise of Abstract Algebra" Archive for History of Exact Sciences. Vol 8. No 3, p 155–242. 1971

Hadamard J. [1924]: "Le principe de Huygens". Bulletin de la Société Mathématique de France. vol 52, pp 610–642. 1924. [1932]:Le Problème de Cauchy et les équations aux dérivées partielles hyperboliques. Hermann et Cie Paris 1932

Lagrange [1781/1783]: "Mémoire sur la Théorie du mouvement des Fluides" Nouv. Mém. de Berlin, année 1781. Berlin 1783. O. L., t IV p 695–748.

[1786/1788]: "Sur la manière de rectifier deux endroits des Principes de Newton relatifs à la propagation du son, et au mouvement des ondes"Nouv. Mém. de Berlin, année 1786. Berlin 1788. O. L., t V p 591–609.

[1788]: Méchanique Analitique. Paris. Vve Desaint..., 1788. 2ème édition 1811 et 1815. O.L. t XI et XII.

Laplace[1808]:"Mémoire sur divers points d'Analyse" Journal de l'Ecole Polytechnique T 8. 1808. p 229–265.

Lamb [1932]: Hydrodynamics. Cambridge University Press. 6°ed. 1932.

Poisson [1815/1818]:"Mémoire sur la Théorie des Ondes" lu le 18 Octobre et le 18 Décembre 1815. Mém. de l'Acad. des Sc. (1816) 2è s. t 1. 1818 p 70–186.

[1819/1820]:"Sur l'Intégration de quelques équations linéaires aux différentielles partielles et particulièrement de l'équation générale du mouvement des fluides élastiques", lu le 19 Juillet 1819. Mém. de l'Acad. des Sc. (1818) 2è s. t3. 1820 p 121–176.

Petrova S. [1987]: "Heaviside and the Development of the Symbolic Calculus" Archive for History of Exact Sciences. Vol 37. No 1, p 1–23.

Notes

1. (Livre II, prop 44, 45, 46)
2. La largeur d'une onde est l'intervalle entre les sommets de 2 ondes successives, l'une saillante, l'autre en creux. La vitesse d'une onde est celle avec laquelle les sommets changent de place et est différente de la vitesse propre aux molécules du fluide.
3. *Mém. de l'Académie* 1776
4. *Mém. de Berlin 1786 et Méchanique Analitique*

4a. "Méchanique analitique tII, 11è section, art 37"

5. Le mémoire de Poisson comprend une centaine de pages alors que celui de Cauchy est nettement plus long (près de 180 pages), et comprend 13 Notes purement mathématiques auxquelles s'ajouteront encore 150 pages à la publication.
6. Le potentiel de vitesse ou impulsion diffère chez Poisson et Cauchy par le coefficient de densité du fluide
7. Quand on fait abstraction d'une dimension horizontale du fluide, celui-ci est censé réduit à un plan. Mais on peut aussi le supposer contenu dans un canal vertical d'une largeur quelconque, pourvu qu'elle soit constante dans toute la longueur du canal, et que les molécules du fluide n'aient aucun mouvement dans le sens de cette largeur. C'est ce que postulent tant Cauchy que Poisson.
8. Cette datation nous a été proposée par Bruno Belhoste qui a touvé un manuscrit de Cauchy, dans un *"Cahier sur la Théorie des Ondes"* qui appartenait à Madame de Pomyers, une des descendantes de la famille Cauchy. A l'exception de quelques pages, le contenu de ce manuscrit est presque identique à la version imprimée des treize premières Notes ; il contient aussi les Notes XIV et XV que Cauchy

a ajoutées en 1821. Le manuscrit de la note XVI se trouve dans un autre *Cahier*, qui se trouve à la Bibliothèque de la Sorbonne, Ms 2057. C'est un véritable deuxième mémoire sur la théorie des ondes.

9. En fait Lagrange avait indiqué dans son "Mémoire sur la Théorie du Mouvement des Fluides" (Nouv. Mém. de Berlin, 1781. O.L. t IV, p 714), puis dans la *Mécanique Analytique*, que si un potentiel de vitesse existe en un instant quelconque, pour une portion finie quelconque de fluide en mouvement (pourvu que la densité du fluide soit constante ou ne dépende que de la pression), alors le potentiel de vitesse existe encore pour la même portion de fluide en tous instants antérieurs ou postérieurs. Pourtant sa démonstration laissait beaucoup à désirer, et la première démonstration rigoureuse est dûe à Cauchy dans ce mémoire, p 38. Une autre démonstration sera donnée par Stokes, avec une introduction critique et historique de la question. ((Math. and Phys. Papers. t I, p 106, 158; t II p 36)
10. O. C. 1ère s. t I Note XIX, p 301
11. idem p 146 à 149.
12. c.f. la Note VI au mémoire (idem p 133 et suivantes) ; la démonstration est d'abord connue par sa publication dans la Note au *Bull. Soc Phil.* en 1817
13. idem Note IX, p 152.
14. On peut vérifier directement, par des considérations de dimensions, que l'effet d'une élévation initiale de niveau concentrée sur une section d'aire A, est de la forme

$$y = \frac{A}{x} f\left(\frac{gt^2}{x}\right).$$

14a. O.C.1°s, tI, p 82.
14b. idem. p 83.
15. O.C.1°s, tI, p 84–85.
16. c.f. Mémoires de l'Académie p 113.
17. Il est difficile de distinguer nettement, dans la version publiée dans les Mémoires de l'Académie [1è série Année 1816 (1818), pp 71–186], les deux parties initiales du mémoire de Poisson ; en effet les lois de propagation pour les deux types d'ondes sont énoncées dans la même section III du mémoire, p 113 et p 120.
18. Dans une partie ultérieure, il traite aussi le cas où la perturbation initiale est dûe à un système de forces "impulsives appliquées dans une certaine portion de la surface libre du fluide". Cauchy traitera également les deux cas.
19. idem. p 103.
19a. idem p 103 et 104.

20. En fait chez Poisson c'est la coordonnée z qui est verticale et la valeur de z à la surface est désignée par z'. Cauchy utilise pour coordonnée verticale y et désigne sa valeur au dessus de la surface par la même lettre.
21. idem p 111.
22. La série définissant η était apparue dans la théorie des intégrales de diffraction de Fresnel.
23. C'est aussi la valeur trouvée par Cauchy (O. C. 1ère s. t I, p 87.)
24. l'équation qui donne les valeurs de k correspondant aux ondes successives n'est valable, elle aussi, que si $gt^2/4x$ est petit par rapport à x/α
25. On démontre facilement que dans le cas où la longueur d'onde est très petite par rapport à la profondeur, $\sigma^2 = gk$. C.f. Lamb [1932, p 365]
26. Poisson utilise aussi les formules suivantes qui étaient connues, et seront utilisées par Fresnel dans son mémoire sur la diffraction:

$$\int_0^\infty \cos\zeta^2 d\zeta = \int_0^\infty \sin\zeta^2 d\zeta = \frac{\sqrt{\pi}}{2\sqrt{2}}$$

Cauchy aussi utilise ces "intégrales de Fresnel" dans la Note XVI, O. C. 1ère s. t I, p 244.
27. C.f. Poisson [1815/1818, p 119]
28. Selon Newton elle aurait dûe être la même que celle des oscillations d'un pendule simple, d'une longueur égale à la demi-largeur des ondes, soit $\pi\sqrt{\frac{\lambda}{2g}}$.
29. C.f. Poisson[1815/1818, p 120]
30. C.f. Poisson [1815/1818, p 122] La vitesse de "l'onde dentelée," ce qu'on nomme aujourd'hui la vitesse de groupe (cinématique), est définie comme la dérivée de la pulsation par rapport au nombre d'onde k, c'est-à-dire $U = \frac{d\sigma}{dk} = \frac{d(kc)}{dk} = c - \lambda\frac{dc}{d\lambda}$ où λ est la longeur d'onde qui vaut $\frac{2\pi}{k}$ et est la vitesse de l'onde (en physique moderne, ce qu'on appelle la vitesse de phase). Une première approche de cette notion est donneé par Stokes [Papers, V, 362]. Le fait que cette vitesse de groupe corresponde à la vitesse de transmission de l'énergie, devait être mis en lumière encore plus tard, par Reynolds dans "On the Rate of Progression of Groups of Waves, and the Rate at which Energy is transmitted by Waves" *Nature*, XVI. (1877) 343 [Papers, I, 198]

31. idem p 123. Poisson montre aussi que les deux premiers maximas sont plus grands que tous les autres, et ceux-ci forment une suite décroissante, quand on se rapproche de l'origine des ondes.
32. Cauchy a nommé "sillons" les différentes dents ou ondelettes d'un groupe d'ondes. C.f. infra et [Cauchy 1815/27, p 237]
33. On poura se reporter à J-M. Lévy-Leblond, F. Balibar: *Quantique*. Interéditions C.N.R.S. 1984, p 105 pour trouver ces résultats sur la vitesse de groupe, obtenus en application des inégalités spectrales classiques.
34. Nous utilisons ici les graphiques de Lamb dans *Hydrodynamics* [1932, p 387–388]
35. Nous avons gardé les notations de Cauchy mais rappelons que y est l'élévation au dessus de la surface libre, que nous avons notée auparavant η
36. O. C. 1ès. t I, p 191.
37. Note XVI, p 193
38. Une telle phrase chez Mandelbrot, théoricien des fractals, à la fin du XXème siècle, ne nous surprendrait pas... ce qui n'est pas le cas chez les physico-mathématiciens du début du XIXème.
39. idem p 198 La fonction $F(x)$ représente l'élévation initiale arbitraire. On voit que comme Fourier, quand il discutait les conditions initiales des équations de diffusion (chapitre IX de la Théorie Analytique de la Chaleur) Cauchy a rompu clairement avec la conception eulérienne des fonctions.
40. Note XVI, p 239.
41. O.C. 1 ère s. t I, p 122.
42. Dans la Note II (idem, p 125).
43. O. C. 1 ère s. t I, p 286.
44. Notons que Cauchy a vérifié, dans la note V, que le premier développement asymptotique est tel que K/k est majoré par 1.

44a. O. C. 1ère s. t I, p 288.

45. O. C. 1ère s. t I, p 185–186.
46. 1822 c.f. O.C. 2° s. t II, p 276–82.
47. C.R. de l'Académie t XV, p 534, 673.
48. Il est étonnant que ce mémoire de Poisson ne soit pas souvent cité dans les articles portant sur l'histoire du calcul symbolique comme par exemple ceux, fort intéressants par ailleurs, de E. Koppelman [1971] et de S. Petrova [1987].
49. La notation $\phi_{,t}$ désigne la dérivée de ϕ par rapport à la variable t.
50. Parmi les noms de ceux qui ont contribué de façon importante à cette élucidation, citons Kirchoff, Helmoltz, Poincaré, Volterra et Hadamard.

51. Le cas d'abord traité par Poisson est celui d'une perturbation unique, confinée dans un sphère centrée à l'origine et de petit rayon r.
52. J. Hadamard: *"Le problème de Cauchy et les équations aux dérivées partielles linéaires hyperboliques."* Paris. Hermann. 1932, p 75. C. f. aussi "Le principe de Huygens" *Bulletin de la Société mathématique de France.* vol 52 [1924], pp 610–642.
53. Hadamard parle ici des perturbations lumineuses, mais le principe d'Huygens est valable pour les ondes sonores et toute la question posée est de savoir pour quel type de perturbation ondulatoire et donc pour quel type d'équation aux dérivées partielles linéaires il s'applique.

53a. Hadamard [1932, p 77].

54. Dans son mémoire de [1818/1819], Poisson a intégré l'équation des ondes à deux variables d'espace, par ce qu'Hadamard a dénommé la "méthode de descente", qui consiste simplement à remarquer que qui peut le plus, peut le moins: si on peut intégrer les équations à 3 variables, en imposant aux conditions initiales de ne faire intervenir que deux variables, alors on peut en faire autant pour celles où n'interviennent que 2 variables.
55. On a pu dire qu'il est bien heureux que l'espace dans lequel nous vivons ait 3 dimensions: car s'il n'en avait que deux, et que par conséquent la propagation du son y soit régie par l'équation des ondes cylindriques, nous entendrions encore - très étouffés certes- tous les bruits qui se sont produits depuis la création du monde!
56. On peut tirer mathématiquement de la formule de Poisson que si une onde plane peut se propager à vitesse $\phi_{,t}$ de signe constant, en revanche une onde sphérique correspond nécessairement à des régions de "condensation" et de "dilatation". Elle est oscillante. Ceci a été remarqué pour la première fois par Stokes en 1849, et développé par Rayleigh, Theory of sound, 2 (1878), p 99.
57. En particulier dans la polémique qui s'est déroulée en 1823 entre Poisson et Fresnel sur le point de savoir si et comment il était possible que les effets des ondes lumineuses se détruisent sur le front interne (arrière), la réponse était contenue dans la formule intégrale de Poisson mais celui-ci n'y a jamais fait allusion. C.f. *Annales de Chimie et de Physique* (2) t XXII, p 250, 270.
58. Cauchy retrouve la forme intégrale de Poisson dans la Note XV au mémoire sur la théorie des ondes (O. C. 1ère s. t I) mais il y retravaille dans ses mémoires de [1821] et [1823]
59. L'analogie entre l'équation de Laplace et l'équation des ondes dans le plan, par l'intermédiaire de la variable imaginaire, sera largement exploitée plus tard par Volterra, Hadamard, Riesz...

60. NoteII O. C. 1 ère s. t I, p 117–118
61. Oeuvres de Fourier. t I, p 429.
62. Cette équation avait été trouvée d'abord par Lagrange en 1811, quand il avait été examinateur du premier mémoire de Sophie Germain pour le Prix des Surfaces élastiques de l'Académie, puis elle avait été retrouvée par la jeune femme elle-même, lors du deuxième Concours sur le même sujet en 1813.
63. C.f. *Bulletin de la Société Philomatique 1818. Août et Septembre.*
64. Cauchy y prend date de ses résultats sur la transformation de Fourier, et la formule de réciprocité; il les applique à des calculs d'intégrales définies et des sommations de séries; enfin il obtient la formule sommatoire dite de Poisson: si a, b sont deux constantes telles que

$$ab = 2\pi,$$

et si f et F deux "fonctions réciproques de 1ère espèce (f est paire et F sa transformée pour le cosinus), alors soient

$$S_1 = (1/2)f(0) + f(a) + f(2a) + f(3a) + \dots \text{ etc.}$$
$$S_2 = (1/2)F(0) + F(b) + F(2b) + F(3b) + \dots \text{ etc.}$$

On a $\sqrt{a}S_1 = \sqrt{b}S_2$ et S_1 sommable $\Leftrightarrow$ S_2 sommable. Cauchy retrouve ainsi des exemples de sommation de séries d'Euler dans *l'Introductio...*
65. C.f. A. Dahan Dalmedico:"Aspects de la mathématisation: la voie d'Augustin-Louis Cauchy" à paraître.

Pure versus Applied Mathematics In Late 19th-Century Germany

Klein 22, L, 1

1.

Erlanger Antrittsrede
November 1872.
(gehalten 7.XII.72).

Prorector magnifice!
Collegen, Commilitonen!
Hochgeehrte Versammlung!

Wenn es auch wohl Sitte ist, dass der neu Angekommene
Ihnen von hiesiger Stelle eine Schilderung entwirft von den
neuesten Errungenschaften, die seiner Wissenschaft zu Theil
geworden sind, oder weiter ausholend von dem Entwickelungs-
gange, der zu den heutigen Auffassungen hingeleitet hat,
so habe ich geglaubt, bei dem wenig zugaenglichen Character meines Faches den Gegenstand fuer meinen heutigen
Vortrag in einer etwas anderen Richtung suchen zu sollen.
Ist doch bei der eigenthuemlichen Schwierigkeit, welche jede
ungewohnte mathematische Gedankenoperation mit
sich fuehrt, eine einmalige mathematische Vorlesung
nur zu leicht selbst engeren Fachkreisen unverstaend-
lich! Um wie viel mehr muesste das bei einer Ge-
legenheit, wie der heutigen, der Fall sein! Es ist ja
doch so, dass selbst geringe mathematische Kenntnisse
nur wenig verbreitet sind, dass die einfachsten ma-

2

From the Manuscript of Felix Klein's Unpublished
"Erlanger Antrittsrede," Delivered December 7, 1872
Courtesy of Niedersächsische Staats- und Universitätsbibliothek, Göttingen

Pure and Applied Mathematics In Divergent Institutional Settings in Germany: The Role and Impact of Felix Klein

Gert Schubring

During the last decades, historiography of the sciences has markedly changed: formerly largely "monolithic," concentrating on the history of ideas, it has become rather multi-faceted. In the aftermath of Thomas Kuhn's *Structure of Scientific Revolutions* (1962), the focus has shifted—for almost all disciplines—from isolated scientific "heroes" to the larger communities that produce and communicate knowledge. Rather than assuming that a continuous increase of knowledge is self-evident, today's historians of science are more interested in studying ruptures and periods of discontinuity in order to understand better the conditions that shape general scientific development. By means of a plurality of approaches, recent researches have attempted to reveal the processes characteristic for the growth of specific disciplines by tracing their emergence and evolution. The historiography of mathematics has also participated in this broadening of perspectives. Symptomatic of this trend has been a growing interest in "styles" of mathematics ("Denkstile") peculiar to certain periods, cultures or scientific schools. Even the term "discipline" itself can be used as an analytical tool: since their "discipline" connects scientific actors with the institutions in which they are working, the investigation of ideas thus becomes tied to that of the institutional context.[1] Of particular value for refining the analysis of disciplinary development is the recent sociological theory of systems. This theory provides a framework for studying the relations of subsystems to their environment as a characteristic pattern within modern, functionally differentiated societies (cf. Luhmann 1980-81; Stichweh 1984).

For the field of mathematics, the first to have taken up seriously questions of disciplinary development seems to have been Felix Klein. As the *spiritus rector* of Wilhelm Lorey's classical history of mathematics at the German universities in the 19th century (Lorey 1916), Klein constantly urged and advised its author to investigate the reasons behind the dramatic dissolution of the former unity between pure and applied mathematics, as represented by Gauss, into an absolutely one-sided cultivation of pure mathematics. A typical expression of Klein's interest in

ISBN 0-12-599662-4.

these developments is the following query to Lorey: "We find in Crelle a wonderful congruence between pure and applied mathematics, which becomes distorted later in Berlin into its opposite. How was this effected?"[2] Whereas Klein was searching for the historical roots underlying the dominance of pure mathematics at the end of the 19th century—his struggle to reverse this trend was the core of his political agenda throughout most of his life—historical research has begun to investigate the motivation behind Klein's program for reorienting mathematics and what effects it really had. This reassessment has evolved in a dialectical fashion: The first serious account of Klein's professional policy on higher education, K.-H. Manegold's *Universität, Technische Hochschule und Industrie* (Manegold 1970), presented his efforts to integrate the technical colleges (*Technische Hochschulen*) into the universities as a far-sighted conception. Klein, a mathematician, was depicted as a protagonist fighting for the true interests of the engineers' movement whose own narrow-minded leadership opposed Klein's views for political reasons and because his ideas did not conform to their own reformist agenda. In brief, Manegold presented Klein's position as that of a visionary who, amid the growing conflict between mathematicians and engineers, recognized the essential congruence and harmony between their implicity-understood interests. L. Pyenson's *The Persistence of Pure Mathematics in Wilhelmian Germany* (Pyenson 1983) focuses on parallel developments in secondary school education. It marks the anti-thesis of Manegold's study, by claiming that Klein's principal intention was to establish a new kind of supremacy for pure mathematics that would extend from the secondary schools to the highest levels of university science education. Pyenson's thesis is based largely on the contention that Klein distorted an already existing reform movement among mathematics teachers aimed at reviving applications within school mathematics.

One problem with these divergent judgements is that they regard Klein's positions and opinions as essentially consistent and unchanging throughout his lifetime. Since then, a more balanced view has evolved regarding this matter; D. Rowe, for example, has shown that Klein changed his mind with regard to one of the most basic questions of school structure (Rowe 1985a, 127). Other important contributions include S. Hensel's analysis of the anti-mathematical movement within the engineers (Hensel 1987), and studies by R. Tobies on Klein's journal policy (Tobies 1986/87), his appointment policy (Tobies 1987), and his conception of applied mathematics (Tobies 1988).

II. Professional Careers and Disciplinary Orientations

Klein's efforts to establish applied mathematics within the universities and thereby reorient pure mathematics present a telling case study for the interaction between the institutional context of an academic discipline and its conceptual development. That the type of institution in which mathematicians teach their subject determines to a certain degree the style and substance of their mathematics would appear to be undeniable. An important example is provided by French mathematics during the first half of the 19th century: while the *facultés des sciences* had no independent educational function, the *École Polytechnique*—which trained engineers and administrators—was the dominant institution. As is well known, mathematical productivity during this period centered on "applied" fields like mathematical physics rather than "pure" branches like algebra, number theory, etc. (cf. Grattan-Guinness 1981). On the other hand, despite its manifest significance, it is a highly complex problem to identify the mechanisms and agents within the institutional context that shape and condition conceptual productivity.

One reason for these difficulties can be ascribed to a shortcoming of the dominant type of institutional histories. These are often in the form of "Festschriften" or similar publications which emphasize the contributions of the institution's leading researchers. Occasionally certain details about their teaching will be mentioned, but the existence of such mathematical institutions, their *raison d'être*, is largely taken for granted. If this question is addressed at all, it will generally be confined to problems the institution encountered in its founding phase. Larger issues concerned with the way in which the particular institution is imbedded within the general patterns of education and training are almost always ignored. An underlying assumption characteristic of this traditional historiographical approach is that the autonomy of mathematics as an academic discipline is regarded as self-evident. Yet, for most fields, this autonomy only emerged over the course of the 19th century, during which time their subsequent development was rarely what could be described as continuously stable. Indeed, it has often been the case that new developments in the relationship between culture and technology in a given society have decisively affected the level of autonomy attained.

In this essay, I propose to apply some of the new, aforementioned approaches in history of science to mathematics in order to arrive at a deeper understanding of the interaction between the discipline and its institutional context. To do so, I start from the system-theoretic conception that modern societies (or national states) are segmented into subsystems according to certain functions they exercise. These subsystems can enjoy considerable autonomy (so that the mutual relations between the subsystems might be obscured and become visible only in periods of crisis), but there are always functional relations and dependencies between them. Analyzing the institutional context of an academic discipline implies, therefore, investigating its functional relations to the (directly) relevant subsystems. Evidently, the educational system of a society constitutes such a subsystem; it splits up, however, into several semi-autonomous subsystems, the most important for us being the secondary school system and the system of higher education. Their functional relationship is obvious, and in certain cases has been studied in detail (cf. Schubring 1983, Eccarius 1987). A second system that is relevant here is the employment system of the respective society—the functional relationship of interest being due to the fact that higher education trains those who will later enter the marketplace of the academic professions.

My basic hypothesis is that the functional relationship between the educational system and the employment system—or to put it more concretely: the types of professional careers accessible to graduates by virtue of the training they receive in academic institutions—largely mold the institutionalization of the related disciplines. The autonomy of an academic discipline depends upon the existence of specialized professional careers for the discipline's graduates. As a consequence, discipline and profession are functionally related. Changes that take place at one of these two poles will induce changes in the other. However obvious this may seem, it is important to stress this interaction since recent sociology of science has tended to strictly separate the notions of discipline and profession.[3]

This essay studies the institutional development of mathematics in Germany by exploring the dynamic between two different pairs of opposite poles: universities vs. technical colleges on the institutional level and, on the geographical-cultural level, northern German/ Protestant conceptions of higher learning vs. southern German/Catholic ones. Section III discusses how these opposing conceptions affected the institionalization of mathematics within different educational subsystems. Section IV traces the rise of polytechnical schools from their beginnings as secondary level schools to the period when they became technical colleges. As a consequence of this rise, the fundamental role of mathematics within their curriculum became questioned, a topic considered in section V. Section VI discusses the policy of Felix Klein who attempted to master this growing crisis in mathematics education. The evolution and change of his program for mathematics education are examined, leading up to a key document of 1900 in which he formulated a new policy that encompassed the different secondary schools as well as the universities and technical colleges. This culminates in the realization of a reorientation program, in particular, through the creation of national and international organizations aimed at addressing problems in mathematics education. The final section analyses the notion of applied mathematics underlying this reorientation program. Two appendices follow: the first gives some hitherto unknown background information on Klein's formative years of study at Bonn; the second reproduces the important document of 1900 mentioned above.

III. Mathematics: Autonomy via Teacher Education

One can roughly characterize the position of mathematics in German higher education during the second half of the 19th century as follows: within the universities, mathematics had succeeded in attaining autonomy; it lacked such an autonomy, however, at the technical colleges. The mathematical courses in the philosophical faculties thus prepared students for a professional career in mathematics, whereas those offered at technical colleges did not—they were solely designed as preparatory courses for engineers and technicians. The opposition between these two functions of mathematics can be found in many national education systems.[4] The underlying tension between its cultural and technological functions is intrinsic to mathematics. Yet these two aspects can be studied with advantage in 19th century Germany, since its educational systems reflected this dichotomy between educational and vocational functions, and since the various German states (there were 36 between 1815 and 1866) had applied several divergent conceptions and epistemologies of education and science to this differentiated system.[5]

This advantage is reinforced by the fact that the German states differed in particular by providing either a Protestant educational system or a Catholic one. The differences between Catholic and Protestant systems represent entirely different functional relations between the respective social subsystems, but these cannot be discussed here.[6] Whereas for other European countries, the effect of religion on learning was interwoven in the nation's culture, Germany presents a case where—within a largely common culture—the existence of independent states (with different religions) facilitates a separation of variables, as it were.

The structural and functional differences between Catholic and Protestant education became particularly sensible after the 17th century in the case of the lower faculties (originally named *Artistenfakultät*) within the universities. Both had their roots in the classical curriculum of the *quadrivium*, which included mathematics but not the sciences. The grades obtained by students of this faculty had, in general, no real signifance for their subsequent careers: they were simply part of the requirements for admission to the three higher, professional faculties. But from about the

17th century onwards, the pattern diverges: in Catholic states and, in particular, in France, the lower faculty led an increasingly marginal existence, until they almost vanished. Most of their curriculum was taken over by the *collèges* (i.e., secondary schools)—which left mathematics even more marginal than before, being taught, if at all, in the last classes when most students had already left the schools (cf. Julia 1986, and for the case of Münster, Schubring 1985).

In the Protestant states, on the other hand, the lower faculties were spared from suffering a similar fate.[7] As philosophical faculties, they established a broad range of courses required of all students before progressing to their professional studies. Within this structure, mathematics held a stable and firm position. Of course, mathematics was still far from constituting an autonomous discipline; at this point of development, mathematics, physics, etc. referred essentially to categories for the classification of knowledge. Not only were there no students who specialized in studying these branches of knowledge, neither were there professors specialized as teachers and researchers for these subjects in the universities. Their professors were themselves *generalists*, who held combined professorships or "ascended" to chairs in the higher faculties. Practitioners of mathematics in 17th and 18th century universities embraced several branches of knowledge, and they united theory as well as applications of mathematics in their teaching and research.[8] The character of the applications, however, altered immensely over the next century, and by the 19th century what were formerly fields of application had now developed into their own largely independent technical disciplines.

The contrast between Protestant and Catholic conceptions of higher learning in Germany can best be studied by comparing Prussia with Bavaria. These two states had six and three universities respectively,[9] while most of the other states had either one or none at all. Their differences are clearly reflected in the respective structure and status of their philosophical faculties. One of the major results of the fundamental educational reforms instituted in predominantly Protestant Prussia after 1806/10 was the secularization of teacher education. This function was removed from the theological faculty and handed over to the philosophical faculty, which thereby won its independent status. Instead of being regarded as a lower faculty, it achieved equal status with the other three, since it now provided direct professional careers to the graduates of its disciplines: careers as teachers at secondary schools. Their scientific authority, based on the new research orientation of the faculty's professors ensured them high social ranking as "scholars" ("Gelehrte"), gave them the power and status that helped make teaching become one of the first modern professions (cf. Schubring 1983, pp. 10 ff.).

Mathematics became one of the three principal subjects taught in the Prussian secondary schools. At the universities, mathematics attained therefore its autonomy as a discipline by preparing students to become mathematics teachers in these schools. This close connection between mathematicians at the universities and the secondary schools, supported by shared values regarding the structure and epistemology of their subject, is what I call a "disciplinary-professional complex." Up until around 1940, the teaching profession remained the almost exclusive basis for the disciplinary autonomy of mathematics at the universities.[10]

In predominantly Catholic Bavaria, on the other hand, neohumanist-minded reformers succeeded after 1800 in establishing well-endowed philosophical faculties, but these were without independent functions; the disciplines taught there were regarded as "general sciences" (*allgemeine Wissenschaften*). University regulations, imposed in an increasingly severe manner (the 1837 version was the strictest enacted), gave these studies an obligatory character. Students were required to study them for two years before they could go on to the "special" (professional) sciences offered by the upper faculties (cf. Dickerhof 1975, U. Huber 1987).

Within this system, mathematics enjoyed no autonomy. The regulations for teacher examinations enacted in 1833 made mathematical studies at the university mandatory for future mathematics teachers (cf. Neuerer 1978, 111). However, the number of such students was much too small to allow for a specialized development of mathematics within the philosophical faculties. Eventually, the obligatory propedeutic courses in the philosophical faculty were abolished in Bavaria in 1848, but the suppression of their traditional function left a conceptual hole since a disciplinary development based on research had not yet been prepared. It took several decades and the "importation" of Prussian professors before a research orientation was finally established.

These problems were particularly acute for mathematics as its status within the school system, too, differed markedly from that in Prussia. In 1808, Bavarian neohumanist reformers had established a bifurcated system of secondary schools: the "Gymnasial-Institutes," which taught classical languages and mathematics, and the "Real-Institutes," teaching mathematics, science and technology. In 1816, as part of the political reaction, this entire structure was abolished: the modernist branch was dissolved, the mathematics teachers in the classical schools were dismissed, and the weekly hours for mathematics reduced to one. Though the number of hours increased somewhat after 1822, along with reemployment of teachers for mathematics, instruction in the subject at the *Gymnasien* remained at an elementary level and was oriented entirely towards the humanities,[11]. The demand to fill vacant positions also stayed fairly low.

The vacuum in modern realistic education was filled in 1833 by the founding of a system of trade and agricultural schools. This system, however, did not run parallel with the classical schools; its status was rather very low. Since these schools aimed at vocational training, not general education, they were not administered by the Ministry of Instruction, but rather by the Commerce Ministry. As a consequence, mathematics teaching at these institutions had an entirely practical character. In 1833, a technical college was founded in Munich and attached to the university. Its purpose was to provide a "top" in this system's hierarchy; the college failed to flourish, however, and was eventually closed in 1842.[12] Despite its problematic status, the new vocational school system nevertheless gradually developed, and the polytechnical school at Munich was reestablished in 1868—this time as independent institution. It provided the final rung in a ladder of institutions that was independent of the system of universities and *Gymnasien*. This autarky meant,in particular, that the polytechnical schools trained those mathematics teachers who were to teach at the trade and technical schools.

The existence of a second independent system of secondary schools in Bavaria, together with its own institution of higher education, led to a weighting of the mathematics and science curriculum that differed remarkably from the better known Prussian example. In 1867 the Minister of Commerce proposed to his colleague for instructional matters that new regulations for the examination of mathematics teachers be enacted which would give complete control of the their administration to the Commerce Ministry. His colleague protested by pointing out that there were in fact a number of positions for mathematics teachers in the general school system which he was entitled to fill. But since he was unable to come up with an alternative proposition, he ended up ceding control over the testing of mathematics and science teachers to the agency in charge of commercial affairs. In September 1869, the Commerce Ministry then issued examination regulations for the "Lehramt an technischen Unterrichtsanstalten" (teaching profession at technical schools) that operationalized the autarky of the vocational school system even over the testing of its teaching candidates.[13]

One can conclude from this that the later proposals of Felix Klein, which charged the technical colleges with the additional task of educating prospective mathematics teachers, were not only inspired by his personal experiences at Munich from 1875-1880, but they were also meant to extend existing Bavarian practices to all of Germany.[14] Likewise, the introduction in 1898 of a teaching license for applied mathematics in Prussia largely followed the Bavarian model, where in fact applications had been predominant in the examinations of teachers for the trade/technical schools.

IV. The Rise of the Technical Colleges

Before discussing the role of mathematics within the German technical colleges, it is necessary to say a few words about the character of these institutions. Unlike the universities, with their long traditions and relatively stable structures, most of the technical colleges were founded during the first half of the 19th century. Although they drew much of their legitimation from the fame of the *École Polytechnique* in Paris, their resources and general level of instruction were incomparably poorer. Still, their status gradually improved during the 1850s, and in the wake of Germany's industrial expansion in the 1870s and 1880s they grew rapidly in size and importance.

This institutional development is not well documented in the literature which focusses more on the later levels than on the poor beginnings. Likewise, there is no equivalent for the technical colleges of Lorey's classical account of the development of mathematics at the German universities (Lorey 1916), since Stäckel's parallel study (Stäckel 1915) is very sketchy in its historical account and concentrates instead on the contemporaneous debates over the role of mathematics in the technical colleges.[15] The following table lists the *Technische Hochschulen* in Germany (excluding Austria), along with the dates of their founding:

Berlin	1821	Hanover	1831
Karlsruhe	1825	Brunswick	1835
Darmstadt	1826	Munich	1868
Dresden	1828	Aachen	1870
Stuttgart	1829		

Almost all of these colleges began at a very low level. Their character was not of an institution of higher education, but rather that of a secondary school, as is evident from the usual age of its students upon entering—between 12 and 15 years years old. Sometimes they functioned primarily as evening schools for artisans, as was the case at Dresden. Apart from the obvious exceptions—the foundings that took place after 1860, i.e., already in the new era of college status—only the schools at Berlin and Hanover operated at a somewhat higher level, since they accepted only graduates of the technical schools also founded in these states.[16] A pattern common to all these technical colleges, and which marks them as schools rather than institutions of higher education, was the lack of *Lernfreiheit*: students were organized in classes and had to follow a strictly prescribed curriculum. It was not until the very end of the 19th century that the technical colleges finally outgrew their inferior status and lost all vestiges of their former role as secondary schools. In

1899 they were officially recognized as institutions of higher education for which the *Abitur* (the final examination at the nine-year secondary schools) became obligatory for admission.

V. The Function of Mathematics Within the Technical Colleges

Over the course of the 19th century, mathematics at the technical colleges was undergoing a process of transformation that was in a certain sense similar to that which it already experienced at the universities. Originally, mathematics had the same preparatory function: it had to be studied *before* one began specialized studies in the professional (i.e., here: technical) discipline. The place of mathematics within the curriculum was characterized by its designation as an *allgemeine Wissenschaft* (general science). This conception harkens back to the traditional polytechnical function of mathematics, and its unanimous adoption within the German polytechnical schools was in fact largely the effect of the transmission of the model of the *École Polytechnique.* One must stress, however, that this conception also had roots in German epistemological thought.

These roots lay in the rationalist view of science, according to which there exists a clear-cut hierarchy of disciplines. Mathematics, as a model of logical strength and systematic reasoning, stood at the pinnacle of this hierarchy. The natural sciences as well as technology, on the other hand, were subordinate disciplines, and were required to deduce their particular rules from mathematical laws. In Germany, this rationalist view enjoyed widespread acceptance and appreciation; it was often identified as a Kantian doctrine since Kant had espoused his famous dictum regarding the necessary mathematization of the sciences.[17]

Although this rationalist view appreciated the technical disciplines only insofar as they were capable of being mathematized, one must stress that throughout the first half of the 19th century the technicians themselves pleaded most fervently for the deductive character of applications. They, too, promoted the idea that improvements in applications went hand in hand with progress in the most abstract realms of pure mathematics. Among the influential technicians who adhered to this position was August Leopold Crelle:

> So it is also important that pure mathematics should be explained in the first instance without regard to its applications and without being interrupted by them. It should develop purely from within itself and for itself. For only in this way can it be free to move and evolve in all directions. In teaching the applications of mathematics it is results in particular that people

> look for. They will by extremely easy to find for the person who is trained in the science itself and who has adopted its spirit.[18]

The unique and privileged role of mathematics as the fundamental scientific basis for technology was increasingly questioned during the 1870s and 1880s. This was a result of growing self-awareness among the ranks of engineers and technological practitioners that accompanied the rapid transformation of the polytechnical schools into institutions of higher learning. As time went on, other subjects from the humanities and sciences were joined with mathematics as part of the preparatory curriculum. At the same time, the rising status of the colleges created a demand for scientifically educated staff, and this meant hiring graduates of the universities. The young university mathematicians, however, only reinforced the growing "anti-mathematics" movement among the engineers. For they were educated in the purist spirit of rigorous mathematical theories, and were thus altogether unprepared or unwilling to adapt to the needs of the technical colleges that employed them.[19] By 1900 the "antimathematics" movement succeeded in noticeably reducing the hours for mathematics courses (Stäckel 1915, pp. 37–41). Under these circumstances, most mathematicians did not care to search for a constructive policy that would overcome this rift between mathematicians and engineers.

VI. The development of Klein's policy regarding applications and his *Gutachten* of May 1900

Felix Klein and Paul Stäckel were among the few mathematicians who realized that the engineers' movement represented an imposing challenge to the mathematical community. In particular, Klein, strove to reorient mathematical research and teaching so as to confront this challenge. He understood that the enormous expansion of enrollment at the technical colleges was symptomatic of a general change in higher education that was bound to have an impact on the future course of mathematics at the universities. In fact, the data shown in figure 1 indicate that the technical colleges rapidly attained an enrollment that was competitive with that of the universities. In particular, during the decade 1890-1900, a rather stagnant period for the universities, enrollment at the colleges more than doubled—in Prussia as well as throughout the entire German *Reich*. After the technical colleges obtained equal status with the universities in 1899, the role of mathematics in this expanding subsystem of technical education became even more crucial. The importance of the *Technische*

a. NUMBER OF STUDENTS/PRUSSIA

	Total	Philos. Fac.	Math./ Sciences	Technical Colleges Total	Technical Colleges "general sciences"
1869/70	7,750	3,014	581	n. d.	—
1890	13,280	3,826	1,072	1,910	1
1900	15,960	6,105	2,208 (Math.: 760)	4,702	20
1910	24,830	14,220	2,870 +Chem. Math.: 1,900)	4,144	84
1930	57,130		Math.: 3,749 (Career options for: Univ. 21, Schools 3,359, others 11)	8,032	Math.: 185 (for Schools: 167)

b. NUMBER OF STUDENTS/GERMAN REICH

	Total	Philos. Fac.	Math./ Sciences	Technical Colleges Total	Technical Colleges "general sciences"
1869/70	14,000	4,865	855	2,928	—
1890	28,900	7,730	2,318	4,209	n. d.
1900	33,000	12,665	4,749	10,472	n. d.
1910	53,380	27,800	5,851 +Chem	10,593	380
1930	99,580		Math.: 5,813 (Career options for: Univ. 26, Schools 5,250, others 12)	22,032	Math.: 660 (for Schools: 602)

Source: Gross/Knauer 1987

Figure 1.

Hochschulen for the discipline is already obvious from the number of positions it provided for mathematicians. In 1914 there were 47 positions for mathematics professors (*Ordinarien* and *Extraordinarien*) at the 21 universities, whereas the 11 technical colleges provided 35 professorships. Moreover, the latter institutions offered many more positions for younger faculty as *Assistenten*.

a. The development of Klein's agenda

In order to understand Klein's policy in this critical situation, let us briefly summarize how he became involved in matters of mathematics education, and the development of his views on the structure of the educational system. Educated at a Prussian Gymnasium in Düsseldorf, Klein attended Bonn University from 1865 to 1868. There he studied not only mathematics, but all the natural sciences in a universalist manner that was in keeping with the regulations of the Bonn natural sciences seminar, then still in vigour (Schubring 1986), (cf. Appendix I). Most formative for him was his contact with J. Plücker (1801-1868), professor of mathematics and physics, under whom Klein served as a laboratory assistant in physics.

Klein seems to have internalized Plücker's own self-assessment that his position within the German mathematical community was a marginal one. Plücker's program for combining geometry with algebra had been ostracized by the Berlin school of synthetic geometers headed by Jacob Steiner. Klein however, did not accept this situation as irremediable and set out to establish another mathematical program for geometry in Germany. Moreover, he gained the chance to realize such an alternative program through an extraordinary career as an influential mathematics professor at several universities and technical colleges. He received his first call in 1872 when he was only 23 years old. The appointment was an *Ordinarius* position at the University of Erlangen in Bavaria. In his famous "Erlanger Program," which was published as part of a traditional inaugural ceremony for newly appointed professors at Erlangen, Klein outlined the leading principles behind his alternative program for geometry. In his parallel inaugural address ("Antrittsrede"), he discussed his guiding principles for the organization of mathematics education (cf. Rowe 1985a).

This parallel elaboration of two programs clearly underscores the importance Klein attached to fundamental reforms in the orientation of mathematical research and its institutional structures. At this stage in his career, he advocated a traditional Prussian version of neohumanism.[20] He deplored the "fateful division" of the educational system, "the division between humanistic and scientific education," and expressed the hope "that in the not too distant future these oppositions will once again be balanced out, and that a unified education will come into being in which the presently polarized elements will be harmoniously joined together."[21]

It was precisely this search for a unified structure for secondary and higher education that led Klein to propose the idea of integrating the technical colleges into the universities in 1888.[22] This proposal, however, and others met with violent opposition from the engineering community. Even compromise initiatives, like attaching technical departments to the universities, were met with rejection (cf. Manegold 1970). The failure of this integration program for higher education in the 1890s made it clear to Klein that a fundamental reform of mathematics institutions in Germany could not be achieved by mere organizational reshuffling, as was implied in his earlier "Denkschriften" to the Ministry. Klein therefore initiated a first phase of complementary activities in the mid-90s that aimed to build a local nucleus in Göttingen for developing a fruitful interactive network of research in mathematics, physics, and technology. These efforts led to the founding of the "Göttinger Vereinigung zur Förderung der angewandten Physik" (later: "und Mathematik") in 1898—a private association of scientists and industrialists that succeeded in providing Göttingen with well-endowed technological institutions.

With the second phase of his activities in the 1890s, Klein followed through on his insight that a reform of university mathematics necessitated taking an even more extensive system into consideration: the school system as the basic foundation for higher education. He began, therefore, to interest himself in the improvement of teacher education. By so doing, he hoped to reverse the trend toward one-sidedly formal, abstract approaches to mathematics instruction by promoting practical instruction and the development of spatial intuition. An early breakthrough in this pursuit, although still realized by the traditional method of administrative decrees from "on high," took place in 1898. In that year new regulations for the examination of prospective teachers in Prussia were introduced. These created a second teaching license for applied mathematics, thus complementing the traditional one, which was thereafter referred to as the examination in "pure" mathematics. The new regulations also enabled the Prussian technical colleges to introduce disciplinary studies in mathematics for the first time. Henceforth, future mathematics teachers were allowed to spend three terms studying at these colleges. At the same time, the enlarged university curriculum made it possible to establish positions for applied mathematics at some of these institutions, thereby breaking up the monopoly that pure mathematics once enjoyed there. There are remarkable differences in the degree how the various universities responded to this possibility: not everywhere mathematicians were willing to enlarge the disciplinary spectrum of the positions (cf. Schubring 1985, 1985a). Particularly reluctant were the mathematicians at Berlin University.

The new university-like status attained by the technical colleges in 1899 as well as the imminent recognition of all three types of nine-year secondary schools as entitled to have their graduates admitted to institutions of higher education posed new structural problems, however, for these institutions. Should the graduates of the modernist schools (*Realgymnasien* and *Oberrealschulen*) be directed only towards the *Technische Hochschulen* or to the universities as well? If the former course were to be adopted, this would imply acceptance of the Bavarian model with its bifurcated educational system and consequent split approach to mathematics instruction.

Klein's mature views on these matters reflect the fact that between the time he delivered his "Erlanger Antrittsrede" in 1872 and the year 1900 he changed his mind profoundly with regard to the structure and function of both secondary and higher education. By the turn of the century, Klein not only acknowledged the equivalence of three parallel types of secondary schools, he also recognized that there should be a second type of technical institution for higher education. We may leave the question open, whether this change of viewpoint was prompted by inner conviction or by political pragmatism. In any event, in his own clear-sighted way Klein realized that such a multi-faceted educational system could pose a danger to mathematics if the discipline became too closely attached to a particular track within the schools. Klein believed that this risk could be minimized if a free and flexible transition from each type of secondary school to the two institutions of higher education was provided for.

Klein's change of position carried with it two far-reaching consequences. Instead of assimilating and integrating the technical schools (*Realschulen*) into the *Gymnasien* and the technical colleges into the universities, as he had proposed earlier, Klein now favored the full and independent development of the *Realschulen* (which had undergone a remarkable process of transformation throughout the 19th century since their beginnings as vocational schools of low status), and the *Technische Hochschulen* (whose transformation had been almost as profound). The second consequence was that the mathematical curricula in all the various secondary and tertiary institutions would have to be redefined in order to make the free transition from each school type to both higher education types. Thus by 1900, Klein's views had evolved to the point that he now recognized that an effective and stable reform of the subsystem of higher education required changes that went beyond this institutional plane. A new approach to teacher education was needed; but beyond this the link to the underlying school system had to be reformed, in particular, by reforming the mathematical curricula in the secondary schools.

b. A new agenda

A hitherto unknown document of May 1900 reveals the new stage of evolution and transformation in Klein's thinking. It contains in a nutshell the intentions underlying the famous curriculum reform movement in mathematics, which found its first public expression five years later at a congress held in Meran. The document is a memorandum ("Gutachten") prepared by Klein at the request of Friedrich Althoff, the leading figure in the Prussian Ministry of Education, in preparation for the Berlin *Schulkonferenz* of June 6-8, 1900. It was this conference that decided to place classical and scientific secondary schools on an equal footing.[23] On May 8th, 1900 Althoff asked Klein to state his expert opinion with regard to two questions: (1) Which advances in mathematics and sciences instruction had been made since the 1890 school conference? (2) What kind of secondary school preparation should be preferred for studies at technical colleges?[24]

Klein's answer, reproduced here as Appendix II, is a document of prime importance for the history of the mathematical reform movement, as in it Klein outlined his program for the reconstruction of secondary and higher education.[25] Moreover, the conclusions he expresses were drawn in part from his debates with representatives of the engineers' movement in the 1890s. The main point in Klein's answer to the first question was his assertion that the three types of secondary schools should be regarded as equivalent. He underscored further that all three offered sufficient opportunities for successful mathematics instruction. With respect to the second question, Klein described the profound crisis confronting mathematics education in the technical colleges. In these schools, following the famous model of the *École Polytechnique,* mathematics had been the main element of the general studies all students had to complete before commencing their work in special technical disciplines. This "polytechnical" tradition of mathematics instruction had, however, become obsolete. On the one hand, mathematical research and teaching had come to emphasize more than ever the logical grounding of the discipline, advancing towards a program of "arithmetization."[26] The technical disciplines, on the other hand, had not only developed and specialized in a highly dynamic manner; they also had changed in character. Having no interest in the complexities of higher mathematics, they stressed instead the role of "ingenious intuition," as Klein called it. He characterized the consequences for mathematics of this change in orientation within the technical colleges as simply disastrous. The great majority of the students at these colleges, he indicated, simply refused to follow the mathematical courses taught there: "they use the simple means of playing truant." As for the mathematics professors at these colleges, their efforts were met with indifference not only from their students but from their colleagues in engineering as well.

To overcome this crisis would not be easy, as Klein well knew, but he was prepared to propose a radical remedy. Since the mathematical courses at the technical colleges consisted of a basic preparatory "general" part and an advanced or higher part, he recommended that the basic studies be transferred to the preparatory schools, i.e. the secondary schools, and that only the advanced studies should remain as part of the college curriculum. However, they would be reformed and taught not as independent branches of knowledge, but rather "in permanent touch with the requirements of the students of mechanical and construction engineering.... Mathematics at the colleges can thus become once again what it ought to be: a specific power pervading the whole." Transferring the preparatory courses to secondary school would help create a firm mathematical basis for the pursuit of college studies, due, in particular, to the "benevolent coercion of school," as opposed to a university environment where academic freedom reigned. But which subjects were to be considered as part of the basic core of this new secondary mathematical curriculum? According to Klein, this core would include analytical geometry, but primarily differential and integral calculus!

In private letters to close friends and co-workers, Klein in fact always stressed that the key to his reform plans was to solve the question of mathematical preparation for the technical colleges.[27] It is evident that the most complicated part of his proposal concerned the changes that would be required in school mathematics, particularly since the secondary schools had dropped analytical geometry and calculus from their curriculum more than a half-century earlier. In 1900 and 1901, Klein carried on negotiations with the Ministry of Education in an attempt to persuade it to proceed in the traditional manner of curriculum reform, i.e. by decreeing changes in the mathematical curriculum from above. In fact, Klein engaged his friend, Wilhelm Lexis (a professor of political economy in Göttingen who, at the time, was an assistant in the Ministry), to make sallies in this direction. In April 1902, Lexis informed him of the decision made by the Prussian officials, which was unexpected and certainly most unusual. Although they approved of Klein's ideas, the Ministry refused to decree the desired curricular changes from above. Instead, they advised him to organize the introduction of these curricular changes "from below" by enlisting the support of appropriately trained teachers who could act as agents for the implementation of the reforms in selected schools. Thus, by following the principle of methodological freedom (*Lehrfreiheit*)—one of the characteristic neohumanist doctrines of the teaching profession in Prussia—these reformist teachers would be entitled to promote Klein's curricular changes without requiring the prior approval of school authorities[28].

In the wake of this decision, Klein began an intense and thorough study of the state of mathematics education in order to discover the pivotal issues that might activate mathematics teachers[29]. Having familiarized himself with some of the main problems facing mathematics teachers in the schools, he proceeded to coin the key phrase that would hereinafter serve as the slogan for his reform program, but which also helped to convey the impression this program was motivated exclusively by a desire to improve education within the schools. This was the famous notion of *functional reasoning*, or—to say it more concretely—the idea that the function concept should pervade all parts of the mathematics curriculum. Walther Lietzmann (1880-1959), Klein's principal assistant in organizing the reform movement (at least after R. Schimmack's sudden death in 1912) later recalled the role this idea played as an ingenious strategical device in vivid terms. Lietzmann noted that the success of the reform movement depended on finding a fundamental idea that would serve as a rallying point and which at the same time would automatically carry the calculus into the *Gymnasium* curriculum. That pivotal rallying point was the concept of function, which according to Klein's program would already be introduced in the lower grades (Lietzmann 1930, 255).

c. The beginnings of the reform movement

This slogan of functional thinking in hand, Klein began gathering support for this reform movement from below[30]. Indeed, its success was in large part a reflection of his efforts. Beginning with his own former students, he went on to persuade a substantial number of teachers and teachers' associations to join the ranks of his cause. The movement received added impetus from a "coalition" of mathematics and natural sciences teachers who sought better representation and more influence for their disciplines (cf. Gutzmer 1908). In the course of this campaign, Klein expended an enormous amount of energy maintaining an extensive correspondence and communicating with those who voiced their opinion on these matters in public. He sometimes even visited adversaries in their homes in hopes of winning them over[31] His goal was clearly to forge an extraordinarily broad and powerful alliance of teachers, scientists and engineers that would advocate a series of reforms for mathematics and sciences curricula (cf. Schimmack 1911).

The Prussian Ministry supported this movement from below as promised: five secondary schools, each a different type, were assigned the status of experimental institutions. At a number of other schools, activist teachers also introduced curricular changes that incorporated the spirit of the function concept. In fact, at all these schools the goal of the teachers was to introduce the elements of differential and integral calculus. It must be stressed, however, that one of the most complicated tasks within Klein's program was to extend these reforms not only to the "realist" schools (*Realschulen*) but also to the "humanist" ones (*Gymnasien*). Only then could the problem of transition be settled in the desired manner.

d. The work of the IMUK

Further evidence that Klein's report of May 1900 did, in fact, contain the heart of his future program for reforming mathematics education can be found in the work subsequently undertaken by the IMUK (*Die Internationale Mathematische Unterrichtskommission*). This organization was established in 1908 at the Fourth International Congress of Mathematicians, where it was assigned the task of preparing reports on the state of mathematics instruction in the industrialized countries for the next Congress, which was to convene in 1912. At this initial meeting, Klein was elected president and H. Fehr of Geneva was made general secretary[32]. Not surprisingly, Klein saw this as a welcome opportunity to enlarge the reform movement and to strengthen, through the support of an international body, the curricular reforms already underway in Germany. In particular, he saw this as a way to promote his program for the technical colleges, which still posed a major problem. The compilitory task of the IMUK thus became linked with a reformist agenda and the organization began to function as an agent for curricular change. At the Fifth ICM meeting in 1912, it was decided that because the statistical work of IMUK was still in progress, the Commission should be prolonged until 1916.

This work led to an enormous quantity of published material in the form of reports by national committees: 294 of them had appeared by 1920. Its reformist goals, however, required a second type of publication, topic-oriented comparisons, which clearly reveal the motivation and goals of the leading IMUK figures, and in particular its President. Of the seven subjects dealt with before 1914, only three were concerned with the Commission's official assignment, methodological and educational reforms within the school system. The other four were concerned with the relation between secondary and higher education or even with higher education exclusively (cf. Schubring 1987, pp. 9-10). It is quite remarkable that the culmination of the IMUK work was reached at the 1914 IMUK session in Paris where the two topics presented corresponded to the two cornerstones outlined in Klein's report to Althoff in 1900.

In Paris, the subject that attracted the most attention and participation was "The Evaluation of the Introduction of Differential and Integral

Calculus into the Secondary Schools." It was hotly debated, and the report on it was the most voluminous of all the international IMUK-reports (cf. *L'Enseignement Mathematique*, 16 (1914), 245–306). This was also the topic that Klein prepared more carefully than any other. He not only helped design the international questionnaire that dealt with this matter, but he also chose its coodinator and reporter, Emanuel Beke, a Hungarian scholar and one of the most fervent adherents of Klein's program.[33] Beke had transmitted the reform ideas directly to Hungary and he was highly successful in initiating an analogous movement there[34].

International comparisons were also made with respect to a second topic at the Paris IMUK session, another cornerstone of Klein's reform program: the mathematical preparation of engineers. Here as well, Klein chose as chairman a trustworthy person, Paul Stäckel, who had worked intensively since 1907 in the *Deutscher Ausschuss für technisches Schulwesen (DATSch)* in order to overcome the anti-mathematical movement within the engineering community[35]. Klein, who did not participate in the Paris session, was immediately informed about the course of the debates in letters from his assistants and co-workers. The points they stress were, of course, those that most interested Klein.

The most important of these letters came from Lietzmann and Stäckel at the close of the conference. With regard to the calculus, Lietzmann lamented the fact that no "palpable results" had emerged from the meeting due to the resistance of "the Italians."[36] Concerning the mathematical training of engineers, however, Lietzmann was able to report, much pleased, that "the engineers want—this was the general opinion—to get their mathematics from the mathematician, not from the engineer."[37] Stäckel also reported that he was quite satisfied with the international response (cf. *L'Enseignement Mathematique*, 16 (1914), 307–356), particularly from the French engineers, the overwhelming majority of whom had stressed "the necessity of a *culture générale* for the engineers."

He complained bitterly, however, about the lack of interest shown by German mathematicians who—although they were professors at technical colleges—preferred continuing their exclusive attachment to pure mathematics, rather than adapting their teaching to the needs of the students. Klein wrote marginal comments on this letter that reveal how central these questions were for his entire activity: "It is sad how short-sighted our colleagues are with us [Klein: right]. It was also sad that the German technical colleges were represented only by [Walther von] Dyck." Although Stäckel had sent invitations to all the mathematicians at these colleges, only one had responded, and this individual wrote to inform the organizers that he would be unable to participate: "I do not wish to become sharp, but we must nevertheless state that if this situation does not change, the day will come (maybe sooner than imagined) when these gentlemen will have shaken the position of mathematics so badly that nothing remains to be saved. And the universities are also in danger! [Klein: Exactly my opinion. The reason for all my activity]."[38]

e. The impact of the program

What were the results of these diverse yet carefully orchestrated activities after 1900? Most remarkable were the effects with regard to the technical colleges.[39] In 1907, the engineers' association (*Verein Deutscher Ingenieure (VDI)* established a special body, the *Deutscher Ausschuss für technisches Schulwesen (DATSch)*, in order to study the entire range of technical education. Mathematics was represented by Paul Stäckel, then professor of mathematics at the technical college in Karlsruhe and a strong advocate of Klein's policy. The deliberations with regard to the mathematical curriculum were closed in December 1913 and the following resolutions, proposed by Stäckel, met with no resistance:

1) engineers should obtain their mathematical education in the technical colleges;
2) their mathematics courses should complement those given in the secondary schools, and they should begin at a fairly advanced level;
3) the appropriate courses should be taught by mathematics professors who are favorably disposed towards technology;
4) mathematics should constitute a separate department in the technical colleges, and should be granted the same rights ("gleichberechtigt") as the other departments;
5) teachers of mathematics and the sciences should be given the opportunity to study within these mathematics departments;
6) the mathematics departments should be allowed to award the *Habilitation* to qualified candidates (Stäckel 1915, pp. 63–65).

The concluding general report of the DATSch, approved in March 1914, confirmed this stablilized function of mathematics (*Abhandlungen* ... 1914, pp. 159 ff.). The subsequent development of the technical colleges, particularly after World War I, saw the realization of the essence of these proposals. The most important result to emerge from this was that, besides its traditional service function, mathematics took on an independent status through the training of secondary school teachers. This general trend is clearly reflected in the considerable number of mathematics students who enrolled at the technical colleges by 1930—and not just in Prussia but throughout Germany (cf. Fig. 1). Most of these students had the intention of becoming teachers, as alternative careers outside the educational system were few. The academic degree of *Diplom-Mathematiker*, which was designed for careers in industry and civil service seems to have first been established in the 1920s at the technical College in Danzig (then not part of Germany) (cf. Wangerin 1979, p. 68). This degree was subsequently introduced throughout Germany in 1942. Up until then, all teacher examination regulations had maintained two separate teaching licenses in pure mathematics and in applied mathematics ever since the latter option was established in 1898.

In this connection, however, we should note a serious research *desideratum*. It is not known how many students obtained a teaching license in applied mathematics nor is it known how this number evolved in relation to the license in pure mathematics.[40] These data are necessary in order to judge the real effect of the differentiation in the mathematical teaching license on teacher training. Since the yearly detailed reports of the examination boards to the Prussian *Kultusminister* are extant in the archival files, it is in principle possible to obtain this information. Much more complicated would be a prosopographical study of the professional careers of the holders of an applied mathematics license: did they really become teachers and, if so, in which types of schools?

In general, one has to consider that the teaching license for applied mathematics constituted an element that was somewhat foreign to the system of liberal arts schools. The inclusion of applications in the form of a teaching license can be regarded as an expression of the fact that the discipline of mathematics was still exclusively oriented towards the teaching profession and that it did not provide effective alternative careers. The substitutive function of the license is confirmed by its redefinition after World War I. By a decree of 5 July 1921, the Prussian *Kultusminister* changed the content of the examinations in applied mathematics profoundly. The three traditional subjects for testing were replaced by a much greater number covering a variety of subjects (see below). This enlargement of the examinations by subjects that were rarely ever taught in the schools was criticized by traditionalists, particularly the Bonn mathematicians, as a "Göttingen dictate."[41]

VII. The evolvement of the notion of applied mathematics

We are thus led to the question: what were the guiding conceptions of applied mathematics that underlay the initiatives of the reform protagonists? It is not easy to arrive at a balanced assessment of the development of applied mathematics during the 19th century. Yet only through a clear understanding of this development—which evidently molded the views of the protagonists at the end of the century—can one give a fair account of the latters' views.[42] If one excepts the contemporaneous reports in the *Encyclopädie der mathematischen Wissenschaften*, related historical investigations are still largely lacking.[43]

As Lorey observed, at the universities the "applied mathematics of old had perished" due to its "encyclopedic flatness" (Lorey 1916, p. 260). Therefore, there existed no tradition which could be built upon and continued. Astronomy had evolved into an isolated specialty, and geodesy and descriptive geometry (as applied to technical mechanics) had been relegated to the technical schools and colleges, where they were mainly subsumed by technological disciplines. Thus, a new beginning had to be made in order to open mathematics towards applications. Neither Stäckel nor Klein were practitioners of applications. Obviously they could not base their views on epistemological reflections stemming from their own activity; rather they had to rely on common philosophical notions. Indeed, the only widespread conception of applications was the traditional one associated with the name of Kant. In Kant's view, pure and applied disciplines were strictly separated: due to their empirical character, the applied sciences were clearly subordinated to the theoretical fields, which provided them with the laws necessary for studying nature. One has to consider, moreover, that space intuition—in F. Klein's view an essential source of mathematical productivity—constituted, according to Kant, the supreme instantiation of pure mathematics. The impact of mathematical technology, on the other hand, such as calculating machinery, took place only after Klein's time. It was not only in our times that algorithmic aspects of concepts and methods have led to a profound change in mathematical epistemology.

For the most part, Klein operated with intuitive and pragmatic notions about applied mathematics. On occasion, he altered his views after discussions with colleagues, in particular, the first professional practitioner of applied mathematics at a German university, Carl Runge.[44] In his 1872 "Erlanger Antrittsrede," Klein made no explicit references to applied mathematics; he emphasized rather the indirect value of theory for applications (cf. Rowe 1985, p. 132). Some twenty years later, in his Evanston Colloquium Lectures of 1893, he stressed "the heuristic value of the applied sciences as an aid to discovering new truths in mathematics" (Klein 1894, p. 46). But here he was still playing the part of a conventional Kantian, differentiating between "exact" and "approximate" mathematics: "The ordinary mathematical treatment of any applied science substitutes exact axioms for the approximate results of experience, and deduces from these axioms the rigid mathematical conclusions" (ibid., p. 47). He even alluded to applications as "inexact mathematics" (ibid., p. 51). Klein was therefore prepared to adopt a notion proposed in 1900 by K. Heun (cf. Richenhagen 1985, p. 246). In his lectures of 1901 on the application of differential and integral calculus to geometry, later republished as volume 3 of his *Elementarmathematik vom höheren Standpunkte aus,* he characterized the relation between pure and applied mathematics as that of *precise* and *approximative* mathematics ("Präzisions- und Approximationsmathematik"). These terms did not express, however, Klein's definitive view of applications. Stäckel critized this definition in 1904 by claiming that applied mathematics is not approximative mathematics but rather a mathematics of approximate relations which places the same exigencies on rigor and uses the same instruments for proof as does precise mathematics (Stäckel 1904 a, p. 317). Klein then remarked in the preface to the second edition of his book, which appeared in 1907, that no pejorative meaning was to be attached to the term "approximative mathematics" and he affirmed in Stäckel's words that it is indeed nothing but the "the precise mathematics of approximate relations" (Klein 1928, p. 5).

Despite this somewhat strained terminology, Klein was clearly convinced of the essential role of applications for the future progress of mathematics. In his memorandum to Althoff of 1900, he expressed his "confidence in the motive power of scientific technics" for mathematical developments (app. II, fol. 13). And he stressed that mathematical "laws" are not immutable but are subject to changes arising from applications:

"In scientific technics, and in physics and other disciplines as well, a return to experimentation has taken place together with a simultaneous rejection of theory. In fact, under closer investigation, nature proves to be much more complex than the earlier theoreticians had previously thought. Therefore it is quite necessary to make use of experiments" (ibid., fol. 9–10).

Later, under the influence of Runge's original work in applied mathematics, Klein regarded the distinction between pure and applied mathematics as rooted in their respective functions as the "legislative" and "executive" branches of the discipline: "... Our colleague Runge has made us understand that applied mathematics is the doctrine of the mathematical executive, i.e. the numerical and graphical methods of which descriptive geometry constitutes a particular branch" (quoted from Tobies 1988, p. 266). Due to the still embryonic character of applied mathematics, questions surrounding its definition continued to be debated for some time. Horst von Sanden, a disciple of Runge and Klein, proposed an alternative differentiation: that of "systematic mathematics," providing the *general* theory, on the one hand, and "practical mathematics", on the other, the latter understood as the "methodology of numerical computation of *particular* solutions" (Sanden 1927, p. iii; my emphasis).

What subject matter actually accompanied these varied conceptions of applied mathematics? The teacher examination regulations of 1898 referred to three subdisciplines which were largely understood as the "field of applied mathematics": (1) descriptive geometry; (2) mathematical methods of technical mechanics, in particular graphical statics; (3) lower geodesy and elements of higher geodesy, together with the theory of compensation for observational errors. These disciplines might look somewhat scant when compared with the actual scope of applied mathematics. One must consider, however, that they were meant to comply with the limitations of teacher education. Moreover, this was the content largely approved by German mathematicians as the effective scope of applied mathematics. Apart from critical remarks made by certain mathematicians who preferred to understand descriptive geometry as part of pure mathematics, these three disciplines were generally viewed as constituting the proper core of applied mathematics, and one mathematician even lauded them as a new "Gaussian program" (Gutzmer 1904, p. 521).

This general consensus is also evidenced by a series of standard textbooks edited by H. Timerding in 1914 under the serial title *Handbuch der angewandten Mathematik.* Its three volumes were "Praktische Analysis" by H. v. Sanden, "Darstellende Geometrie" by J. Hjelmslev, and "Grundzüge der Geodäsie und Astronomie" by M. Näbauer. Taken as a whole, this series shows that the conventional view of applied mathematics as laid down in 1898 had not changed dramatically during the fifteen years that had passed since the enactment of a teaching license in the field. At the same time, it indicates that certain changes in the content of applied mathematics were indeed taking place. The volume on "Praktische Analysis," for example, underscores the important new role of numerical analysis as developed by Runge. The volume on geodesy and astronomy, on the other hand, reveals the still problematic status of probability theory. Back in 1897, Klein had proposed to Althoff that probability theory be included as one of the subjects for the new teaching license in applied mathematics. And in a letter written to the Ministry eleven years later, he criticized the narrowly-defined scope of the subjects in the teaching license, which included the "Theorie der Ausgleichung der Beobachtungsfehler" but nothing more on probability theory. Finally, by 1914 technical mechanics had lost most of its former importance. There had been debates in 1903/04 between H. Lorenz, Stäckel, and Klein over whether technical mechanics should be entirely subsumed under applied mathematics, handed over to the physicists, or simply restricted to its mathematical methods (cf. Stäckel 1904, p. 296 and 1904a, pp. 328–329).

In his book on mathematics at the German universities, Lorey undertook to expound "the modern view of applied mathematics" (Lorey 1916, p. 250). Considering his close supervision of the book's contents, this presentation may be presumed to be identical with Klein's mature view of this subject. Following Lorey, technical mechanics no longer played a central role in applied mathematics, and graphical statics, which had now become aligned with descriptive geometry, also possessed only marginal importance. In the case of graphical statics, the reason for this decline from the preeminent position it had enjoyed fifty years earlier through Culmann's influence was largely because its over-simplifying assumptions gave it little practical value in modern construction technics.

In fact, these changes in viewpoint regarding the content of applied mathematics were largely due to the influence of Runge's work, particularly his research on numerical analysis. Runge in effect replaced the former notion of applied mathematics as a series of loosely related subdisciplines with a methodological notion. For him, applied mathematics was the study of theoretical principles underlying the numerical and graphical methods employed by astronomers, physicists, engineers, etc., and their generalization to even broader principles of potential application (cf. Stäckel 1915, p. 74). In 1907 this methodological definition was approved by a conference of mathematicians interested in applications. Their key resolution was formulated as follows: "The essence of applied mathematics is constituted by the establishment and practice of methods for the numerical and graphical solution of mathematical problems."[45]

In 1914 the committee for the reform of mathematics curricula (the *DAMNU*), which was established in 1908 and decisively influenced by Klein, proposed a revision of the 1898 decree for teacher examinations in applied mathematics. In so doing, it followed exactly the lines set out by Runge and formalized at the 1907 conference that had adopted his methodological philosophy rather than the conventional view of applied mathematics. The teaching candidate should henceforth show his mastery of graphical and numerical methods and his ability to apply them in at least one relevant field. The committee listed six potential fields of application: astronomy, geodesy, meteorology and geophysics, applied mechanics, applied physics, mathematical statistics and actuary science (cf. Lorey 1916, pp. 255–256). This new approach was adopted for the teacher examinations in July 1921 by the Prussian *Kultusminister*, who even enlarged the number of fields of application beyond the six recommended by the *DAMNU* (see above, chapter VI).

Still, even in 1915 Stäckel continued to plead for a *Hochschulmathematik* that would be molded by an applied style suitable to the technical colleges (Stäckel 1915, p. 73). During the course of the Weimar Republic, however, the clamor to continue Klein's emphasis on applied mathematics gradually faded. In the meantime, the relationship between the universities and technical colleges stabilized, the universities maintaining their leading role, although mathematics was well represented in the general departments of the technical colleges. As a result, some of the positions that had been specially created for applied mathematics between 1900 and 1910 were now being filled with pure mathematicians.[46] It was only in the 1950s that technological innovations and new conceptual developments in the field of applied mathematics led to a rapid increase in the institutionalization of this subdiscipline. In the meantime, the legacy of Klein's agenda for introducing applied issues into university mathematics was confined to and cultivated in the realm of teacher education.

NOTES

1. Not only has the traditional dichotomy between "internal" and "external" history lost much of its character (cf. Schubring 1986), but one now also finds mutual support of these approaches in the "social history of ideas" (Brockliss 1987).

2. Quoted from: Schubring 1983, p. 184. The correspondence between Klein and Lorey will be presented in more detail in a forthcoming biography of Lorey that will also include unpublished manuscripts from his *Nachlass*.

3. Cf. Stichweh 1987 for a thorough discussion of the notions of discipline and profession within the system-theoretic approach of sociology.

4. An example is the Italian educational system around 1900. Despite the existence of independent "mathematics faculties," they functioned largely as propedeutic for technical high schools. Very few students—and in general not the brightest—continued their mathematical studies in order to obtain the degree *laurea di matematica* (cf. Pincherle 1911).

5. Felix Klein experienced these divergent systems first hand when he left Erlangen University, where he usually had three or four students attending his lectures, and never more than eight, to accept a position at the polytechnical school in Munich, where he often had 100 auditors and sometimes even more than 200.

6. A famous approach to this question was introduced by the sociologist Max Weber. His notion of a "Protestant ethics" provides evidence for the different impact of Protestant and Catholic Religion on economic structures as well as science.

7. Curiously enough, the general history of universities has not dwelled upon this split and the subsequent divergence in development. I first became aware of this in studying the origins of the various forms of institutionalization of mathematics, but so far, I can only describe it, rather than account for it or give precise answers about the reasons for this split. One complex of reasons seems to lie, however, in the policy of the Jesuits. Since the corporate universities were unwilling to render them one or more of their faculties, they established a system of *collèges*

that ran parallel with the arts' faculties, thereby siphoning off the latter's resources.

8. The new trend toward specialization first became visible at the two (Protestant) "reform" universities, Halle and Göttingen, founded in 1694 and 1737, respectively. During the 18th-century their professors, e.g. A. G. Kästner at Göttingen, were characteristic representatives of the beginning stages of this transition process. Those, like Felix Klein, but also Paul Stäckel (cf. his 1904 paper), who invoked the pre-19th-century unity of theory and practice as an ideal point of reference tended to overlook the decisive difference between the stage of merely classificatory delimitations within knowledge during the 18th-century and the level of modern disciplinary development thereafter. Gauss, who was Klein's favorite model figure upholding such a unity, was a representative of the universalistic ideals of the 18th-century academies rather than those of modern university professorships.

9. The original Prussian universities were: Berlin, Bonn, Breslau, Greifswald, Halle, and Königsberg; they were complemented after the 1866 war by Göttingen, Kiel, and Marburg. The Bavarian universities were Erlangen, Munich,and Würzburg.

10. F. Richelot attempted after 1869 to separate somewhat the formation of researchers from the formation of teachers (Schubring 1983, p. 111 and 289).

11. Only in 1854 was physics admitted in the upper grades as the sole representative of the natural sciences.

12. A *polytechnische Zentralschule* had existed before, from 1827 to 1833.

13. Cf. *Hauptstaatsarchiv München*, files of the *Ministerium des Kultus*, section: *Studienwesen in genere, MK 14873, Unterricht in der Mathematik und Physik*, vol. II, 1866–1934 (not paginated).

14. This, for instance, is confirmed by the resolutions of a conference on problems of secondary and higher education, organized on September 12 and 13, 1904 by the *Verein Deutscher Ingenieure (VDI)*. Contrary to its former position, the *VDI* now backed F. Klein's policy. An essential element was the decision to "recommend a further enlargement of the teacher education provisions at the technical colleges in *northern* Germany" (*Zeitschrift des VDI* 1905, p. 1976; my emphasis).

15. An exact description of the evolution of mathematical teaching within these schools/colleges is therefore not yet possible. In particular, what is missing is a prosopographical study of the teachers/professors who worked there. At the present time we know almost nothing about

their educational and personal background, teaching role, and research activities. A first step in this direction is the thesis by S. Hensel (1987), which lists the mathematics professors at the technical colleges since 1870 and describes their educational background. An exceptionally careful study concerning the origins of a polytechnical school was conducted by Viefhaus on Darmstadt (Viefhaus 1977).

16. After 1834, when *Realschulen* were founded in the state of Baden, the polytechnical school at Karlsruhe attained the same character as a central school.

17. "Ich behaupte aber, dass in jeder besonderen Naturlehre nur so viel *eigentliche* Wissenschaft angetroffen werden könne, als darin *Mathematik* anzutreffen ist" (Kant, *Metaphysische Anfangsgründe der Naturwissenschaft, Vorrede,* A IX).

18. Quoted from a memorandum of Crelle concerning the curriculum for the projected polytechnical institute at Berlin (1828), from: *Zentrales Staatsarchiv der DDR*, files of the former Prussian *Kultusministerium,* Rep. 76 V c, Sekt. 2, Tit. 23, Lit. A, Nr. 17, fol. 33.

19. Only a few mathematicians attempted to find a conciliatory that was suitable for the needs of technicians, Runge being the most prominent example (Richenhagen 1985, p. 85).

20. The consequence of this notion, a unified structure for secondary education with just one type of school, had long been advocated, in particular by mathematics teachers (cf. Schubring 1983, pp. 80–83).

21. Quoted from Rowe 1985a, 135–136.

22. Although such an approach was not, in principle, unrealizable, as examples from other countries show (USA, England), it can be regarded as naive with respect to the German situation. In particular, it overestimated the reach of administrational policy with regard to existing institutions and their clientele.

23. The changes within the secondary school system, due to the steady rise of "realist" school types, led to several approaches in Prussia to establish a clearly defined new structure. The most important were the two School Conferences of 1890 and 1900. The first conference proposed to establish a segregated system by abolishing the intermediate type schools *Realgymnasien* (with Latin, but no Greek) and granting the realist type *Oberrealschule* (without any classical languages) only limited access to the universities. These proposals were not realized, however, and the School Conference of 1900 instead granted equal access to higher education to the graduates of all three types of schools: *Gymnasium* (Greek and Latin), *Realgymnasium*, and *Oberrealschule* (cf. Albisetti 1983).

24. Nachlass Klein, no. 30; and Lietzmann 1930, p. 289.

25. With the exception of a short mention by Lietzmann (1930, 289), Klein's memorandum of 15./16. May 1900 has never been discussed in the literature. Although Lietzmann did publish several highly important documents in his 1930 article, he gave no hint at all as to the content of this document, evidently because of its frank and strategic character. Klein's participation at the conference itself lacked this strategic character; he focused instead on issues connected with certain parts of his answer to the first question, and there were only some hidden allusions to his new plans (Klein 1900, 153–155).

26. A telling example of the refusal of many mathematicians at technical colleges to change their purist attitudes can be found in certain pronouncements made by H. Timerding. In a report for the *IMUK* on mathematical education of science students, Timerding not only deplored the"amathematical" or "antimathematical tendency" within scientific and technological circles, but also that "one was obliged to adapt mathematics to the vocations" so that mathematics professors could not simply pursue pure mathematics qua "general" knowledge (Timerding 1911, 485–487).

27. Particularly revealing is Klein's letter of 8 December 1913 to Walther Lietzmann with regard to a meeting of the commission on technical colleges (*DATSch*): "Die Sache hat für ihren an Beke gerichteten IMUK-Bericht [report on the introduction of calculus, 1914 Paris] grosse Bedeutung. Denn die Herren stellen an die Vorbildung der Hochschulstudierenden genau die Anforderungen, die unsere Reform erstrebt." (Nachlass Lietzmann, UB Göttingen, I, 269). On the significance of that *DATSch* meeting, cf. Stäckel 1915, pp. 65–70. After Klein's death, Lietzmann took over the responsibility for the German *IMUK* section. Lietzmann's Nachlass at the Universitätsbibliothek Göttingen consists essentially of Klein's letters to him. Another part of Lietzmann's Nachlass (actually the greater part, documenting his scientific and educational activities) is located in the Stadtarchiv Göttingen; this also contains material concerning work of the IMUK, in particular the 1914 IMUK session at Paris.

28. This remarkable document appears in (Lietzmann 1930, 294).

29. Numerous notices and excerpts from the literature document these studies.

30. It is evident that Klein could not "divert" a reform movement (Pyenson 1983, 52) since there existed no such movement beforehand. One has to consider that historians of the mathematical reform movement, all of them followers of Kiein, presented his efforts in a line of continuity and therefore overemphasized the relatively isolated positions

of "reformist" mathematics teachers before 1900 in order to show Klein as a culminating point within a longer tradition.

31. Even G. Holzmüller, who was wholeheartedly opposed to the introduction of calculus, was won over by Klein, in 1904. A great deal of Klein's campaign after 1902 is well documented in his Nachlass, nos. 32–40 (the Holzmüller letters are in no. 33).

32. D.E. Smith of Columbia first proposed in 1905 that such a descriptive international comparison be made. In particular, he helped Klein establish relations with national experts, thereby establishing the international network of scholars needed to undertake such a project (cf. the correspondence between Klein and Smith in the Butler Library, Columbia University, Rare Book Room, D.E. Smith Professional Correspondence).

33. Klein's preparatory work is documented in: Nachlass Klein, no. 51, and in: Nachlass Lietzmann, I, 255 and passim (in particular: 269–270, 277).

34. Beke had been one of Klein's students at Göttingen in the 1890s. His attachment to Klein and his program is clearly documented by his letter of protestation against the dissolution of the IMUK in 1920 (25.9.1920, copy to Klein, Nachlass Klein, No. 51).

35. As the correspondence between Klein and Stäckel shows, both were close allies with regard to professional mathematical policy. In fact, Stäckel was one of Klein's key co-workers in several pivotal projects involving history and applications of mathematics, as well as mathematics education (the edition of Gauss' works, a report on the history of non-Euclidean geometry (with F. Engel), and articles on applied mathematics for the *Encyklopädie der mathematischen Wissenschaften*, the German IMUK-reports, and the IMUK-international comparisons). Although Stäckel was an independent personality, his career was decisively promoted by Klein.

36. The refusal of the majority of Italian mathematicians to support Klein's curricular reforms constituted a major threat to his plans (cf. Schubring 1987).

37. Lietzmann to Klein, 4 April 1914 (Paris). Nachlass Klein, no. 51.

38. Stäckel to Klein, 9 April 1913 [sic! 1914] (Heidelberg). Nachlass Klein, no. 51.

39. The realization of the intended curricular reforms in the secondary schools constituted an even more complex process. The famous "Meraner Lehrplan" of 1905, recommended by a committee of several associations, did not propose calculus as an obligatory subject, but rather only as an optional one. However, the mobilization of the teachers proved

successful, and by 1910 a majority of them were already teaching calculus. This *de facto* practice was officially acknowledged in 1925, when the elements of differential and integral calculus became part of the new curriculum in Prussia for all types of secondary schools, including the *humanistisches Gymnasium*.

It should be noted that the foregoing analysis of the context of Klein's curricular reform plans can be used to explain a feature of them that has until now remained enigmatic. For although the development of functional thinking has been highly touted as the very essence of the "Kleinian program" for secondary school mathematics, it was never clear how this key concept was to be implemented concretely, in particular with regard to curricular links between algebraic and geometric concepts. Such concrete matters, however, were of no principal concern to Klein. His interests were rather focused on the redefinition of the curricular transition between secondary and higher education.

40. For the years 1901 to 1904 these numbers are given by Stäckel (1904, 293).

41. See HStA II, fol. 296–297. Cf. Schubring 1983a, p. 34.

42. Recently, M. Otte has criticized, from a philosophical standpoint, Klein's view of applied mathematics as a reductionist psychologism (Otte 1987, p. 38).

43. Important steps in this direction are the studies by Erhard Scholz on graphical statics (cf. his article in this volume) and several studies directed by Walter Purkert.

44. Cf. Richenhagen (1985). Richenhagen's doctoral dissertation of 1984 contained a chapter on the institutionalization of applied mathematics. It is regretable that the author has not maintained this chapter in the published version.

45. "Besprechung von Vertretern der angewandten Mathematik in Göttingen am 22. und 23. März 1907," *Jahresberichte der Deutschen Mathematiker-Vereinigung*, 16 (1907), 496–519, on p. 518.

46. Courant argued, for instance, in his obituary to Runge that it was no longer necessary to have special labels for mathematical professorships (Courant 1927, p. 231), but evidently wished also to legitimate the circumstance that Runge's successor, Gustav Herglotz, was not an applied researcher (cf. Neuenschwander, Burmann 1987, p. 25).

47. Complementary note: "Eine von vielen Seiten erhobene Forderung: (und von meiner Universitätserfahrung)"

48. Marginal note: "durch die Verhältnisse des Studiums a.d. Universitäten erläutern"

49. Marginal note: "Immer das Bekannte als bekannt hinstellen. Dann das Eigene heranbringen."

50. Due to the change in the text, Klein did not rework the grammar of the phrase.

51. Inserted but not fully elaborated note (not entirely legible): "umgekehrt wird die Hauptaufgabe".

52. Note above 'keine': "nicht etwa an die [?]"

53. Insert: "sind bekannt genug"

54. Marginal note: "dies teile [?] ich nun den Herren mit"

55. The last line is underlined in a manner such as to indicate that a change is to be made.

56. F. Klein/E. Riecke (eds.): *Allgemeines über angewandte Mathematik.* Teubner: Leipzig 1900.

57. Insert: "dies meine persönliche Stellungnahme, die übrigens nichts Individuelles ist, sondern von ausgedehnten Kreisen getheilt wird."

58. Above the line: "nichts"

59. Between the lines: "enthalten einen richtigen Kern, werden aber in der Form vielfach übertrieben."

60. Note: "so ist das eine unvors.[ichtige] Formulierung, die ... [illegible word] H. nicht würdig [?] ist, mehr u conditionell einführen: Missverstehen der math.[ematischen] Bildung überhaupt"

Bibliography

A. Unpublished Sources

1. Universitätsarchiv Bonn: Exmatrikel.

2. Bayerisches Hauptstaatsarchiv München. Files of the Ministerium des Kultus, section: Studienwesen in genere, MK 14873, Unterricht in der Mathematik und Physik, vol. II, 1866–1934 (ByHaA).

3. Zentrales Staatsarchiv der DDR, Abteilung Merseburg. Files of the former Prussian Kultusministerium, Rep. 76 V c, Sekt. 2, Tit. 23, Lit. A, Nr. 17 (ZStA).

4. Hauptstaatsarchiv für Nordrhein-Westfalen in Düsseldorf. Files of the former Prussian Kultusministerium (Bestand NW5): Nr. 483, Acta betreffend das naturwissenschaftliche Seminarium der Universität zu Bonn, vol. III, 1859–1888 (HStA I); Nr. 558, Acta betreffend den mathematischen Apparat der Universität zu Bonn, vol. I, 1835–1929 (HStA II).

5. Niedersächsische Staats- und Universitätsbibliothek Göttingen, Handschriftenabteilung: Cod. Ms. Felix Klein, Nr. 11, 30, 32–40, 51 (Nachlass Klein); Cod. Ms. Lietzmann (Nachlass W. Lietzmann).

6. Stadtarchiv Göttingen: Dep. 89, Nachlass Walther Lietzmann.
7. Butler Library, Columbia University, New York, Rare Book Collection: D.E. Smith Collection, Professional Correspondence.
8. Nachlass Julius Plücker, CISTI, Ottawa, Canada.

B. Published Sources

Abhandlungen und Berichte über Technisches Schulwesen, Band V: Arbeiten auf dem Gebiete des Technischen Hochschulwesens, (Hrsg.: Deutscher Ausschuß für Technisches Schulwesen). Leipzig/Berlin: Teubner, 1914.

Albisetti, J.C., 1983: *Secondary School Reform in Imperial Germany.* Princeton: Princeton University Press.

Beke, E., 1914: Les resultats obtenus dans l'introduction du calcul différential et intégral dans les classes supérieures des établissements secondaires. Rapport Général. *L'Enseignement Mathématique,* 16, 245–284.

Courant, R., 1922: Carl Runge als Mathematiker. *Die Naturwissenschaften,* 10, 88–93.

Dickerhof, H., 1975: *Dokumente zur Studiengesetzgebung in Bayern in der ersten Hälfte des 19. Jahrhunderts.* Berlin: Duncker und Humblot.

Eccarius, W., 1987: *Mathematik und Mathematikunterricht im Thüringen des 19. Jahrhunderts.* Dissertation B, Pädagogische Hochschule Erfurt-Mühlhausen.

Grattan-Guinness, I., 1981: Mathematical Physics in France, 1800–1840: Knowledge, Activity, and Historiography. J.W. Dauben, ed., *Mathematical Perspectives.* New York: Academic Press, pp. 95–138.

Gross, H.-E. & Knauer, U., 1987: *Der Beruf des Mathematikers. Materialien zur Entstehung.* Oldenburg.

Gutzmer, A., 1904: Über die auf die Anwendungen gerichteten Bestrebungen im mathematischen Unterricht auf den deutschen Universitäten. *Jahresbericht der Deutschen Mathematiker-Vereinigung,* 13, 517–523.

__________, 1908, (Hrsg.): *Die Tätigkeit der Unterrichtskommission der Gesellschaft Deutscher Naturforscher und Ärzte. Gesamtbericht.* Leipzig/Berlin: Teubner.

Hensel, S., 1987: *Die Auseinandersetzungen um die mathematische Ausbildung der Ingenieure an den Technischen Hochschulen Deutschlands Ende des 19. Jahrhunderts.* Dissertation A, Universität Leipzig.

Huber, U., 1987: *Universität und Ministerialverwaltung. Die hochschulpolitische Situation der Universität München (1832–1847).* Berlin: Duncker u. Humblot.

Julia, D., 1986: Universités et Colléges à l'Epoche moderne. Les institutions et les hommes. J. Verger, ed., *Histoire des Universités en France*. Paris: éditions privat, pp. 141–197.

Klein, F., 1894: *Lectures on Mathematics. The Evanston Colloquium*. New York: Macmillan.

_______, 1901: [Beitrag zur Frage 5 über mathematischen Unterricht]. *Verhandlungen über Fragen des höheren Unterrichts, Berlin 6. bis 8. Juni 1900*. Halle: Waisenhaus-Buchhandlung, pp. 153–155.

_______, 1902: *Rathschläge betreffend das Studium der angewandten Mathematik*. Göttingen.

_______, 1923: Göttinger Professoren. Lebensbilder von eigener Hand. 4. Felix Klein. *Mitteilungen des Universitätsbundes Göttingen*, 5(1), 11–36.

_______, 1928: *Elementarmathematik vom höheren Standpunkte aus*. Bd. 3, Präzisions- und Approximationsmathematik (ausgearbeitet von C.H. Müller, für den Druck fertig gemacht und mit Zusätzen versehen von Fr. Seyfarth). Dritte Auflage. Berlin: Springer.

Kuhn, Th., 1962: *The Structure of Scientific Revolutions*. Chicago: University of Chicago Press.

Lietzmann, W., 1930: 25 Jahre Meraner Vorschläge. *Zeitschrift für mathematischen und naturwissenschaftlichen Unterricht*, 61, 289–300.

Lorey, W., 1916: *Das Studium der Mathematik an den deutschen Universitäten im 19. Jahrhundert*. Leipzig/Berlin: Teubner.

Luhmann, N., 1980–81: *Gesellschaftstheorie und Semantik. Studien zur Wissenssoziologie der modernen Gesellschaft*. 2 volumes. Frankfurt: Suhrkamp.

Manegold, K.-H., 1970: *Universität, Technische Hochschule und Industrie*. Berlin: Duncker u. Humblot.

Neuerer, K., 1978: *Das höhere Lehramt in Bayern im 19. Jahrhundert*. Berlin: Duncker u. Humblot.

Neuenschwander, E. & Burmann, H.-W., 1987: Die Entwicklung der Mathematik an der Universität Göttingen. *Georgia Augusta. Nachrichten der Universität Göttingen*, Nov. 1987, 17–28.

Otte, M., 1987: Zum Verhältnis von Mathematik und Technik im 19. Jahrhundert (Teil I). *Occasional Paper des IDM der Universität Bielefeld*, no. 87.

Pincherle, 1911: Sugli studi per la laurea di matematica e sulla sezione di matematica delle scuole di magistero. *Atti della sotto-commissione Italiana* (IMUK). Roma.

Pyenson, L., 1983: *Neohumanism and the Persistence of Pure Mathematics in Wilhelmian Germany*. Philadelphia: American Mathematical Society.

Richenhagen, G., 1985: *Carl Runge (1856–1927): Von der reinen Mathematik zur Numerik*. Göttingen: Vandenhoeck & Ruprecht.

Rowe, D., 1985a: Felix Klein's "Erlanger Antrittsrede," a Transcription with English Translation and Commentary. *Historia Mathematica*, 12, 123–141.

________, 1985b: Essay Review. Felix Klein (Renate Tobies with Fritz König; Karl-Heinz Manegold; Lewis Pyenson). *Historia Mathematica*, 12, 278–291.

________, 1989: Felix Klein, David Hilbert, and the Göttingen Mathematical Tradition. To appear in K.M. Olesko, ed. *Osiris*, vol. 5: Science in Germany, Problems at the Intersection of Institutional and Intellectual History.

Sanden, H.v., 1927: *Mathematisches Praktikum I*. Leipzig/Berlin: Teubner.

Schimmack, R., 1911: *Die Entwicklung der mathematischen Unterrichtsreform in Deutschland*. Leipzig/Berlin: Teubner.

Schubring, G., 1983a: *Die Entstehung des Mathematiklehrerberufs im 19. Jahrhundert. Studien und Materialen zum Prozess der Professionalisierung in Preussen (1810–1870)*. Weinheim/Basel: Beltz.

____________, 1983b: Seminar-Institut-Fakultät: Die Entwicklung der Ausbildungsformen und ihrer Institutionen in der Mathematik. *Reihe Diskussionsbeiträge zur Ausbildungsforschung und Studienreform, IZHD der Universität Bielefeld*, Heft 1, 1–44.

____________, 1985: Das mathematische Seminar der Universität Münster, 1831/75 bis 1951. *Sudhoff's Archiv*, 69, 154–191.

____________, 1985a: Die Entwicklung des Mathematischen Seminars der Universität Bonn 1864-1929. *Jahresbericht der Deutschen Mathematiker-Vereinigung*, 87, 139–163.

____________, 1986: Wilhelm Lorey (1873–1955) und die Methoden mathematikgeschichtlicher Forschung. *mathematica didactica*, 9, 75–87.

____________, 1987: The Cross-Cultural 'Transmission' of Concepts—the First International Mathematics Curricular Reform around 1900. *Occasional Paper des IDM der Universität Bielefeld*, no. 92.

____________, 1989a: Warum K. Weierstrass beinahe in der Lehrerprüfung gescheitert wäre (1841). *Der Mathematikunterricht*, 35, 13–29.

____________, 1989b: The Rise and Decline of the Bonn Naturwissenschaften Seminar: Conflicts between Teacher Education and Disciplinary Differentiation. To appear in K.M. Olesko, ed. *Osiris*, vol. 5: Science in Germany, Problems at the Intersection of Institutional and Intellectual History.

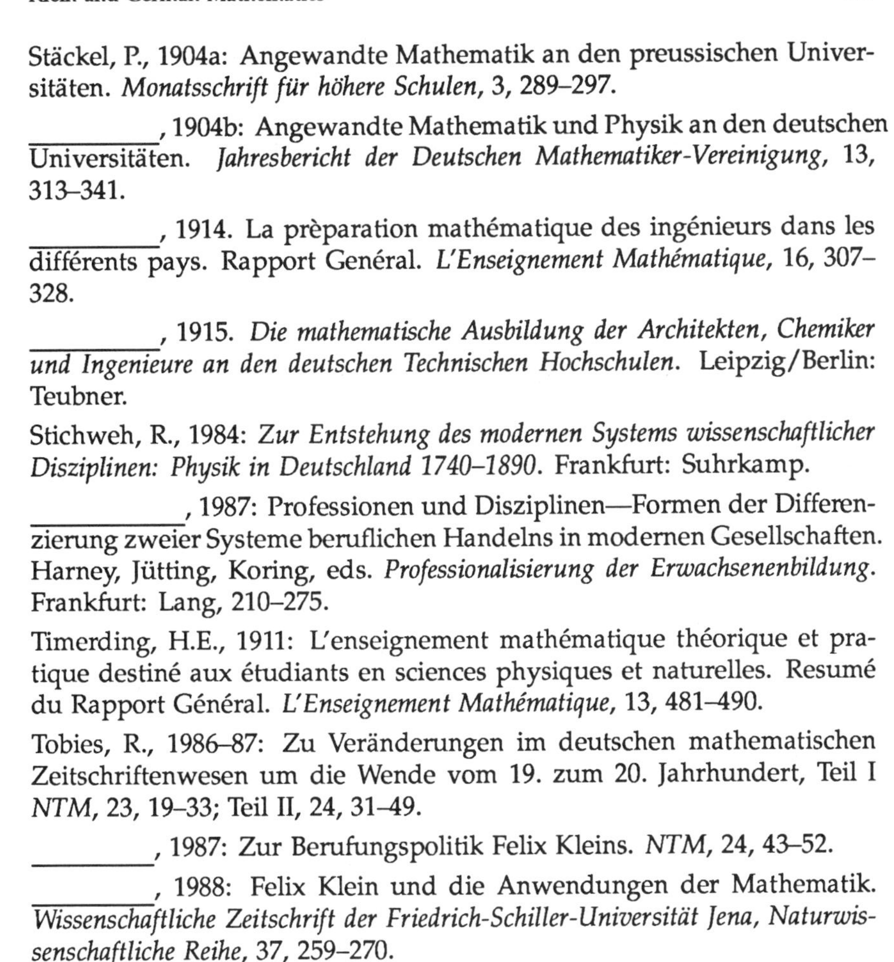

Stäckel, P., 1904a: Angewandte Mathematik an den preussischen Universitäten. *Monatsschrift für höhere Schulen*, 3, 289–297.

———, 1904b: Angewandte Mathematik und Physik an den deutschen Universitäten. *Jahresbericht der Deutschen Mathematiker-Vereinigung*, 13, 313–341.

———, 1914. La prèparation mathématique des ingénieurs dans les différents pays. Rapport Genéral. *L'Enseignement Mathématique*, 16, 307–328.

———, 1915. *Die mathematische Ausbildung der Architekten, Chemiker und Ingenieure an den deutschen Technischen Hochschulen*. Leipzig/Berlin: Teubner.

Stichweh, R., 1984: *Zur Entstehung des modernen Systems wissenschaftlicher Disziplinen: Physik in Deutschland 1740–1890*. Frankfurt: Suhrkamp.

———, 1987: Professionen und Disziplinen—Formen der Differenzierung zweier Systeme beruflichen Handelns in modernen Gesellschaften. Harney, Jütting, Koring, eds. *Professionalisierung der Erwachsenenbildung*. Frankfurt: Lang, 210–275.

Timerding, H.E., 1911: L'enseignement mathématique théorique et pratique destiné aux étudiants en sciences physiques et naturelles. Resumé du Rapport Général. *L'Enseignement Mathématique*, 13, 481–490.

Tobies, R., 1986–87: Zu Veränderungen im deutschen mathematischen Zeitschriftenwesen um die Wende vom 19. zum 20. Jahrhundert, Teil I *NTM*, 23, 19–33; Teil II, 24, 31–49.

———, 1987: Zur Berufungspolitik Felix Kleins. *NTM*, 24, 43–52.

———, 1988: Felix Klein und die Anwendungen der Mathematik. *Wissenschaftliche Zeitschrift der Friedrich-Schiller-Universität Jena, Naturwissenschaftliche Reihe*, 37, 259–270.

Viefhaus, E., 1977: Hochschule—Staat—Gesellschaft. *100 Jahre Technische Hochschule Darmstadt*. Darmstadt, pp. 57–112.

Wangerin, A., 1979: Abteilung für Mathematik und Physik. *Beiträge und Dokumente zur Geschichte der Technischen Hochschule Danzig, 1904–1945*. Hannover, pp. 67–69.

APPENDIX I

On the Biography of Felix Klein

In view of the opposition between the "Berlin school" and the "Göttingen school" of mathematics, it is interesting to compare the educational backgrounds of two of their chief protagonists, Karl Weierstrass (1815–1897) amd Felix Klein (1849–1925). In fact, it would be hard to imagine two mathematicians whose training differed more sharpely. For whereas Klein's university studies decisively shaped his subsequent universalist approach to science and mathematics, Weierstrass undertook only specialized mathematical studies. As I have shown elsewhere, Weierstrass was not really an autodidact in mathematics, as is commonly reported: he studied for at least three semesters with Gudermann in Münster. On the other hand, he acquired almost no knowledge of the natural sciences during the course of his entire education, and, as a consequence of his ignorance of these disciplines, he nearly failed his teacher examination in 1841 (cf. Schubring, 1989a). Since Klein's education at Bonn University from 1865–1868 has never been studied in detail, historians have had to depend on secondary sources from a later period, in particular (Klein, 1923). Here I should like to document what is to be found in various archival sources regarding his university education. These documents reveal that Klein undertook an extraordinarily broad range of studies in mathematics *and* the natural sciences. Furthermore, despite some negative remarks by Klein regarding his thesis advisor, Rudolf Lipschitz (Klein 1923, 14), these sources reveal that Klein attended nearly as many of Lipschitz's courses and seminars with Lipschitz as he did those of his preferred mentor Julius Plücker.

An Overview of Klein's Studies at Bonn

This information can be found in Klein's "Exmatrikel," the official document listing the courses a student attended before leaving a university. Klein's "Exmatrikel" shows that he began with an intensive study of mathematics, but that from the second semester on the natural sciences equalled or even superseded his mathematical studies.

Felix Klein's "Exmatrikel"

WS 1865/66
Logik (Prof. Neuhaeuser)
Analytische Geometrie (Prof. Dr. Lipschitz)
Experimentalphysik (Prof. Plücker)
Differentialrechnung (Prof. Gehring)
Mathematische Übungen (Prof. Lipschitz)
Methode der kleinsten Quadrate (Prof. Argelander)

Göthe (Prof. Springer)

SS 1866

Elektizitätslehre (Prof. Plücker)
Anorganische Experimentalchemie (Prof. Landolt)
Allgemeine Botanik (Prof. Hanstein)
Zahlenlehre (Prof. Lipschitz)
Mechanisches Practikum (Prof. Lipschitz)
Mathematische Übungen (Prof. Plücker)

WS 1866/67

Analytische Mechanik (Prof. Lipschitz)
Organische Experimentalchemie (Prof. Landolt)
Ausgewählte Kapitel der Chemie (Prof. Landolt)
Fortpflanzung und Entwicklung der Pflanzen (Prof. Hanstein)
Mathematische Übungen (Prof. Lipschitz)
Mathematische Übungen (Prof. Plücker)
Naturwissenschaftliches Seminar (Prof. Troschel)

SS 1867

Theorie der Kräfte, die nach dem Newton'schen Gesetze
 wirken (Prof. Lipschitz)
Zoologie (Prof. Troschel)
Mineralogie (Prof. Nöggerath)
Mathematisches Seminar (Prof. Lipschitz)
Mathematisches Seminar (Prof. Plücker)
Das natürliche Pflanzensystem (Prof. Hanstein)
Naturwissenschaftlicihes Seminar (Prof. Troschel)

WS 1867/68

Chemisches Praktikum (Prof. Landolt)
Mathematisches Seminar (Prof. Lipschitz)
Mathematisches Seminar (Prof. Plücker)
Interferenzerscheinungen (Prof. Ketteler)

SS 1868

Differentialgleichungen (Prof. Lipschitz)
Mathematisches Seminar (Prof. Lipschitz)
Mathematisches Seminar (Prof. Plücker)
Variationsrechnung mit Beispielen aus der analytischen
 Dynamik (Dr. Gehring)

WS 1868/69

Zahlentheorie (Prof. Lipschitz)
Analytische Statik (Prof. Radicke)
Über parasitische Pilze (Dr. Pfitzer)

Klein as a Participant in the Natural Sciences Seminar

The Bonn natural sciences seminar was a unique institution because of the fact that it encouraged students to study the entire scope of the sciences. The history of this seminar is examined in (Schubring, 1989b). Felix Klein was quite remarkable for his active participation in all five sections of the seminar. Since students had to enroll in these sections with the director of the seminar, their "Exmatrikel" generally does not reflect their participation in them. To obtain this information it is therefore necessary to analyze the seminar files. These reveal that Klein was a member of the seminar from his first semester (WS 1865/66) until the summer of 1867. His participation in its five sections is given in the following chart:

	WS 1865/66	SS 1866	WS 1866/67	SS 1867
physics	X	X	X	X
chemistry	X	X	X	X
minerology	X	X	X	X
zoology	X	X	X	
botany	X	X	X	X

Klein's high degree of participation in all sections of the seminar was unusual as the *Reglement* that required students to take part in all of them was dropped beginning in the winter semester 1866/67. During his first year, Klein was very active in the physics section, where Plücker praised his performance with these words: "Durch Talent, Kenntnisse und Fleiss ragte vor allen anderen Mitgliedern des Seminars Klein hervor." The following year after he won an award (*Prämie*) for a paper on magnetic curves, Plücker wrote: "Ein eminentes Talent für mathematische Physik sowohl, als für Experimentalphysik legte Klein an den Tag." That same year Klein made presentations in each of the other four sections as well. During the winter semester 1866/67, he spoke on "Kristallsysteme" in mineralogy, on "Gefäss-Kryptogamen" in botany (rated "sehr gut"), and on "Schmetterlinge" in zoology (rated "vorzüglich). In zoology, too, the section head praised Klein: "...dessen hervorragendes Talent sich kund gab." The following semester Klein spoke about natrium in chemistry ("frei und gut"), and on "Generationswechsel" in botany ("recht gut"). When he left the seminar after the summer semester of 1867 to concentrate on mathematics, the seminar heads expressed their regret over his departure (HStA I, fol. 91 and 116–118).

Klein as a Participant in the Mathematics Seminar

Due to the activities of the science professors, mathematics was of rather marginal importance at Bonn University, particularly since the mathematics professor Plücker held a chair in physics as well. From the early 1840s, Plücker had devoted himself primarily to physics. Only after Lipschitz was appointed to a position at Bonn in 1864 was there an impulse to establish a mathematics seminar. The *Kultusministerium* hesitated, however, due to a feud between Plücker and Lipschitz. Eventually, the seminar was inaugurated in the winter semester 1866/67 with two separate sections run by the two rival professors (cf. Schubring, 1985a). Felix Klein joined it immediately as one of its initial members. Indeed, as his "Exmatrikel" shows, he even participated in both sections. Not surprisingly, both directors praised his excellent performance, but unfortunately the files of the seminar provide no further evidence of his precise activities in its two sections during the four semesters he was a member. There is, however, indirect evidence that strongly suggests he was an active member in both sections. A traditional means of rewarding active participation by students in seminars was by confering a *Prämie*), an amount equivalent to a small scholarship. Klein received a *Prämie* from the mathematics seminar semesters he was a member. Furthermore, Lipschitz explicitly stated that he was awarded a *Prämie* for the summer semester 1868 (HStA II, fol. 56). Since Plücker had died in May 1868, the second section of the seminar was suspended that summer. So it is clear that at least on this occasion Klein received a *Prämie* for his participation in Lipschitz's section of the seminar.

APPENDIX II

Thanks to the kind permission of the *Handschriftenabteilung, Niedersächsische Staats- und Universitätsbibliothek Göttingen*, Klein's *Gutachten* written in two parts on May 15-16, 1900 for Friedrich Althoff is published here from the Klein *Nachlass* (Cod. Ms. Klein, 30). The text represents Klein's draft, which documents his systematic understanding of the interrelationships between secondary and higher education. Several notes between the lines and in the margins indicate that the version actually submitted to Althoff differs in some passages. The draft with its changes gives insight into Klein's thinking processes regarding critical issues for mathematics education in Germany. It is likely that the final version that was sent to Althoff is still extant in his papers on the 1900 school conference or in the voluminous ministerial files dealing with the debates over school reform.

(Accompanying letter)

Ew. Excellenz

beehrt sich der ehrerbietigst Unterzeichnete in Erledigung der ihm durch das Schreiben U II 1398 vom 8. Mai 1900 erhaltenen Auftrag beiliegende zwei Gutachten zu überreichen. –F. Klein.

Gutachten I.

Frage: Welche Fortschritte sind seit der Schulconferenz vom Jahre 1890 auf dem Gebiet des mathematischen und naturwissenschaftlichen Unterrichts an den höheren Schulen, insbesondere auch nach der angewandten und technischen Seite hin, zu verzeichnen und was kann in dieser Beziehung noch weiter geschehen?

1. *Umgränzung meiner Aufgabe*

Es hat keinen Zweck, dass ich mich über Einzelheiten des Schulbetriebs äussere, die ich nur mangelhaft kenne und die Andere jedenfalls viel besser kennen als ich. Es kann sich einzig darum handeln, dass ich Gesichtspunkte hervorkehre, die mir in meiner Praxis als Universitätslehrer und Mathematiker lebendig entgegentreten sind und über die ich glaube, einiges Nützliche sagen zu können. Ich verstehe dabei unter höheren Schulen Unterrichtsanstalten von *allgemeinem* Charakter. Höhere Fachschulen sind allerdings auch etwas sehr wesentliches, und ich glaube, dass dieselben in Preussen lange noch nicht genug entwickelt sind. Sie entgehen der Schulconferenz, weil [sie] nicht dem Cultusmin.[isterium] unterstellt [sind]. Aber es würde zu weit führen, hier darauf einzugehen.

2. *Einteilung der in Betracht kommenden Disciplinen.*

Es findet eine natürliche Trennung zwischen den biologischen (oder wie man früher sagte: den beschreibenden) Naturwissenschaften statt und den mathematisch-physicalischen Disciplinen. Chemie und Mineralogie, resp. Geologie stehen in der Mitte und können zum Teil der einen Hälfte, zum Teil der anderen zugerechnet werden.

3. *Die biologischen Naturwissenschaften.*[47]

Dieselben sind seither auf den höheren Schulen unbillig vernachlässigt. Sie sollten bis zur Oberstufe durchgeführt werden. Lebendige Beobachtung der umgebenden Natur bildet den naturgemässen Anfang. Auf der Oberstufe aber sollte einiges Verständnis für Physiologie, Hygiene etc. erzielt werden. Alles Dogmatische (Hypothetische) ist dabei wegzulassen. Für Einzelheiten wird keine Zeit sein.

Durchaus wesentlich scheint mir, dass die biologischen Naturwissenschaften auf allen Anstalten ihre eigenen Lehrer erhalten. Es geht auf die Dauer nicht mehr an, dass dieselben beim Studium unserer Lehramtscandidaten mit den mathematisch-physikalischen Disziplinen verquickt werden.[48] Es handelt sich auf beiden Seiten nicht nur um einen ungeheuren Complex von Erscheinungen, sondern auch um eine durchaus unterschiedene Geistesrichtung. Die so formulierten Forderungen haben vermutlich zunächst wenig Aussicht auf Berücksichtigung. Sie müssen um so energischer gestellt und, wenn es nicht anders ist, als Programm für die Zukunft festgehalten werden.

4. *Mathematik und Physik und ihre Bedeutung für die Schule.*[49]

Im Betrieb der Mathematik sind von je zwei entgegenlaufende Tendenzen hervorgetreten. Wir haben als das eine Extrem den rein scholastischen Betrieb. Die Mathematik wird nur als Mittel angesehen "den Verstand zu schärfen", dementsprechend das logische Element einseitig betont. Nach der anderen Seite haben wir den rein utilitarischen Betrieb. Es werden bestimmte Formeln oder Constructionsmethoden, die den Anwendungen nützlich sind, ohne strengeren Beweis autoritativ übermittelt und vor allen Dingen auf Gewandtheit in der Durchführung der äusseren [?] Operationen gesehen. Wir haben denselben Gegensatz bei anderen Fächern, z.B. beim Sprachunterricht. Auf der einen Seite ausschliessliches Hervorheben der Grammatik, auf der anderen Seite ebenso ausschliessliche Betonung der Phonetik und Uebung der Gewandtheit im wirklichen Sprechen. Hiervon huldige ich folgendem Grundsatze. *Ein rationeller Schulunterricht soll sich in der Mitte halten,* d.h. die wesentlichsten Momente von beiden Seiten zur Geltung bringen. Im einzelnen wird der Individualität des Lehrers viel Freiheit zu belassen sein. Denn die Persönlichkeit des Lehrers muss als solche zur Wirkung kommen.

Als Ziel des mathematischen Unterrichts für die höhere Schule hat [möchte ich vor allem] die Überzeugung hervorzurufen,[50] dass *richtiges Nachdenken aufgrund richtiger Prämissen* die Mittel zur Beherrschung der Aussenwelt an die Hand gibt". Hier berührt sie sich mit[51] der Physik, bei welcher das Herbeischaffen der richtigen Voraussetzungen, d.h. das *inductive Element* ein Hauptgeschäft ist. Übrigens ist die Mathematik keine [52] Naturwissenschaft; sie hat ihr eigenes Gebiet, welches nach vielen anderen Seiten ausgreift (cf. politische Arithmetik).

Ich befürworte durchaus, dass der mathematische Unterricht vom Sinnfälligen beginnt: der Anschauung, dem Zeichnen, Messen, numerischen Rechnen. Die Verbindung soll auch auf der Oberstufe festgehalten werden, aber durchaus das logische Element zur Geltung kommen. Der Schüler muss ein sicheres Urteil darüber bekommen, ob ein Beweis richtig oder falsch ist; dies ist ein unerlässlicher Prüfstein eines erfolgreichen mathematischen Unterrichts.

5. *Neuere Tendenzen beim mathematischen Unterricht.*

Die angesprochenen Grundsätze decken sich in der Hauptsache mit den sog.[enannten] *Braunschweiger Beschlüssen* des Vereins zur Förderung des mathematisch-naturwissenschaftlichen Unterrichts. (Man vergleiche auch die Referate der Herren *Holzmüller und Schwalbe* in den Berichten des Vereins 1896.) Inzwischen sind andere Einflüsse an der Arbeit, den mathematischen Unterricht nach dem einen oder anderen Extrem hin zu verschieben.

Die lebhaften Bestrebungen der *Ingenieure*, beim mathematischen Unterrichte der Anschauung ihr Recht zu sichern[53] verdienen so lange alle Anerkennung und Unterstützung, als sie den logischen Kern des Unterrichts bestehen lassen.[54] Im wissenschaftlichen Betrieb der reinen Mathematik tritt umgekehrt in den letzten Jahrzehnten fortschreitend immer mehr die rein logische Seite mit Ausschliesslichkeit in den Vordergrund. Allgemeine Auffassung, math.[ematische] Phantasie und dergleichen kommen neben der Tendenz auf Verschärfung der Grundbegriffe und Beweismethoden nur noch beiläufig zur Geltung. Das ist, was ich die *Arithmetisierung* der Mathematik genannt habe. Wieweit dieselbe fortgeschritten ist, bzw. das Terrain einseitig beherrscht, tritt in überraschender Weise in den bis jetzt erschienenen Teilen der mathematischen Encyklopädie hervor. Es handelt sich um eine Bewegung, welche übergreift. Es fehlt nicht an Bestrebungen, welche die Arithmetisierung auch an der Schule in weitgehender Weise durchführen wollen.

Die Ingenieure berufen sich bei ihren Bestrebungen vielfach auf amerikanische Universitätseinrichtungen. Demgegenüber bildet es einen merkwürdigen Gegensatz, dass zahlreiche amerikanische Mathematiker z.Z. damit beschäftigt sind, die arithmetisierte Wissenschaft, die sie an europäischen Universitäten gelernt haben, in ihrem Vaterlande zur Geltung zu bringen. Die Gegensätze, um die es sich hier handelt, greifen eben durchaus über das einzelne Land auf die ganze Culturwelt über. Auch Frankreich ist der einseitigen Arithmetisierung verfallen, seit die Ausbildung der gelehrten Mathematiker von der Ecole Polytechnique, welche dieselbe früher besorgte, an die Ecole Normale Supérieure übergegangen ist.[55]

6. *Wertbestätigung der verschiedenen Gattungen höherer Schulen und ihres mathematischen Unterrichts*

Ein erfolgreicher Unterricht ist ohne eine gewisse Concentration nicht möglich, bloss encyclopädischer Vortrag verflacht. Nach welcher Richtung die Concentration gewählt wird, ist vom allgemeinen Standpuncte aus gleichgültig; im speciellen wird man sie so wählen, wie es dem späteren Berufe am meisten zu gute kommt. Hiermit soll natürlich einem engen Fachbetriebe nicht das Wort geredet sein, wohl aber dem Nebeneinanderbestehen verschiedener Gattungen unter sich gleichwertiger höheren Schulen.

Welche der bestehenden Schulgattungen wird hiernach für den späteren Mathematiker (nicht nur den gelehrten Forscher sondern den Lehrer der Mathematik) am meisten zu empfehlen sein? Die allgemeine Antwort geht vermuthlich dahin, dass dies die Anstalten realistischen Characters sein möchten. An den Oberrealschulen bez. Realgymnasien wird in der That die Mathematik im allgemeinen in weitergehendem Umfange gepflegt als an den humanistischen Anstalten. Ausserdem werden die mathematischen Fertigkeiten: das Zeichnen etc. sehr viel mehr gepflegt. Ich habe aber Erfahrung genug gesammelt (und die grosse Mehrzahl der F.[Fachkollegen?] wird mir darin beistimmen), dass vielfach das logische Element, ja die Fähigkeit sich klar im Zusammenhange auszudrücken, zurückbleibt, und dass in dieser Hinsicht—gerade für den Mathematiker–das humanistische Gymnasium den Vorzug verdient. *Wenn ich keiner besonderen Schulgattung hier den Vorzug zuerkenne, so habe ich andererseits die Überzeugung, dass an sämtlichen Arten höherer Schulen ein tüchtiger Lehrer der Mathematik bei einiger Bewegungsfreiheit ausgezeichnete Resultate erzielen kann.* Wir Mathematiker werden uns überall zurechtfinden. Um so wichtiger erscheint mir die Frage der geeigneten Ausbildung unserer mathematischen Lehramtscandidaten.

7. *Lehrerbildung*

Die neue Prüfungsordnung ermöglicht nach verschiedenen Richtungen hin ein sehr viel rationelleres Studium der Mathematik und Physik, als früher. Indem sie neben reiner Mathematik angewandte Mathematik als besondere Lehrbefähigung einführt, gibt sie den mathemat.[isch-]physikalischen Studien nicht nur die lang vermisste Ergänzung nach der praktischen Seite, sondern sie ermöglicht auch eine Concentration auf Mathematik und Physik allein (also ein Beiseitelassen der beschreibenden Naturwissenschaften, wie es unter 1) empfohlen wurde). Indem sie andererseits die angewandte Mathematik an die reine Mathematik bindet, betont sie, dass mit der Ausdehnung der Studien nach der praktischen Seite kein Verfall der logischen Grundlegung verbunden sein soll. Übrigens betont die Prüfungsordnung auch bei der Physik die Anwendungen. Sie gestattet den Lehramtscandidaten der Mathematik und Physik überdies ein facultatives Studium an der Technischen Hochschule in der Ausdehnung, dass nur drei Universitätssemester nachgewiesen werden müssen. Also auch nach dieser Hinsicht eine weitgehende Freiheit. Hierzu ist allerdings zu bemerken, dass an den Technischen Hochschulen Preussens keine Lehramtsabteilungen bestehen, dass also der Lehramtscandidat an ihnen zwar allgemeine Anregung durch die Fühlung mit den technischen

Kreisen aber nicht einen besonderen für ihn umittelbar passenden Unterricht findet. Umso nothwendiger erscheint es, dass die Universitäten ihre Unterrichtseinrichtungen nach Seiten der angewandten Mathematik (und der angewandten Physik) ergänzen.

8. *Neue Unterrichtseinrichtungen an den Universitäten*

Wir haben in Göttingen die besondere Gunst der Umstände benutzen können und Einrichtungen der angedeuteten Art in den letzten Jahren zur Entwicklung gebracht. Auf mathematischer Seite handelt es sich um den Unterricht in darstellender Geometrie, technischer Mechanik und Geodäsie, auf physikalischer Seite um Electrotechnik und allgemeine theoretische Physik. *Hierüber soll in allernächster Zeit eine Schrift erscheinen* (die sich schon unter der Presse befindet!);[56] dieselbe ist aus den Vorträgen erwachsen, welche bei unserem letzten Feriencurs für Lehrer der Mathematik und Physik gehalten wurden. Diese Schrift soll die Gesammtheit der hier vorliegenden Fragen klarstellen. *Ich werde dieselbe Schrift in einer grösseren Zahl von Exemplaren einsenden.* Unser besonderer Wunsch ist, damit einen Anstoss zu geben, dass ähnliche Einrichtungen wie bei uns, jedenfalls an einer Mehrzahl anderer Universitäten getroffen werden mögen. Nur so lässt sich der allgemeine Fortschritt für die Ausbildung unserer Lehramtscandidaten erzielen, auf den es ankommt. Und noch ein Anderes sollte generalisiert werden. Feriencurse für Lehrer höherer Lehranstalten (Oberlehrer) finden ja z.Z. viele in regelmässiger Wiederholung statt. (an preussischen Univ.[ersitäten] mehrere) Es ist dies eine vorzügliche Einrichtung, die sich ebenfalls erst in den letzten Jahren herausgebildet hat (deren allgemeine Bedeutung ich hier nicht darzulegen brauche). Aber gerade Mathematik ist bis jetzt was Preussen angeht, nur erst bei uns in Göttingen in den Bereich der Feriencurse einbezogen worden. Es ist dringend wünschenswerth, dass dies ebenfalls allgemeiner (auch an anderen Orten) geschieht, dass jedenfalls auch an einem Platze in den *östlichen* Provinzen Curse in Mathematik eingerichtet werden.

Göttingen, 15. Mai 1900. F. Klein

Gutachten II.

Frage: Welche Art der Vorbildung auf höheren Schulen verdient für das Studium der Technischen Hochschulen den Vorzug?

1. *Präcise Antwort, nebst Erläuterungen.*

Auf die vorstehende Frage glaube ich kurz antworten zu sollen: "eine 9 jährige Vorbereitungsanstalt allgemeinen Characters, welche die mathematisch-naturwissenschaftlichen Disciplinen voll zur Geltung bringt und die neueren Sprachen nicht vernachlässigt".

Die Anstalt soll *allgemeinen* Characters sein, keine Fachschule. Ich will mich aber allerdings nicht dagegen aussprechen, dass besonders tüchtige Absolventen mittlerer Fachschulen ausnahmsweise zum Hochschulstudium zugelassen werden. Andererseits will ich Absolventen *humanistischer* Gymnasien nicht principiell vom Hochschulstudium ausschliessen. In der That erscheint es mir nützlich, dass verschiedene Berufskreise, bez. Auffassungsweisen immer wieder zu gegenseitiger Einwirkung gebracht werden. Könnte man die Studienzeit unbeschränkt verlängern, so wäre man versucht im Interesse allseitiger Bildung den Technikern eine humanistische, den Philologen und Theologen eine technisch-naturwissenschaftliche Bildung vorzuschreiben. Doch das sind Utopieen.

Das Realgymnasium hat den Vorzug, durch Beibehaltung des Latein eine Brücke zur traditionellen Bildung aufrechtzuerhalten. Die Oberrealschule vertritt ein eher in sich geschlossenes System, doch liegt hier die Gefahr wohl besonders nahe, dass der Unterricht encyclopädisch verflacht. An sich scheinen mir beide Schulsysteme zur Vorbereitung für den technischen Beruf gleich geeignet (indem sich Vortheile und Nachtheile die Waage halten). Genau dasselbe würde ich auch betr.[effs] die Vorbereitung zum *medicinischen* Berufe sagen.

2. Excurs, betr.[effs] die Mathematik an den Technischen Hochschulen

Die unter 1) zusammengestellten Aussagen[57] können kaum etwas[58] Eigenartiges bieten. Ich will aber umso lieber im Folgenden einen einzelnen Punct herausgreifen, betreffs dessen ich einen Vorschlag zu befürworten habe, der mir in vielen Richtungen beachtenswerth scheint. Allerdings sind viele praktische Bedenken zu überwinden.

Ehe ich den Vorschlag begründe, muss ich einige Worte über den Betrieb der Mathematik an den Technischen Hochschulen sagen (der den Fernerstehenden orientiert). Es dürfte bekannt sein, dass in den letzten Jahren zwischen den Mathematikern und den Vertretern der technischen Fächer über den Umfang, den man der Mathematik an den Hochschulen geben soll, noch mehr aber über die *Art* des mathematischen Unterrichts grosse Meinungsverschiedenheiten hervorgetreten sind: Diese Meinungsverschiedenheiten wurzeln im Grunde in der Verschiedenartigkeit der Entwicklung, welche die an der Hochschule in Contact kommenden Wissenschaften als solche während der letzten Jahrzehnte genommen haben. Auf Seiten der Mathematik ist ganz besonders die logische Seite, die " Arithmetisierung" (wie ich in Gutachten I,4 ausführte) zur Entwicklung gekommen. Andererseits hat in der wissenschaftlichen Technik, wie auch in der Physik und anderen Gebieten, eine Rückkehr zum Experiment stattgefunden, unter Abwendung von der Theorie. In der That erweist sich die Natur bei näherem Zusehen als sehr viel complicierter, als die früheren Theoretiker vorausgesetzt hatten, es ist also sehr nothwendig, Experimente zu Rathe zu ziehen. Andererseits ist es so bequem, sich des Rüstzeuges der Mathematik zu entschlagen und stattdessen die geniale Intuition hervorzuheben. Nichtmathematiker werden den ganzen Gegensatz verstehen, wenn sie die modernen Richtungen der bildenden Kunst mit dem classischen Kunstbetrieb vergleichen, der mit Perspective und Anatomie begann. Ich bin immer für Verständigung auf mittlerer Linie eingetreten. Die Forderungen, welche in Folge dieser Umstände von den Technikern an den Hochschulmathematiker gestellt werden, lassen sich unter folgende drei Puncte fassen:[59]

a) *Der Mathematiker soll sich in den Gesammtzweck der Anstalt einordnen*— hiergegen wird kaum etwas zu sagen sein; sofern die Anstalt nicht gerade mit einer Lehramtsabteilung versehen ist, hat der Mathematiker darauf zu verzichten, seine Wissenschaft um ihrer selbst willen zu entwickeln— b) *beim Unterrichte soll mehr das anschauungsmässige Element als das logische hervortreten*—(soll auf das Milieu abgestimmt sein.) Hierin liegt viele Wahrheit, nur darf die Sache nicht übertrieben werden; eine Mathematik ohne logischen Kern ist keine Mathematik mehr (wie in Gutachten I noch ausführlicher dargelegt).[60]

c) *Die Studierenden sollen nur solche Mathematik lernen, welche sie später beim technischen Betriebe* (womöglich in den Vorlesungen der gerade an der Anstalt wirkenden technischen Docenten) *direct gebrauchen.* Diese Forderung (die in der That vielfach gestellt wird), ist natürlich in dieser Form durchaus zu verwerfen, weil mit dem Character einer Hochschule unverträglich, sie widerspricht sich überdies selbst, denn die Mathematik, welche der künftige Bauingenieur oder Vermessungsingenieur nach dem heutigen Stand ihrer Disciplinen unmittelbar gebrauchen, ist eine ganz andere als die Mathematik der Maschineningenieure oder Electrotechnicer. (Ich habe nicht den Streit abzugleichen, sondern nur zu schildern).

Der Streit, den ich hiermit schildere, hat sich inzwischen nicht auf theoretische Erörterungen beschränkt, sondern praktische Consequenzen gezogen. Zusammenziehen der früheren mathematischen Vortragsstunden ist das Mindeste. Die Studierenden der technischen Hochschulen lehnen in Mehrzahl den höheren mathematischen Unterricht, der ihnen geboten wird, überhaupt ab; um zu grosse Anstrengung beim Lernen zu vermeiden benutzen sie das einfache Mittel des Schwänzens (welches nebenbei bemerkt, in keinem Fache so unangebracht ist, als gerade im Fach der Mathematik) und erlassen [?] gelegentlich, was Examensforderungen angeht, Massenprotest gegen zu hohe Anforderungen. *Von wenigen Ausnahmen abgesehen leiden die Mathematiker an den Hochschulen gleichzeitig unter dem Misstrauen iherer Collegen und ihrer Zuhörer; es ist eine wahre Calamität.* Wie gering unter diesen Umständen der Lehrerfolg ist, und wie wenig dadurch der Studierende an mathematischer Vorbildung in seinen späteren Beruf hinübernimmt, braucht nicht erst ausgeführt zu werden.

3. *Vorschlag betr.[effs] Verlegung eines Teils der höheren Mathematik an die Vorbereitungsschule.*

Die Mathematik, welche an technischen Hochschulen (in dem ersten Semester) gelehrt wird, umfasst im wesentlichen analytische Geometrie, Differential- und Integralrechnung, sowie darstellende Geometrie. Dabei kann man auf allen diesen Gebieten eine Unterstufe und eine Oberstufe unterscheiden. Während letztere den Maschinen- und Bau- Ingenieuren vorbehalten bleibt, soll die Unterstufe beispielsweise auch von den Architecten und den Chemikern (sofern man denselben nicht thatsächlich alle höhere Mathematik erlässt) erworben werden. Der Vorschlag, den ich hier befürworten will oder doch zur Discussion stehen will, ist nun der, *besagte Unterstufe der Mathematik von der Hochschule weg an die Vorbereitungsschule zu verlegen.*

Für die technische Hochschule und auch, wenn es richtig gemacht wird, für die *Mathematik* an den technischen Hochschulen würde hieraus nach verschiedenen Richtungen ein Vorteil entstehen. Die Hochschule würde einen Vorbereitungsunterricht abgeben, der so, wie er jetzt getrieben wird, doch seinen Zweck verfehlt. Unter dem wohlthätigen Zwange der Schule wird derselbe sehr viel erfolgreicher gedeihen als jetzt unter dem Schatten der akademischen Freiheit. Es wäre eine wahre mathematische Grundlage für die Hochschulstudenten gewonnen. Die Hochschule könnte dann gleich im ersten Semester mit specifischen Fächern wie technischer Mechanik und Vermessungswesen beginnen.

Der Mathematiker aber müsste die Aufgabe bekommen, die technischen Vorlesungen für Maschinen- und Bauingenieure mit höheren mathematischen Vorlesungen zu begleiten, welche sich in stetem Contact mit den Bedürfnissen der heranwachsenden Zuhörer von Semester zu Semester steigern. Die Mathematik würde solcher Gestalt an der Hochschule wieder werden, was sie sein soll, eine specifische Potenz, die das Ganze durchdringt. Endlich wäre hier die aus allgemeinen Gründen so sehr erwünschte Möglichkeit gegeben, die durchschnittliche Dauer der Hochschulstudien etwas herabzumindern.

4. *Bedenken und Einwände gegen die vorgeschlagene Maassregel*

Ich setze voraus, dass nicht etwa Verfall der mathematischen Studien an den Hochschulen eintritt. Vertrauen in die Triebkraft der *wissenschaftlichen* Technik! *Caveant consules.* Ob es möglich ist, den genannten mathematischen Lehrstoff, der ja zum Teil schon auf unseren höheren Schulen behandelt wird, auf denselben vollständig zur Geltung zu bringen, das ist eine schultechnische Frage, welche sich meiner Competenz entzieht und über welche die Sachverständigen entscheiden müssen.

Sollte es beispielsweise an den humanistischen Gymnasien nicht angehen, so müsste man zu dem Hülfsmittel greifen, welches z.B. in Würtemberg eingeführt wird, wo die Absolventen der humanistischen Gymansien, welche die Hochschule besuchen wollen, zunächst in einen Vorbereitungscurs eintreten müssen (den "Mathématiques spéciales" der Franzosen entsprechend). Ein derartiger Candidat hat eben dafür, dass er sich Alles in Allem einen vielseitigere Bildung erwirbt, ein oder zwei Semester zuzusetzen.

Ein anderer denkbarer Einwand wäre der, dass durch die vorgeschlagene Maassregel die Vorbereitungsschulen einseitig auf das Bedürfnis der technischen Hochschulen zugeschnitten würden. Diesen Einwand kann ich nicht gelten lassen, weil es nämlich eine sehr grosse Zahl von Studierenden der Universitäten gibt, die zwar thatsächlich an den Universitäten bislang meistens keine mathematischen Studien treiben, denen mathematische Studien in dem bezeichneten Umfange ausserordentlich nützlich wären. Dies sind die Studierenden der Naturwissenschaften und Chemie, ganz besonders aber auch, wegen der mathematischen Hülfsmittel, die auf ihrem eigenen Gebiete immer mehr benutzt werden, die Mediciner. Die Interessen der Universität sind hier mit denjenigen der Hochschule durchaus übereinstimmend. Also ich meine, dass der Vorschlag ernste Erwägung verdiene. Er setzt allerdings eines voraus, nämlich

5. *eine verbesserte Ausbildung der mathematischen Lehramtscandidaten.*

Hiermit mündet dieses zweite Gutachten in dieselben Betrachtungen, die ich zum Schluss von Gutachten I auseinandergesetzt habe. Andererseits Maassregeln zu geeigneterer Vorbildung der späteren Hochschullehrer. Einzelheiten gehören nicht hierher.

Göttingen 16. Mai 1900. F. Klein

The Founding Members of the Deutsche Mathemtiker-Vereinigung, Bremen, 1890

On the Contribution of Mathematical Societies To Promoting Applications of Mathematics in Germany

Renate Tobies

This paper will concentrate on particular issues related to applied mathematics in Germany which I consider to be most important and relevant for this Symposium.[1] I begin with the following thesis: "applied mathematics" is an historical category and it is therefore necessary to determine what we understand by it. The formation of a special field called "applied mathematics" was the result of disciplinary differentiation within mathematics and increasing division of labour among mathematicians. Attempts to define the term applied mathematics are as diverse as its history. Ultimately, the definition depends on the historical stage of development of the sciences, but the standpoint of the observer may also lead to subtly diversified opinion.

One of this century's leading applied mathematicians, Richard von Mises (1883-1953), expressed the opinion that:

> The mathematician concerned with the conceptual development of infinitesimal analysis speaks of 'applications' of differential calculus to geometry when he draws even the most trivial geometric conclusions from his theorems. Someone who works in an area of theoretical mechanics, say elasticity theory, regards this differential geometry as the mathematical or theoretical tool he 'applies' to exhibit or analyze his equations. The scientifically-oriented engineer, on the other hand, uses elasticity principles as 'theory' in order to arrive at concrete calculations of solidity, which for him are the only real 'applied' mathematics. Finally there is the practical design engineer, for whom such calculations of solidity are still highly theoretical mathematics. He masters the tasks of his profession with completely elementary rules of thumb, and for him these constitute truly 'applied' mathematics. [2]

My presentation here will be confined to mathematical applications to other sciences (natural, technical and social sciences) as well as their

ISBN 0-12-599662-4.

relations to the field of education. Direct applications to material production or for planning and managing industrial processes fall outside the scope of my inquiry.

Let us begin by considering some of the mathematical organizations that promoted applications in Germany before 1870. Of these organizations, I should like to single out the following: the first subject-specific mathematical society, the *Kunstrechnungs lieb- und übende Societät* (founded in Hamburg in 1690), the Mathematical Society of Jena (founded in 1850), the "Mathematical Circle" in Karlsruhe (1862), and the mathematics and astronomy section (1843) of the Society of German Natural Scientists and Physicians (*Gesellschaft Deutscher Naturforscher und Ärzte*, or *GDNA*). The activities of these bodies through lectures and publications reveal a broad interest in applications of mathematical knowledge to other sciences—the main emphasis being on mechanics, geometrical optics, goedesy—as well as the solution of practical problems, for example, in navigation or the compilation of tables of measurements and weights. Further evidence of the strong connection between mathematics and other fields can be seen in: 1) the renaming of the Hamburg Society, which from 1790 to 1870 became known as the "Society for the Dissemination of Mathematical Knowledge in Hamburg;" 2) the growing participation of representatives from other disciplines in various mathematical organizations; and 3) the detailed discussions on the relationship between mathematics and the natural sciences that preceded the admission of mathematics as a special section within the Society of German Natural Scientists and Physicians (the initiative here was taken primarily by representatives of the other sciences rather than by the mathematicians themselves).

The perception that mathematics was a valuable tool for the consolidation of other sciences was widely shared during this period and served as a basis for the institutionalization of mathematics. A typical expression of this prevalent philosophy came from the astronomer Julius Zech (1821-1864), who wrote:

> . . . I believe that one cannot deny at least the possibility that everything that is comprehensible to the senses is capable of being mathematically calculated. The reason this has not yet occured lies not with the state of mathematics, but rather the natural sciences, which in large part have not yet progressed beyond the phenomena to the causes behind them. If the latter were known, then the possibility arises of applying mathematics. In this respect, the extent to which mathematics has already been applied to a branch of the natural

> sciences can serve as a criterion for the general degree of its utility.[3]

These views conform with the well-known assertion of Immanuel Kant "...that in any field of general science there can be found only as much science proper as can be found mathematics."[4] It was in this sense that the high appreciation of mathematics in the technical sciences was generally understood, and it was this philosophy that brought together the mathematician Alfred Clebsch and the engineer Franz Grashof, two of the leading figures associated with the "Mathematical Circle" at the Technical College in Karlsruhe.[5]

By the beginning of the last third of the nineteenth century, the activities of German mathematical societies were directed toward establishing mathematics as an autonomous discipline based on its own fundamental principles. At the same time the view that mathematics should serve to enrich the other sciences—especially physics and astronomy—was never openly disputed. The growing number of people working on mathematics in a professional capacity, increasing mathematical productivity, and the fragmentation of mathematics into isolated, specialized subdisciplines were factors that prompted mathematicians to seek an independent organization on a national scale. The attempts during the 1870s to arrange regular conferences for German mathematicians should be regarded as an initial stage in this pursuit, even though these failed to achieve their purpose.[6] The impulse to found a separate mathematical organization was further accentuated by the fact that other fields, which had hitherto considered the existing mathematical organizations as their home base, were also beginning to organize independently. Among these were the astronomers, surveyors, engineers and technicians who formed the following organizations: the Astronomical Society (1863); the German Association of Geometers (1871); and the Association of German Engineers with their respective district organizations (1856).

By the 1870s the majority of mathematicians saw their principal task in the elaboration of the fundamental theoretical principles of mathematics, that is the implementation of what came to be called the "arithmetization of mathematics." They wished to expand the scope of research in mathematics by systematizing its principles and and their mutual penetration. The lecture programs organized by the local mathematical societies reflect the dominant role of these activities. But, at the same time, there were always mathematicians who also engaged in solving problems posed by physics and astronomy. However, a certain arrogant attitude spread that looked upon applications of mathematics to concrete practical problems as a somewhat primitive exercise. This attitude is expressed in the remark by E. E. Kummer (1810-1893) that "applied mathematics is dirty mathematics." And in his inaugural address delivered in Tübingen in 1874, Paul duBois Reymond (1831-1889) proclaimed that mathematicians were by no means "...imbued with the desire to extend our knowledge, to create methods, which are useful for practice."[7] Even the inaugural speech delivered by Felix Klein (1849-1925) in 1872, while expressing an appreciation for practical applications, subtly implied that it was somehow inappropriate for university mathematicians to tackle problems that went beyond the realm of the purely theoretical. The 23-year-old Klein declared on that occasion:

> When I speak here of applications of mathematics, I am thinking less of applications in the usual sense, such as those one might care to bring forward before circles that are somewhat removed from the academic outlook in order to demonstrate the utility and significance of one's discipline. So far as mathematics is concerned, for example, one mentions the predictive calculations of the astronomer, one praises the precision of geodetic measurements, and one admires the accomplishments of the engineering art. Nothing could be further from my mind than to dismiss such applications as these. To be sure, so far as the general interests of humanity are concerned, these applications are of the highest and most far-reaching significance by virtue of the attainments they make possible. However, it is not my intention ... to follow these considerations any further. By the word "applications" I am thinking much more of the theoretical services performed by mathematics in the development of other sciences.[8]

With the founding of the *Deutsche Mathematiker-Vereinigung* (*DMV*, or German Mathematical Society) on September 18, 1890, the first organization emerged that brought together German mathematicians on a national scale. Here it is important to highlight the role of Georg Cantor (1845 -1918), who succeeded in uniting the majority of mathematicians despite their existing differences. On October 19, 1890, Klein wrote to Adolf Hurwitz (1859-1919): "Cantor knows how to arrange things so that Weierstrass, Kronecker, and now Neumann, too, equally favour it!"[9] This letter shows how Klein deliberately remained in the background so as not to run the risk of alienating his rivals, thereby endangering the whole effort: "The thing that matters now is to give the days in Halle [where the initial meeting was to take place] an enormous richness and diversity. Then I will be back on the scene again, whereas for now I restrain myself in the interest of the venture so that it will not take on a partisan character."[10] On the other hand—and this is the essential feature for us here—Klein succeeded in influencing the programme, the composition of the executive committee, and the criteria for membership, among other things, as becomes obvious from correspondence with his friends and students. His efforts in this regard should be looked upon as one of many endeavours to support all areas of mathematics and their applications equally. Thus I do not subscribe to the opinion that Klein spoke up only for applications of mathematics: he simply wanted to emphasize them without ignoring other fields.[11]

Klein was one of the few who realized the new demands that would soon be made on mathematics by industry, the natural sciences, and technology as well as by modern financial institutions, particularly the insurance industry. He was also one of the few who tried to prepare mathematicians for these new roles and novel activities. Above all, it was largely due to Klein's influence that the German Mathematical Society was not founded in isolation of other disciplines. While preparing for its inaugural meeting, he wrote to Paul Gordan (1837-1912) on June 20, 1890: "At present I disagree with Cantor. He wants to disengage the mathematical society from the natural scientists right from the beginning, which I regard to be nonsense; I only want the separation to take place if we are strong enough or feel overly restricted."[12]

Right from the beginning, the *DMV* stressed the goal of bringing together all mathematicians and all areas of mathematics while promoting ties with important fields of application. The "Bremen accords," which were enacted at the Bremen *Naturforscher* meeting of 1890 and served as the founding document for the *DMV*, explicitly stated that: "the mathematics-astronomy section of the Society [of German Natural Scientists and Physicians] shall... support a more extensive circle [of mathematicians], whose activities include the entire spectrum of scientific interest in mathematics."[13] It was further resolved that the mathematics-astronomy section should no longer bring mathematicians together on a merely sporadic basis, and that its program should be prepared by the *DMV*.

These early plans reveal that alongside pure mathematics applications of mathematics were to play a conspicuously significant role. Thus, for the 1892 meeting planned for Nuremberg (but eventually cancelled due to a cholera epidemic), the organizers wished to present reports on recent mathematical work related to technical mechanics, astronomy, and modern physics, as well as a comprehensive collection of mathematical models and physical instruments. Through the efforts of Klein's former student, Walther von Dyck, this exhibit was displayed at the *DMV* meeting held in Munich the following year.

Although Klein held no post in the *DMV's* executive committee at this time, he was nevertheless largely responsible for the society's early orientation toward applied mathematics. He made his influence felt through Dyck, who was on the committee. Their correspondence reveals that they conferred with one another over nearly every major question involving the scientific program and the *DMV* membership. Additional correspondence further reveals that Klein solicited several reports on developments in applied mathematics that were presented at *DMV* meetings. He received considerable support in this regard from Ernst Lebrecht Henneberg (1850-1933), a professor of technical mechanics at the technical college in Darmstadt. Through Henneberg, Klein sought to find suitable authorities who could prepare reports on various parts of technical mechanics, including hydrodynamics and hydraulics.[14]

Klein considered the formation of the German Mathematical Society to be an essential opportunity for preserving the relations between mathematicians and natural scientists, and he suggested that it was important to win physicists as members. In 1894 he delivered a lecture at a plenary session of the meeting of the *GDNA* held in Vienna. There he spoke about the importance of Bernhard Riemann (1826-1866) for the development of modern mathematics. As he wrote to Gordon in August of 1894, approximately one month before the lecture, his principal aim was "...to secure the position of mathematics within the sphere of allied natural sciences."[15] And in the lecture itself, he flatly stated that: "The close relationship between mathematics and the theoretical sciences which has existed for the benefit of both fields since modern analysis came to the fore is in danger of disintegrating... As members of the Mathematical Society, we want to do everything in our power to counteract this. It was in this spirit that we joined the *Naturforscherversammlung*."[16]

As Gert Schubring has indicated above, Klein's views on these matters were in part a reaction to the challenge posed by the ever-growing strength of the "engineer's movement." This led him to the realization that mathematics should seek to maintain close ties with developments in the physical sciences and technology. An early expression of this opinion came in October 1880 with his inaugural address as professor of geometry in Leipzig. Klein addressed this issue again in more concrete terms in a memorandum of 6 October 1888 prepared for the Prussian Ministry of Culture. It was not until the 1890s, however, that he turned to this question in earnest and began to tackle it by pursuing a variety of organizational projects.

Klein's efforts to shape and influence the scientific programs at *DMV* meetings was only one of these numerous projects. His influence in this regard, however, was significant and became even more so after he was elected to the executive committee in 1896 and as president in 1897. Following his recommendation, the *DMV* decided to choose a theme that would serve as a focal point for the annual meetings. The theme chosen for the 1897 meeting to be held in Brunswick in 1897 was mechanics. Through Klein's efforts, an impressive number of lectures were given by prominent authorities on technical mechanics, including Föppl, Cranz, Lorenz, Mehmke, Finsterwalder, Schubert, and Stäckel.[17]

Between 1897 and 1913 topics within applied mathematics appeared several times as a special theme for the *DMV* program. These topics ranged from mechanics (1897, 1903, 1908, 1911) to electrodynamics (1898) and electron theory (1902), to the partial differential equations of physics (1905), geometrical kinematics (1909), and numerical analysis (1898). Here, let us devote a few words to the lectures that were held on applications and on the teaching of mathematics up until 1913 (see Table 1). At the turn of the century, the proportion of such contributions within the total number of lectures had been particularly high, amounting to around 38% of the total. This was by no means an accident; it resulted from the fact that the executive committees of the *DMV* responsible for these conferences chose to emphasize these areas.[18] Subjects for long expository articles on relevant topics were assigned and the project of an "Encyclopedia of Mathematics including its Applications" was discussed and promoted. The latter, of course, became the basis for the famous *Encyklopädie der mathematischen Wissenschaften*, eventually published over the course of more than 30 years.

A breakdown into five separate categories of the lectures on applied topics delivered between 1891 and 1913 at the annual meetings of the *DMV* is given in Table 1. Within the first group, descriptive geometry was by far the most popular topic. This field continued to play a central role in mathematics instruction at the German *technische Hochschulen* which were decisively influenced by the model of the *École Polytechnique*. During the 1880s and 1890s, Felix Klein had been instrumental in introducing this subject into the curriculum at some of the leading universities. There were relatively few lectures on vector analysis, although Heinrich Burkhardt (1861-1914) spoke about the importance of vector methods for mathematical physics in 1896.[19] This was two years after A. Föppl first introduced vector analysis in Germany with his *Einleitung in die Maxwellsche Theorie* and a year before the appearance of Föppl's more comprehensive account of the theory in *Geometrie der Wirbelfelder*.

Table 1: Lectures on applied mathematics and mathematical instruction delivered at annual meetings of the German Mathematical Society, 1891–1913.

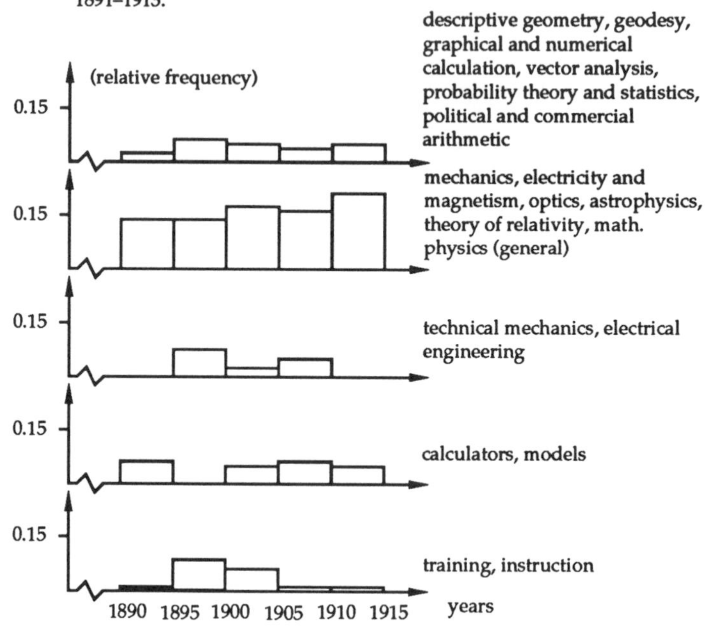

Table1

Contributions to political and commercial mathematics, also few in number, were concerned with calculations undertaken in connection with problems that arose in finance and public administration. One such work, submitted by Andreas Voigt on the "Mathematische Theorie des Tarifwesens,"[20] dealt with a mathematical treatment of certain questions regarding taxation and the regulation of tariffs. Such theories, however, had little impact on praxis. Indeed, on one occasion the *DMV* submitted a written statement to the Finance Ministry expressing its reservations regarding the new income tax law of 29 March 1920 and recommending that mathematicians be allowed to have some input in the formulation of tax proposals.[21]

The second group of lectures in Table 1 deals with mathematical physics and its various subdisciplines, particularly mechanics and electrodynamics. The field of theoretical mechanics was, of course, long considered to be a proper component of mathematics, and it was chosen as a theme for the *DMV* meetings no fewer than four times during this period. Highpoints were the 1903 meeting in Kassel with a series of lectures on "the scientific results and goals of modern mechanics," and the 1908 meeting in Cologne which presented a number of lectures dealing with fundamental problems in mechanics. The 1903 meeting included consideration of the latest results in the field of technical mechanics. It opened with a report on celestial mechanics by the Göttingen astronomer Karl Schwarzschild (1873-1916).[22] A second report on technical mechanics was presented by Arnold Sommerfeld (1868-1951), Klein's former assistant then teaching at the *technische Hochschule* in Aachen.[23]

Sommerfeld emphasized two research trends in technical mechanics: the growing reliability and precision of the experimental findings and the utilization of ever more sophisticated theoretical methods. He pointed out that both aspects were necessary to ensure future progress in the field. His report then went into a detailed discussion of current problems in the field of machine construction. Sommerfeld also reported on Otto Schlick's invention of the "Schiffs- und Schlingerkreisel," a flywheel that acted as a stabilizer by damping out internal vibrations, and whose mathematical analysis was supplied by Hermann Schubert (1848-1911) and Hans Lorenz.[24] At this time Lorenz held an *Extraordinariat* in technical physics in Göttingen.

Another lecture from the 1903 meeting, entitled "Über physiologische Mechanik," was held by Otto Fischer (1861-1916).[25] It dealt with a mathematical treatment of bodily functions: breathing, digestion, and the circulatory system. The mechanical analysis of bodily motions had long been studied by scientists, but Fischer indicated that a variety of mathematical methods from vector analysis to the calculus could potentially be applied to physiological phenomena. That few such problems had been solved till then was, in Fischer's opinion, a result of the fact that the necessary anatomical knowledge for undertaking these studies had only recently become available.

The lectures in the third group in Table 1 dealt with technical mechanics and electro-technology. These works were increasingly concerned with problems in hydro- and aerodynamics. An overview of research in hydrodynamics was given by Richard von Mises (1883-1953) in a lecture held in 1908. [26] Von Mises divided this research into three types: problems that can be treated by use of Euler's theory of ideal fluids, such as out-flow problems; those concerned with friction in lubricants which can be dealt with using the Navier-Stokes theory of viscuous fluids; and a third field where such classical methods cannot be applied, such as in current flows in open rivers, canals, and pipelines. Ludwig Prandtl (1875-1953) made vital contributions to the study of this third class of phenomena after assuming a position at Göttingen in 1904. His work gives a particularly impressive illustration of the successful utilization of mathematics to solve otherwise intractable technical problems. Prandtl often had occasion to express his high opinion of mathematics and argued as follows for the establishment in Göttingen of an institute for hydro- and aerodynamic research:

> To handle the problems of aerodynamics and hydrodynamics requires a considerable amount of mathematics, and to a large degree it is higher mathematics and the highest mathematics that is used. For this reason, considerable value must be placed, on the one hand, on the involvement of trained mathematicians in the institute's work, on the other hand, it would be of the greatest advantage for the institute's researchers to maintain a lively contact with mathematics and, so much as needed, to bring their knowledge of mathematics up to date by attending lectures, etc. For all this, however, Göttingen, as the mathematicians-university *par excellence*, is splendidly suitable.[27]

The fourth group in Table 1 refers to lectures on calculating machines and mathematical models. Such works were promoted by the German Mathematical Society from the time of its founding. One of the most noteworthy undertakings in this direction was the exhibition of mathematical models and scientific instruments and apparatus assembled by Walther von Dyck for the 1893 meeting in Munich. The exhibits were organized in four parts: analysis, geometry, mechanics, and mathematical physics. This effort followed the 1876 exhibition of scientific instruments held in Kensington, but as Dyck pointed out it presented many new models and instruments that had only been developed during the intervening period, particularly models for studying algebraic and differential geometry and apparatus dealing with the numerical and graphical representation of physical phenomena, for example, Sommerfeld's harmonic analyzer. [28]

The fifth group of lectures in Table 1, which dealt with problems in mathematics education, particularly in the field of applied mathematics, will be discussed in more detail below.

Beyond these lectures, what else did the German Mathematical Society do to support applications of mathematics? In this regard, there were two particular activities that should be emphasized. In the first place, the *DMV* tried to bring about changes in the relationship between pure and applied mathematics principally by means of reforms in teacher training. Secondly, these reforms were decisively influenced by the requirements of industry, as expressed in comments made by engineers.

Changes in teacher training around the turn of the century played a key role in the endeavor to reinforce the relations between mathematics and its applications. The examination regulations valid in the various German states served as a common basis for the education of mathematics students at universities, most of whom were prospective teachers. The subjects included within these regulations were an essential part of university education. By introducing new subjects it was possible not only to influence the university curriculum but also to create new teaching posts and research positions in applied areas as well.

Beginning with the Prussian School Conference of 1890, the contents of the examination regulations that had been valid since 1884 in Prussia were subjected to growing scrutiny. By 1893 the German Mathematical Society, which initially had ignored questions about teaching methods, broadened the scope of its interests so as to exert an influence on the structure of the examination regulations under discussion in the various states at the time.[29] Discussions within the *DMV* indicated a widespread desire to promote subject-specific education in the face of demands by school circles for introducing practical and theoretical courses into educational theory. On the other hand, with regard to the structuring of mathematical studies there was a growing body of opinion that wished not only to take into consideration the needs of future teachers but also those of other professions. This idea had already come up at the Vienna *Naturforscher* meeting of 1894. There Ludwig Kiepert (1846-1934), a mathematician at the Technical College in Hanover, delivered a lecture "On the mathematical training of insurance technicians." According to the conference proceedings, Kiepert raised the question: "... whether it would not be useful to compile a report on those lectures dealing with applied mathematics that were to be given at future conferences which would particularly emphasize the possibilities for interplay between the practical interests of engineers and jurists and the demands of theory."[30] At the 1895 *Naturforscherversammlung* held in Lübeck, Alexander Brill (1842-1935), then president of the German Mathematical Society, was assigned this task along with Felix Klein.

It was around this time that Klein's dispute with the engineering community reached its peak over the quantity and kind of mathematical training suitable for prospective engineers. In 1895, in fact, he confronted the Association of German Engineers in their own backyard by joining the organization and proselytizing for his cause at their meetings. In the end, for the sake of a compromise, he chose to limit his efforts to the education of prospective teachers. At the same time, Klein made use of another organization, namely, the Association for the Promotion of Mathematics and Science Instruction (*Verein zur Förderung des Unterrichts in der Mathematik und den Naturwissenschaften*, or *Förderverein*) with the aim of intensifying a training oriented towards applications. At his suggestion, the annual conference of this association was held in Göttingen in 1895, on which occasion Klein delivered the main lecture.[31] The following year the *Förderverein* discussed the relationship between mathematics and engineering at a meeting held in Elberfeld.

In a lecture entitled "On the demands of engineers and the training of prospective mathematics teachers" delivered in Hanover in April of 1896, Klein pointed out that: "...apart from abstract interests those directed toward applications must not wither away, and especially prospective teachers should be given the chance to get to know the elements of descriptive geometry, geodesy, technical mechanics and its graphical methods."[32] This emphasis on the needs of engineering students clearly had an impact on the structuring of examination regulations, as can be seen very clearly from a letter Klein sent to the influential Prussian ministerial officer Friedrich Althoff (1839-1908):

> We are of the opinion, that now is the right time to help do something about the protests of the engineers... over the deficient educational background of mathematics teaching candidates with regard to applications, insofar as their protests appear to be justified. In this regard we have made three additions to our draft, namely: 1) the recognition of a definite number of required semesters at the technical colleges, 2) the placement of professors from the technical colleges on the examination commission, 3) the establishment of an appropriate level of scientific standards.... We have also come upon the idea of introducing a separate faculty for applied mathematics.[33]

Klein then listed the new standards that would be made of teaching candidates, which basically concurred with those eventually adopted in the examination regulations for prospective teachers of secondary schools in Prussia issued on September 12, 1898. These reforms created a special teaching certification in applied mathematics that was later adopted by many of the other German states.[34]

The introduction of this teaching certification, its adoption at the various universities and technical colleges, and the concrete realization of its initiatives through courses in descriptive geometry, technical mechanics, and geodesy raised questions that were discussed in detail by the membership of the *DMV*. Between 1898 and 1907 the *Jahresberichten* published twenty contributions dealing with these matters. At the 1899 meeting in Munich, Paul Stäckel was charged with the task of compiling a report that detailed the progress in implementing this new program for applied mathematics. Stäckel presented the results of his study two years later at the annual meeting in Kiel.[35] They revealed that the necessary prerequisites for teaching descriptive geometry were already in place at nearly all of the German universities with the exception of Berlin, where students had traditionally been referred to the Technical College in Charlottenburg for such instruction. The only institutions, however, with a teaching commission [*Lehrauftrag*] for descriptive geometry were Giessen, Göttingen, and Jena. The introduction of geodesy courses also led to many difficulties, and the instruction in technical mechanics was often confined to the presentation of graphical statics as part of the course on descriptive geometry. The only universities that went beyond this were Giessen, Göttingen, Halle, and Strassburg. As the facilities for instruction in applied mathematics expanded, it often became difficult to find first-rate scholars who could fill the new positions. Indeed, this continued to pose a problem until well into the 1920s.

Influential engineers as well as other authoritative representatives of industry considered a change in teacher training to be a decisive factor that could lead to positive long-term results for industrial production. The fact that the reform of teacher training was a principle concern of the industrialists who supported the Göttingen Association for the Promotion of Applied Physics and Mathematics is rather surprising. Indeed, this policy contrasts sharply with those adopted later by much larger organizations like the Kaiser Wilhelm Society and the Emergency Association of German Science (founded in 1911 and 1920, respectively), which generously funded basic research but deliberately excluded support for the teaching profession. Anton von Rieppel (1852-1926), who was an engineer, industrialist, and one of the founding members of the Göttingen Association, described the following scene from a meeting held on November 30, 1912 (an extended excerpt of Klein's remarks at this meeting is reproduced in the appendix):

> ...in his lecture Klein presented us with the following goals for our organization: 1) above all to effect a better training for future teachers; 2) to promote more research in the direction of the applied sciences; and 3) to redirect university politics along lines that have more to do with practical everyday life than has been the case up till now. We reached agreement above all on the first point, as experience had taught us that young engineers lost considerable time in their graduate studies due to a deficient prior education that was

> divorced from praxis. Often they were forced to recoup knowledge that could easily have been provided to them by the secondary schools. This system, in our opinion, only forces engineers to delay their careers, and the founding idea of the Göttingen Association was to help improve these conditions.[36]

The new curriculum at the universities helped spurn research work in the applied fields. By December of 1900 the Göttingen Association extended its support beyond technical mechanics to other applied mathematics fields which were now part of the new examination regulations. During the decade following, descriptive geometry and graphical statics were taught in Göttingen by Friedrich Schilling (1868-1950), geodesy by Emil Wiechert (1861-1928), and applied mathematics was elevated to a full professorship when Carl Runge (1856-1927) came to Göttingen in 1904.

Until it was dissolved in 1920, the Göttingen Association brought together 50 leading figures from business and finance and 42 professors of science and mathematics who taught at Göttingen University. Among the companies and organizations represented were Bayer, Krupp, the AEG, Siemens, Norddeutscher Lloyd, and the Society of German Engineers, i.e. many of the largest chemical, electrical, steel and metallurgy firms in Germany. The president of the Göttingen Association was Henry Böttinger (1848-1920), a member of the board of directors of the Bayer chemical concern; Klein served as vice-president. In the course of its 23-year existence, the Göttingen Association collected 2,300,000 marks in assets, a relatively modest sum, but enough to make a profound impact on scientific life in Göttingen. After an initial pledge from its founding members in 1896, the following facilities were established there: the institutes for Applied Mechanics (1897), Applied Electricity (1897), geophysics (1898, new building 1901), agricultural bacteriology (1901), inorganic chemistry (1903), applied mathematics (1905), a new building for the physics institute and for the department of applied electricity (1905), an aerodynamics testing center (1907-1908), and a laboratory for testing wireless telegraphy (1909). In the meantime, the Prussian Ministry added new positions at Göttingen, thereby increasing the number of chairs in these fields from five to ten. Böttinger had connections with the Prussian Finance Ministry, and he was able to use his influence there to obtain additional funding for the projects supported by the Göttingen Association. By 1906 the state had granted 185,000 marks compared with 220,900 marks that came from the membership of the Göttingen Association.[37]

A common bond that united the Göttingen professors and the industrialists and engineers who belonged to the Göttingen Association was their concern for providing future teachers with a praxis-oriented education that stressed applications of science and mathematics (for Klein's views on these matters, see the appendix). The membership was in complete agreement with the goals of the school reform movement, about which they received regular reports from Klein. As a member of the lower chamber of the Prussian Parliament from 1891 and the upper chamber from 1908 onward, Böttinger was actively involved in developing new regulations for mathematics and biology instruction in the schools. During the budgetary negotiations of 1907 within the Prussian Ministry of Culture, he argued for additional spending that would expand the role of these subjects in the curriculum.[38] Böttinger also represented the Göttingen Association in the Breslau Education Commission between 1904 and 1907, and in the German Committee for Mathematics and Science Education, founded in 1908. The reforms advocated by these organizations received modest financial support from German industry.[39]

One field that stood out from all others supported by the Göttingen Association was aerodynamics. Research on the technical possibilities for aircraft began in Göttingen only a short time after the Wright brothers' sensational flight of 1903 and the nearly simultaneous discovery of the aerodynamic law for lift made by the German mathematician W. Kutta (1867-1944) and the Russian N. E. Žukovskij (1847-1921). Ludwig Prandtl, who was the first to give a theoretical explanantion for atmospheric friction, one of the principal sources of air resistance, was appointed to an *Extraordinariat* at Göttingen in 1904 and elevated to a *persönliches Ordinariat* three years later. Backed by Klein, the Göttingen Association, and the newly-founded Society for Aviation Studies in Berlin, Prandtl designed an aerodynamics testing center that opened in 1908. Its military significance was not overlooked during the war years when it was expanded and received substantial subsidies from the army. But even before the war, this enterprise commanded more funding than any other project supported by the Göttingen Association. As a means of comparison, Prandtl received a yearly sum of 4,000 marks from the state and an additional 5,000 from the Association to operate the older Institute for Applied Mathematics and Mechanics on the Leine canal. His giant research institute, on the other hand, received 1,500 marks from the *Kaiser-Wilhelm-Gesellschaft*, 6,000 from the Göttingen Association, an additional 60,000 marks from the federal government, an unspecified amount from a private aviation club, and another 5,000 marks for experiments undertaken on behalf of various industrial concerns. Beyond this, Prandtl himself received a guaranteed sum of 6,000 marks in lecture fees from the Göttingen Association.[40]

This generous support of aviation research by the Göttingen Association and other organizations led to the establishment in 1912 of a new national organization, the Scientific Society for Aviation Technology (*Wissenschaftliche Gesellschaft für Flugtechnik*, two years later renamed the *Wissenschaftliche Gesellschaft für Luftfahrt*). The three executive officers chosen by its first board of directors were Böttinger, Prandtl, and August von Parseval (1861-1942); Prince Heinrich of Prussia was named as honorary president. Beyond its scientific interests, this organization—which even more than the Göttingen Association was dominated by leading industrialists—resounded with unmistakable political and military overtones.[41]

At the 1898 meeting of the Society of German Natural Scientists and Physicians held in Düsseldorf a special section was formed for "Applied Mathematics and Physics" (later "Engineering Sciences" were added). This novel organizational development was a direct reflection of growing research activities in this field. For the engineering sciences, it also presented the possibility of establishing a foothold within the framework of the *GDNA*, something that had not always been easy to accomplish. For example, Klein was forced to abandon his plans for introducing an engineering sciences section within the Göttingen Academy of Sciences in the face of steadfast resistance on the part of its membership.

This growing tension between natural scientists and engineers led to open conflict within the new Applied Mathematics and Physics section of the *GDNA*. At the first meeting of this section held in 1898, three mathematicians—L. Kiepert, R. Mehmke, and F. Müller—delivered lectures in it, but this situation was exceptional and contributed papers from mathematicians were thereafter quite rare. The engineers also had misgivings about the lectures on technical subjects given by mathematicians at the meetings of the German Mathematics Society. August Gutzmer (1860-1924) reported to Klein on this ever-widening rift after attending the *Naturforscher* meeting held in Munich in 1899, where the engineers had openly protested against some of the lectures. He recommended that the *DMV* should no longer print reports in the *Jahresberichten* on developments in technology, and that mathematicians should keep their noses out of such matters until such time as they had acquired a better understanding of them.[42] Eventually a compromise was reached, however, that enabled mathematicians and engineers to establish joint meetings within the *GDNA*.

An analysis of the lectures delivered in the Applied Mathematics and Physics (Engineering Sciences) section indicates that preference was given to those developments that evinced substantial mathematical content (see Table 2). Thus fields like electrical engineering, hydrodynamics, aerodynamics, technical mechanics and mechanical engineering occupied the central position.

Table 2: Lectures in the section of "Applied Mathematics and Physics (engineering Sciences)" of the Society of German Naatural Scientists and Physicians, 1898–1913.

subject	1898–1900	1801–1903	1904–1906	1907–1909	1910–1913	Σ	%
general problems	–	2	1	–	–	3	2.9
potential theory	1	–	–	–	–	1	0.9
scientific instruments, models, tables	4	–	2	2	3	11	10.6
mechanics (general)	–	4	1	4	–	9	8.6
theory of elasticity	3	–	–	1	1	5	4.8
theory of plasticity	–	–	–	1	–	1	0.9
hydro-, aerodynamics	3	1	3	1	5	13	12.5
electromagnetic theory	1	2	–	–	1	4	3.8
technical mechanics	4	1	4	–	3	12	11.5
mechnical engineering	5	4	3	3	–	15	14.4
electrical engineering	–	5	2	5	6	18	17.3
cinematography	–	–	–	–	1	1	0.9
economy, hygienics	2	2	1	–	6	11	10.6
total	23	21	17	17	26	104	100.0

Table 2

Table 3: Application of mathematical theories and methods in technology.

algebraic, particularly linear equations	linear differential equations, potential theory, calculus of variations	eigenvalue problems	nonlinear differential equations
construction statics	states of stress in elastic solids	frequency calculation	solid bodies beyond the elastic limit (plasticity, earth pressure)
kinematics of machines	ideal fluid flow	critical states of motion	motion of viscous liquids
	heat conduction and analogous problems	stability of constructions	fluids of finite compressibility
	theory of electrical machines electrical waves		

Source: Theodor von Kármán, *Mathematik und technische Wissenschaften.* In: Die Naturwissenschaften, 18(1930)1, p. 15

Table 3

Attendance at the beginning was fairly sparse, but the fact stands out that Ludwig Prandtl (1857-1953) regularly participated in these conferences and gave more lectures than anyone else at them. His work was highlighted in meetings of the "Society for Applied Mathematics and Mechanics," founded under Prandtl's leadership in 1922. At the same time this society may be regarded as an expression of the closer working relationship between mathematicians and scientists conducting research on the areas bordering mathematics, physics and engineering. According to an investigation by Theodor von Kármán (1881-1963), a large number of modern mathematical theories and methods were applied in technical disciplines by 1930. These findings, which illustrate the degree to which mathematics ultimately succeeded in penetrating technical fields, are summarized in Table 3.

NOTES*

1. For a more detailed analysis see R. Tobies, *Die gesellschaftliche Stellung deutscher mathematischer Organisationen und ihre Funktion bei der Veränderung der gesellschaftlichen Wirksamkeit der Mathematik (1871-1933)*, Dissertation (B), Leipzig University, 1985.
2. R. von Mises, "Über die Aufgaben und Ziele der angewandten Mathematik," *Zeitschrift fur angewandte Mathematik und Mechanik*, 1(1921)1, p. 1.
3. J. Zech, "Über das Verhältnis der Mathematik zu den Naturwissenschaften," *Amtlicher Bericht der 22. Versammlung deutscher Naturforscher und Ärzte in Bremen*, 1844, p. 14.
4. I. Kant, *Metaphysische Anfangsgründe der Naturwissenschaften*, 2nd ed. Riga: 1787, p. VIII.
5. "Mathematisches Kränzchen zu Karlsruhe," *Jahresbericht der Deutschen Mathematiker-Vereinigung*, 21(1912), Abt. 2, pp. 81–82; see also W. Purkert and S. Hensel, "Zur Rolle der Mathematik bei der Entwicklung der Technikwissenschaften," *Dresdener Beiträge zur Geschichte der Technikwissenschaften*, 11(1986), pp. 3–53.
6. See R. Tobies, "Zur Geschichte deutscher mathematischer Gesellschaften," *Mitteilungen der Mathematishen Gesellschaft der DDR*, H. 2/3 (1986), pp. 112–134.
7. Quoted in A. Voss, "Die Beziehungen der Mathematik zur allgemeinen Kultur," *Kultur der Gegenwart. Ihre Entwicklung und ihre Ziele*, Teil 3, Abt. 1, Leipzig/Berlin: 1914, p. 44.

* I wish to thank David E. Rowe for improving the English and translating portions of this paper.

8. F. Klein, "Antrittsrede in Erlangen vom 7.12.1872," quoted from D. E. Rowe, "Felix Klein's 'Erlanger Antrittsrede': A Transcription with English Translation and Commentary," *Historia Mathematica*, 12 (1985), pp. 123–141, on p. 137.
9. Niedersächsische Staats-und Universitätsbibliothek Göttingen, Handschriftenabteilung, Mathematiker - Archiv.
10. Ibid.
11. This viewpoint also appears in D. E. Rowe, "Klein, Hilbert, and the Göttingen Mathematical Tradition," in K. Olesko, ed. *German Natural Science: Problems at the Intersection of Institutional and Intellectual History, Osiris*, 5(1989). The contrary opinion is expressed in M. Otte, "Zum Verhältnis von Mathematik und Technik im 19. Jahrhundert in Deutschland (Teil 1)," *Institut für Didaktik der Mathematik der Universität Bielefeld, Occasional Papers*, 87 (Jan. 1987). Many facts can be cited in support of my position. When Klein expressed his views on this subject, they were not mere rhetoric but rather ideas based on his own experience. In his lecture of 1898 on "Universität und technische Hochschule" Klein stated: "I ask of you at the outset to be convinced that, despite the abstract direction that my own career has taken, no one has a more immediate love for the technical professions or holds them in higher esteem than do I. But, by the same token, my standpoint is definitely not one-sided, and in that very reason, namely my wish to [promote theory] without neglecting [praxis], lies the difficulty I have occasionally encountered in my efforts." (F. Klein and E. Riecke , eds., *Über angewandte Mathematik und Physik in ihrer Bedeutung für den Unterricht an der höheren Schulen*, Leipzig/Berlin: 1900, p. 230.)
12. Niedersächsische Staats- und Universitätsbibliothek Göttingen, Handschriftenabteilung, Cod. Ms. Klein, XII.
13. *Verhandlungen der Gesellschaft deutscher Naturforscher und Ärzte, 63. Versammlung zu Bremen*, Teil 2. Leipzig, 1890, p. 13.
14. See L. Henneberg, "Über die Entwicklung und die Hauptaufgaben der Theorie der einfachen Fachwerke," *Jahresbericht der Deutschen Mathematiker-Vereinigung*, 2(1893), pp. 75–154. See also Hennenberg's correspondence with Klein in Cod. Ms. Klein, IX, Nr. 679.
15. Klein to Gordan, 20 June 1890, Niedersächsische Staats- und Universitätsbibliothek Göttingen, Handschriftenabteilung, Cod. Ms. Klein, XII.
16. F. Klein, "Riemann und seine Bedeutung für die Entwicklung der modernen Mathematik," *Gesammelte Mathematische Abhandlungen*, vol. 3, pp. 482–497, on p. 483.

17. Klein's fruitful efforts to enlist the support of Föppl and Finsterwalder are documented in Cod. Ms. Klein IX, Nrs. 36, 49, Niedersächsische Staats- und Universitätsbibliothek, Göttingen.
18. For example, the preliminary program for the 1892 meeting of the *DMV* planned for Nuremberg (which was postponed and held the following year in Munich) called for reports and summaries of recent work in applied mathematics (technical mechanics, astronomy, physics, etc.) and the exhibition of a comprehensive collection of mathematical models and technical apparatus. See *Jahresbericht der Deutschen Mathematiker-Vereinigung,* 1(1890–91), p. 11.
19. H. Burkhardt, "Über Vektoranalysis," *Jahresbericht der Deutschen Mathematiker-Vereinigung,* 5(1897), pp. 43–52.
20. A. Voigt, "Mathematische Theorie des Tarifwesens," *Jahresbericht der Deutschen Mathematiker-Vereinigung,* 22(1913), pp. 127–135.
21. *Jahresbericht der Deutschen Mathematiker-Vereinigung,* 31/32(1922), p. 35.
22. K. Schwarzschild, "Himmelsmechanik," *Jahresbericht der Deutschen Mathematiker-Vereinigung,* 13(1904), pp. 145–156.
23. A. Sommerfeld, "Technische Mechanik," *Jahresbericht der Deutschen Mathematiker-Vereinigung,* 13(1904), pp. 156–173.
24. H. Lorenz and H. Schubert, "Die Beseitigung von Schiffsvibrationen durch Ausgleichung der Massenwirkungen der Maschinen," *Jahresbericht der Deutschen Mathematiker-Vereinigung,* 6(1898), pp. 120–122.
25. O. Fischer, "Über physiologische Mechanik," *Jahresbericht der Deutschen Mathematiker-Vereinigung,* 13(1904), pp. 173–186.
26. R. von Mises, "Über die Probleme der technischen Hydromechanik," *Jahresbericht der Deutschen Mathematiker-Vereinigung,* 17(1908), pp. 319–325.
27. Zentrales Staatsarchiv Potsdam, RMd I, Nr. 89, 70/1 Bl. 173 ff.
28. W. Dyck, "Einleitender Bericht über die mathematische Ausstellung in München," *Jahresbericht der Deutschen Mathematiker-Vereinigung,* 3 (1894), pp. 39–56.
29. See *Jahresbericht der Deutschen Mathematiker-Vereinigung,* 3(1892-93), pp. 5-6.
30. See "Bericht über die Jahresversammlung in Wien, 24.- 28. Okt. 1894," *Jahresbericht der Deutschen Mathematiker-Vereinigung,* 4(1894-95), p. 5.
31. See F. Klein, "Über den mathematischen Unterricht an der Göttinger Universität im besonderen Hinblicke auf die Ausbildung der Lehramtscandidaten," *Zeitschrift für den mathematischen und naturwissenschaftlichen Unterricht,* 26 (1895).

32. F. Klein, "Die Anforderungen der Ingenieure und die Ausbildung der mathematischen Lehramtscandidaten," F. Klein and E. Riecke, eds., *Über angewandte Mathematik und Physik in ihrer Bedeutung für den Unterricht an den höheren Schulen*, p. 226.
33. F. Klein to F. Althoff, 2 June, 1897, Zentrales Staatsarchiv der DDR, Abt. Merseburg, Rep. 92, Nachlass Althoff B. Nr. 92.
34. See "Ordnung der Prüfung für das Lehramt an höheren Schulen in Preussen, vom 12, Sept. 1898," Berlin: 1898.
35. P. Stäckel, "Über die Entwicklung des Unterrichtsbetriebes in der angewandten Mathematik an den deutschen Universitäten," *Jahresbericht der Deutschen Mathematiker-Vereinigung*, 11(1902), pp. 26–37.
36. Niedersächsische Staats- und Universitätsbibliothek Göttingen, Handschriftenabteilung, Mathematiker-Archiv, Nr. 50^{29}.
37. Zentrales Staatsarchiv der DDR, Abt. Merseburg, Rep. 92, Althoff A I, Nr. 139, Bl. 48.
38. Zentrales Staatsarchiv der DDR, Abt. Merseburg, Rep. 92, Nachlass Schmidt-Ott, C 55, Bl. 125, 146.
39. For example, from 30 October 1909 to 1 November 1910, the Göttingen Association allocated 2,400 marks for courses for workers in mechanical professions; 3,000 marks for technical courses offered to jurists and administrative officials; and 1,700 marks for seminars given to teaching candidates at Göttingen (Zentrales Staatsarchiv der DDR, Abt. Merseburg, Rep. 92, Althoff A I, Nr. 139, Bl. 265).
40. Zentrales Staatsarchiv der DDR, Abt. Merseburg, Rep. 92, Nachlass Schmidt-Ott, B 43; Bl. 16, C 55, Bl. 126.
41. Zentrales Staatsarchiv der DDR, Abt. Merseburg, Rep. 92, Nachlass Schmidt-Ott, Bl. 43.
42. Niedersächsische Staats-und Universitätsbibliothek, Göttingen, Handschriftenabteilung, Cod. Ms. Klein, IX, Nr. 513.

Appendix

Auszug aus dem Protokoll der Versammlung der Göttinger Vereinigung zur Förderung der angewandten Physik und Mathematik vom 30. November 1912. Quelle: Handschriftenabteilung der Niedersächsischen Staats- und Universitätsbibliothek Göttingen, Mathematiker-Archiv 50^{29}.

Geheimrat Klein: An sich ist das Problem der Lehrerbildung ein Thema von ungeheurer Ausdehnung. Für uns kann es sich hier nur handeln um die Ausbildung der Oberlehrer für mathematische und physikalische Fächer, und auch in diesem Gebiete werden wir uns im Wesentlichen auf diejenigen Gesichtspunkte zu beschränken haben, die den Beziehungen auf die Interessen der Göttinger Vereinigung entsprechen.

Es besteht von altersher in unserm Wissenschaftsbetrieb der *Gegensatz zwischen Theorie und Praxis* entsprechend den verschiedenen Seiten menschlichen Begabung, die sich nur selten in einer Persönlichkeit zusammen finden. Nur die ganz Grossen haben beides vereinigt.

Es sei an Archimedes, Newton, Gauss erinnert. Im allgemeinen aber bleiben die Begabungen auf verschiedene Individuen verteilt, und jeder ist geneigt, seine hieraus entspringende besondere Interessenrichtung für allgemein gültig zu halten.

Eine richtige Hochschulpolitik scheint mir doch nun unzweifelhaft die zu sein, *die Interessen und Begabungen möglichst allseitig zur Vertretung kommen zu lassen*, eine allseitige Wirkung zu erzielen durch Koordinierung von Kräften verschiedenster Art.

Das ist, was wir in Göttingen seit einer Reihe von Jahren erreicht haben. Die Göttinger Tradition ist von jeher in dieser Richtung gegangen und darum hatten in älteren Zeiten die angewandten Wissenschaften in Göttingen neben den reinen ausgiebige und z. T. hervorrangende Vertreter. Eine unglückliche Entwicklung hat leider dazu geführt, jene alte Tradition zu unterbrechen. Die Göttinger Vereinigung hat also im Grunde nichts getan, als die alte Tradition in einer der Neuzeit entsprechenden Form wieder neu zu beleben.

Auf pädagogischem Gebiete muss, wie ich glaube, derselbe Standpunkt eingenommen werden: *Man muss verschiedenartig ausgebildete Lehrer hinausschicken, damit die in der Jugend verschlummernden Kräfte möglichst vielseitig geweckt werden können.* Das hierbei die Anwendungen zu ihrem Recht kommen, ist seit Anfang der 90er Jahre in Göttingen beharrlich angestrebt worden. Ein sichtbares Zeichen dafür ist eben die Gründung der Göttinger Vereinigung im Jahre 1898. Ebenso die im Jahre 1898 von Göttingen angeregte Aufnahme der angewandten Mathematik in die preussische Prüfungsordnung für Oberlehrer, übrigens unter Bindung an reine Mathematik, die auch für die angewandte Mathematik immer die Fundierung im logischen Denken bleiben muss. Ferner die Umwandlung der Schlömilch'schen Zeitschrift in eine Zeitschrift für angewandte Mathematik, deren Redaktion neben Professor Mehmke (Stuttgart) unser Kollege Runge besorgt.

Ferner die Aufnahme der angewandten Physik und der angewandten Mathematik in unsere Doctor-Prüfungen; die Einrichtung von Fortbildungskursen für Oberlehrer, die wir seit 20 Jahren alle zwei Jahre für mathematische und physikalische Oberlehrer hier in Göttingen abhalten.

Zwei Bände Veröffentlichungen, die ich mit Kollege Riecke in den Jahren 1900 und 1904 herausgegeben habe, legen Zeugnis ab, dass wir auch in diesen Kursen das Grundprogram der Göttinger Vereinigung zur Geltung zu bringen suchen. Schliesslich folgt im Jahre 1904 das unter unserer Mitwirkung zustande gekommene Eingreifen der Naturforscher-Versammlung mit den sich daran anschliessenden Kommissionen und Ausschüssen, wovon ich schon in meinem Eingangsbericht gesprochen habe.

Welche Entwicklung der *Unterricht in angewandter Mathematik und Physik* im Zusammenhang mit dieser allgemeinen Bewegung bei uns im einzelnen genommen hat, werden die Berichte der einzelnen Fachvertreter nachher erkennen lassen. Ich darf vorweg einiges allgemeine herausheben:

Seit wir Kollegen Runge hier haben, verstehen wir unter angewandter Mathematik die Lehre von der mathematischen Exekutive, d. h. die numerischen und graphischen Methoden, wovon die darstellende Geometrie einen Seitenzweig bildet. Ueber die graphischen Methoden hat letzthin Kollege Runge Vorträge in Amerika gehalten, die in diesen Tagen im Druck erscheinen werden. Geodäsie, Astronomie, angewandte Mechanik, Versicherungswesen etc. sind Bestätigungsgebiete der angewandten Mathematik und ihrer Vertreter sind dank liberaler Auslegung der Prüfungsordnung durch das Ministerium zur Oberlehrerprüfung in der angewandten Mathematik mit zugelassen. (Die mathematische Physik fehlt in dieser Aufzählung, weil sie bei der Prüfung unter "Physik" zur Geltung kommt.) Entsprechend dem oben betonten Princip der freien Wahl hat sich der Kandidat unter den verschiedenen Examinatoren zwei auszuwählen, die das Examen mit ihm abhalten. Aehnlich gliedert sich die angewandte Physik, die sich naturgemäss mit der angewandten Mathematik vielseitig überschneidet, zur Zeit in die Fächer angewandte Mechanik und Thermodynamik, angewandte Elektrizität, Geophysik und Astrophysik. Gern würden wir noch andere hinzunehmen, z. B. angewandte Optik, von der wir bei unserer Jenaer Tagung einen so lebendigen Eindruck gewonnen haben. Auch an angewandte Akustik könnte man denken, die heute in der Phonographentechnik, in der Technik der Musikinstrumente, für Unterwasserschallsignale, Nebelsignale u. dgl. eine ausgedehntere Bedeutung hat, als man im allgemeinen weiss. Ausserdem haben wir Sorge getragen, den Studierenden, angegliedert an die Fachschule für Feinmechanik, einen guten Handfertigkeits-Unterricht zuteil werden zu lassen, damit die künftige Oberlehrer es lernen, sich einfache Apparate selbst herzustellen und fertige Apparate nach Konstruktion und Wert zu beurteilen.

Wenn wir in allen diesen Gebieten den Unterricht organisiert haben, so sind damit der Tradition der Universität entsprechend, überall auch *neue Herde wissenschaftlicher Forschung* entstanden. Anders würde man an der Universität keinen Einfluss gewinnen. Dass der Universitätsprofessor gleichzeitig Lehrer und Forscher sein soll, ist das Lebensprinzip unserer Anstalten. In diesem Zusammenhang möchte ich die Zweifel erneuern, die gegen die Einrichtung der Kaiser Wilhelm Institute erhoben worden sind. Hoffentlich kommt es doch zur Ausbildung des Nachwuchses an Spezialforschern, insofern dort Forschung ohne Unterricht und ohne die Anregung des Unterrichts betrieben werden soll.

Ich komme zur Frage nach dem *Erfolg der geschilderten Einrichtungen und Bestrebungen.* Er äussert sich in den wissenschaftlichen Arbeiten, die aus unsern Instituten hervorgehen, in den Frequenzzahlen der Kurse und Vorlesungen, die wir eingerichtet haben, und in der Zahl der Examina, die in den angewandten Fächern stattfinden.

Ueber den Erfolg der wissenschaftlichen Arbeit wird nicht gestritten. Was die Frequenzzahlen betrifft, so ist ein Erfolg und ein zunehmendes Interesse nicht zu leugnen. Immerhin aber stehen die Zahlen in keinen Verhältnis zu denen, die wir in den Vorlesungen und Kursen über die reinen Wissenschaften haben. Was ist die Ursache? Vor allem bildet in solchen Dingen die studentische Tradition einen gewaltigen Trägheitswiderstand. Noch schickt das ältere Semester das jüngere ausschliesslich in die hergebrachten abstrakten Vorlesungen, weil auch ihm seinerzeit derselbe Rat von seinen älteren Kameraden gegeben worden war. Die Bildung einer neuen Tradition aber vollzieht sich um so langsamer, als wir von anderen Universitäten her nicht genügend Unterstützung erhalten. Denn an andern Universitäten haben unsere Bestrebungen nur erst mangelhaft Fuss gefasst. Am weinigsten noch in Berlin, am meisten in Jena, wo Abbe unser Helfer wurde. An andern Universitäten errichtet man wohl Lehraufträge für angewandte Wissenschaften, vergisst aber gelegentlich die Bereitstellung von Lehr- und Forschungs-Einrichtungen, ohne die es nun einmal nicht geht. Oder aber, man benutzt die bereitgestellten Mittel, um zwar verdienten, für den erwünschten Zweck aber ganz ungeeigneten Dozenten endlich eine kleine Remuneration zu verschaffen; ich möchte dringend wünschen, dass die berufenen Instanzen diesen Verhältnissen mehr Aufmerksamkeit zuweisen möchten, als es seither geschieht.

Dass wir aber doch langsam weiter kommen, und schon Gutes erreicht haben, dafür zeugt die grosse Zahl der von hier ausgegangenen Hochschulprofessoren. Dazu gehören die drei ersten Kandidaten, die das Examen für angewandte Mathematik bei uns bestanden: *Ludwig,* z. Z. Professor der darstellenden Geometrie in Dresden, *Hamel,* z. Z. Professor für Mathematik in Aachen und *Wiegtordt,* z. Z. Professor für Mechanik in Wien. Und von überall kommen Anfragen an uns heran nach jungen Leuten, die in unserm Sinne in angewandter Wissenschaft ausgebildet sind.

Man kann fragen und es wird viel gefragt, warum die *technischen Hochschulen nicht mehr für die Ausbildung der Oberlehrer* herangezogen werden? Nun, die Möglichkeit ist dafür auch in Preussen ziemlich weitgehend vorhanden. Von den 10 Semestern, die das Oberlehrerstudium heute durchschnittlich verlangt, brauchen blos 3 Universitäts-Semester zu sein. Wenn trotzdem in relativ geringem Masse von der Möglichkeit Gebrauch gemacht wird, so liegt dies daran, dass zur Zeit die Einrichtungen der technischen Hochschulen in Preussen diesem Zwecke entsprechend noch nicht genügend ausgestaltet sind. Es würde dies auch nur unter Aufwendung beträchtlicher Geldmittel möglich sein, wie dies u. a. Stäckel im Gesammtbericht der Unterrichts-Kommission der Gesellschaft Deutscher Naturforscher und Aerzte (1908) auseinandergesetzt hat.

Früher hat es zwischen *Universitäten und technischen Hochschulen Gefühle der Rivalität gegeben.* Ich meine, dass heute ernstlich nicht mehr die Rede davon ist. Man kann darüber verschiedener Meinung sein, ob es gut war, dass sich die beiden Hochschulen unabhängig von einander entwickelt haben. Die Entwicklung lässt sich aber nicht zurückschrauben, eine nachträgliche Verschmelzung dürfte kaum möglich und wünschenswert sein. Vielmehr will mir scheinen, als ob es auch hier das Richtige wäre, wenn beide Hochschulen ihre Eigenart recht lebendig zur Entwicklung bringen möchten, dass sie sich aber überall zu einem "Zweckverbande" zusammentun sollten, wo allgemeinen Fragen des nationalen Wohles auf dem Spiele stehen. Man sollte alle die andern Hochschularten, die es gibt, die Forst- und Berg-Akademien, die Handelhochschulen, die neuen Hochschulen für Kommunalverwaltung etc. je nach Lage des Falles einbeziehen.

Die Göttinger Vereinigung darf sich rühmen, in den im vergangenen Jahren mit der technischen Hochschule Hannover gemeinsam abgehaltenen Kursus für Juristen und Verwaltungsbeamte ein bemerkenswertes Beispiel eines solchen Zusammenwirkens gegeben zu haben. Dieser Kurs soll 1913 wiederholt werden, wie heute Nachmittag zusammen mit Vertretern der technischen Hochschule Hannover weiter zu beraten sein wird.

Georg Hamel (1877–1954)
Courtesy of Springer-Verlag Archives

Mathematics at the Berlin Technische Hochschule/Technische Universität Social, Institutional, and Scientific Aspects

Eberhard Knobloch

SUMMARY

This essay attempts to describe the social, institutional, and scientific role of mathematics at the Berlin Technische Hochschule, the leading Prussian technological training institution. It has been divided into six sections with the following headings:

1. Academic rights and duties
2. The era of the independent academies (1770-1879 and 1770-1916)
3. Mathematics at the Technische Hochschule (1879-1945)
4. Mathematics at the Technische Universität after the Second World War
5. The mathematicians of the Berlin TH/TU and the Mathematical Society of Berlin
6. Tables

1. ACADEMIC RIGHTS AND DUTIES

The Berlin Technische Hochschule was founded in 1879 through the union of two considerably older academies, the Building Academy ("Bauakademie," founded in 1799) and the Vocational Training Academy ("Gewerbeinstitut," founded in 1821). It also incorporated the former Mining Academy ("Bergakademie," founded in 1770) in 1916. These three academies were institutions whose teachers were not expected to do scientific research as such, and who offered their training courses avocationally. It was not until 1856 that the first regular professorship of mathematics was established at the Vocational Training Academy, a position first occupied by Karl Weierstrass (1815-1897). This chair still exists to the present day. Only in 1895 was such a professorship created at the Mining Academy, its first occupant being Fritz Kötter (1857-1912). This position, however, was terminated in 1921. The Building Academy, on the other hand, was never provided with a professorship in mathematics, although descriptive geometry was taught there beginning with the painter Karl Pohlke (1810-1876) in 1849.

ISBN 0-12-599662-4.

From the time of its inception, the Berlin Technische Hochschule enjoyed a special prestige. It was the first technical institution in Germany to be planned and inaugurated as a university-like Hochschule rather than as a polytechnical school. The other German Technische Hochschulen—Aachen, Brunswick, Darmstadt, Dresden, Hanover, Karlsruhe, Munich, and Stuttgart— had all been elevated to this status. This transformation process occured for the most part during the decades of the 1870s and 80s [Die deutschen technischen Hochschulen 1941]:

Aachen 1879 Technische Hochschule, since 1.10.1870 Königlich rheinisch-westfälische polytechnische Schule in Aachen

Berlin 1.4.1879 Technische Hochschule

Brunswick 1877 Technische Hochschule, since 5.7.1745 Collegium Carolinum

Breslau (now Wroclaw) 29.11.1910 Technische Hochschule

Clausthal 27.12.1863 Bergakademie, since 21.11.1810 Bergschule

Danzig (now Gdansk) 6.10.1904 Technische Hochschule

Darmstadt 1869 Technische Hochschule, since 1836 Höhere Gewerbeschule, Oktober 1868 Polytechnische Schule

Dresden 3.2.1890 Technische Hochschule, since 1.5.1828 Technische Bildungsanstalt für das Königreich Sachsen, since 23.11.1851 Königliche Polytechnische Schule

Freiberg 13.11.1765 Bergakademie

Hanover 1880 Technische Hochschule, since May 1831 Polytechnische Schule

Karlsruhe 1885 Technische Hochschule, since 7.10.1825 Polytechnische Schule

Munich 27.6.1899 Technische Hochschule, since 12.4.1868 Polytechnische Schule

Stuttgart 1890 Technische Hochschule, since 2.1.1840 Königliche Polytechnische Schule

In fact, there had long been plans to establish a special type of technical institution in Berlin that would be independent of the Building and Vocational Training Academies [Lorey 1916, 40-51]. During the 1820s, Alexander von Humboldt had hoped to entice Gauss away from Göttingen by creating a chair without teaching duties for him at this polytechnical institute. These dreams were, of course, never realized, at least not until 1879.

From the time of its founding, the Berlin Technische Hochschule was not only the largest and most important institution of its kind in Prussia, but throughout all of Germany as well. During its first year of existence it was attended by 1,180 students, and by the academic year 1927/28 the number had increased to 5,205 out of a total of 8,750 studying Prussian Technische Hochschulen. In keeping with its prominent position, the Berlin Technische Hochschule also played the leading role in bringing about the scientific emancipation of all Technische Hochschulen, a movement that gained momentum throughout the 1890s and culminated in 1899 when Emperor Wilhelm II decreed that the higher technical institutions were to receive rights and privileges commensurate with those of the German universities.

When the Danzig Technische Hochschule was founded in 1904 it became a competitor of Berlin in shipbuilding. A second competitor in mining and metallurgy emerged in 1910 when the Breslau Technische Hochschule was founded. As far as mathematics was concerned strong relations developed between the professors at the Berlin institution and the new Hochschulen. This applies, for example, to Ulrich Graf (1908-1954), Wolfgang Haack (*1902), Karl Jaeckel (1908-1984), Arthur Korn (1870-1945), Ernst Mohr (*1910), Ernst Steinitz (1871-1928), and Werner Schmeidler (1890-1969).[1]

Up until 1917 the ordinary professorship ("Ordinariat") at the universities corresponded with the so-called permanent or budgetary professorship at the Technische Hochschulen. Only after 1921 were there extraordinary professors at the Berlin Technische Hochschule, and they were not public officials. Their status was lower than that of a supernumerary professor and should not be confused with extraordinary and ordinary honorary professors. Young scholars thus had to climb up the following rungs of a complicated hierarchy:

private lecturer
title "professor"
non-budgetary lecturer
budgetary lecturer
extraordinary professor (without being public servant)
supernumerary professor ("ausserplanmässiger" professor)

Between 1922 and 1945 there were seven extraordinary professors of mathematics at the Technische Hochschule: Ernst Jacobsthal (1882-1965), Richard Fuchs (1873-1945), Emil Haentzschel (1858-1948), Eugen Stübler (1883-1930), Friedrich-Adolf Willers (1883-1959), Paul Lorenz (1887-1973), and Wilhelm Adolf Cauer (1900-1945).

A similar difference existed between the status of private lecturers at the universities and at the Technische Hochschulen. Originally the unsalaried and salaried lecturers at the Technische Hochschule had different qualifications, legal status, and scientific role than their counterparts at the universities. Of course the Building Academy and Vocational Training Academy already had private lecturers. But their appointment did not require a special habilitation process. Regulations for the conduct of a habilitation at the Berlin Technische Hochschule were only introduced in 1884, and these did not stipulate that the candidate hold a doctoral degree to teach there. This was due to the fact that it was not until 1899 that the Prussian Technische Hochschulen gained the right to confer such degrees themselves. This right, however, was extended only to the engineering sciences and did not apply to auxiliary sciences like mathematics or physics.

Apart from the newly established extraordinary professorships in mathematics, the Technische Hochschule broke new ground in other respects during the early 1920s:

1. In 1920 the mathematician Rudolf Rothe (1873-1942) elaborated a reform program in which the TH Berlin would grant degrees in mathematics and physics and train prospective Gymnasium teachers, and beginning July 5, 1921 it was granted the right to train secondary teachers.
2. In 1922 the seven departments of the Technische Hochschule were transformed into four faculties. On July 1, 1924 mathematics and physics became independent special subjects within the Faculty of General Sciences, and were thereafter entitled to grant diplomas and degrees.
3. In 1924 the first two mathematical dissertations were completed at the Berlin Technische Hochschule by Kurt Lachmann and Walter Kretschmer. Lachmann wrote "Über den Konvergenzbereich des Verfahrens der schrittweisen Verbesserung bei gewöhnlichen linearen Differentialgleichungen zweiter Ordnung, deren Integralkurven durch zwei gegebene Punkte gehen sollen," and Kretschmer's dissertation was entitled "Einige besondere nomographische Verfahren und ihre Anwendung auf technische Formeln." Their thesis advisors names have not been recorded, but they were undoubtedly either Georg Hamel (1877-1954) or Rudolf Rothe.

While all mathematics professors or lecturers at the Berlin Technische Hochschule held doctoral degrees, most of them wrote their dissertations at a university, especially the one in Berlin.[2] Leon Lichtenstein actually published two doctoral dissertations: in 1908 he wrote "Beiträge zur Theorie der Kabel. Untersuchungen über die Kapacitätsverhältnisse der verseilten und konzentrischen Mehrfachkabel" at the Berlin Technische Hochschule, which at the time only had the right to grant degrees in the engineering sciences. Thus in 1909 Lichtenstein completed a mathematical dissertation at Berlin University entitled "Zur Theorie der gewöhnlichen Differentialgleichungen und der partiellen Differentialgleichungen zweiter Ordnung. Die Lösungen als Funktionen der Randwerte und der Parameter." This thesis was accepted by H.A. Schwarz (1843-1921) and F. Schottky (1851-1935).

2. The era of the independent academies (1770-1879 and 1770-1916)

In view of their respective institutional functions, it is clear why mathematics was regarded as an auxiliary science at all three academies. Nevertheless the level of instruction was higher than might otherwise be expected, particularly in those cases where the mathematics teachers were associated with Berlin University which was founded in 1810. Thus Ernst Daniel Chr. L. Lehmus (1780-1863) and Friedrich H. A. Wangerin (1844-1933) taught at the Mining Academy from 1814 to1852 and from 1881 to 1882, respectively, whereas Johann Ph. Gruson (1768-1857), Peter G. L. Dirichlet (1806-1859), Ferdinand Minding (1806-1885), and Martin Ohm (1792-1872) all taught at the Building Academy. Karl Weierstraß taught as a permanent professor at the Vocational Training Academy from 1856 to 1861 while an extraordinary professor at the Berlin university, and Georg Hettner remained extraordinary professor at the Berlin university after he became permanent professor of mathematics at the Technische Hochschule in 1894. It should be emphasized that the famous geometer Jakob Steiner never taught at the Vocational Training Academy, contrary to statements sometimes found in the literature. Steiner taught from 1825 to 1834 at the Friedrich-Werdersche Vocational School which is sometimes confused with the Academy [Knobloch 1978/79, 56]. Just as in Paris, it was not unusual for mathematicians to hold appointments and teach at several of these institutions, especially the three academies. Lehmus taught at the Mining and Building Academies, Ringleb, Karl Pohlke, Siegfried Aronhold (1819-1884), and Ernst Kossak (1839-1892) at the Building and the Vocational Training Academy, and Hugo O. Hertzer (1831-1908) taught at all three (see Tables 1-3).

2.1 The Mining Academy

Up until 1895 only part-time teachers taught mathematics at the Mining Academy. This was the situation, too, when Fritz Kötter became Heinrich Bertram's successor in 1889. Kötter received his doctoral degree at Halle in 1883 and was a private lecturer at the Berlin Technische

Hochschule from 1887 to 1895 before he became the first permanent professor of mathematics at the Mining Academy. As the first eminent mathematician at the Mining Academy, Kötter was known for his successful applications of mathematics to technical problems. In 1900 he assumed the new professorship of technical mechanics at the Berlin Technische Hochschule. His appointment underscored the strong relationship between mathematics and mechanics at the Berlin Technische Hochschule.

Kötter's successor was Adolf Kneser, patriarch of the famous Kneser dynasty in German mathematics. A pupil of K. Weierstraß and L. Kronecker (1823-1891), Kneser wrote his dissertation on algebraic functions in 1884 and passed the habilitation at Marburg in the same year. When he left Dorpat in 1900 to assume the professorship at the Berlin Mining Academy, he had just completed his textbook on the calculus of variations. During his stay in Berlin, he developed the ideas for his famous paper on Sturm-Liouville series [Kneser 1905]. He then accepted a position in Breslau, and another Weierstraß pupil assumed the vacant professorship at the Mining Academy: Kneser's friend Eugen Jahnke (1905-1921), then editor of the Archiv für Mathematik und Physik.

Like Kneser, Jahnke had studied at Berlin University, where he was mainly influenced by Weierstraß and L. Fuchs (1833-1902). He received his doctoral degree at Halle in 1889, and in 1901 he was made a private lecturer at the Berlin Technische Hochschule. When in 1916 the Mining Academy became the Mining Department of the Technische Hochschule, Jahnke did not join the other mathematicians G.W. Scheffers (1866-1945), St. Jolles (1897-1942), E. Lampe, and R. Rothe at the Department of General Sciences, but remained instead at the new Department of Mining because of his growing interest in technical questions. His last field of research was lift technology. In 1919 he was elected rector of the Technische Hochschule, the fifth mathematician upon whom this honour was bestowed. This not only reveals the high esteem in which he was held personally, but it is also an indication of the stature and influence enjoyed by the mathematicians at the Technische Hochschule.

2.2 The Building Academy

The first eminent mathematician to teach at the Building Academy was P.G.L. Dirichlet, who came from Breslau. His profound influence on the course of modern mathematics was, of course, closely tied to his teaching activities at Berlin University rather than at the Building Academy. Dirichlet's worthy successor was his university colleague Ferdinand Minding, who is mainly known today as one of the first mathematicians to make an original contribution to the field of topology.

At the Building Academy great importance was attached to descriptive geometry, which was taught for some time (from 1849 to 1876) by the painter Karl Pohlke (1810-1867), and after him by Guido Hauck (1845-1905), who taught the subject there from 1877 to 1905. Hauck earned his doctoral degree at Tübingen University in 1876 and passed the habilitation shortly thereafter. All his numerous scientific publications concerned descriptive geometry, as he was concerned with elaborating a methodological foundation for the discipline. The Technische Hochschule soon became Berlin's recognized and officially sanctioned centre for teaching and learning descriptive geometry, especially for mathematics students attending Berlin University [Lorey 1916, 340].

In 1864 a first-class mathematician joined the faculty of the Building Academy, and although he never occupied a mathematics professorship, he did become a permanent professor of mechanics in 1874. This was Julius Weingarten (1836-1910), who like Kötter and Jahnke took his doctoral degree from Halle University in 1864 after having completed a dissertation on differential geometry. In the same year he became a private lecturer at the Building Academy in Berlin. Weingarten taught statics, dynamics, and mechanics, but after 1874 his publications were mainly concerned with differential geometry. His work stimulated the leading differential geometers of the day, including Sophus Lie (1842-1899), Luigi Bianchi (1856-1928), and Gaston Darboux (1842-1917). He was the founder of the plane theoretic school at the Berlin Technische Hochschule/Technische Universität, whose members included Georg Scheffers, Gerhard Hessenberg, Eduard Rembs (1890-1964), and Karl Peter Grotemeyer (*1927). Weingarten encouraged Meyer Hamburger (1838-1903) to become a private lecturer at the Academy in 1879, a few months before the Technische Hochschule was established. A friend of the university mathematician Lazarus Fuchs, Hamburger was mainly known for his work on partial differential equations.

2.3 The Vocational Training Academy

The Vocational Training Academy required a somewhat longer time than the Building Academy before its training courses attained a reasonably high scientific level. The first mathematics chair at this Academy was established in 1856. It was occupied by Karl Weierstrass, who was appointed extraordinary professor at the University in the same year. At the Academy he taught six hours of analytic geometry and calculus each week. After four and one-half years, however, he had to give up this lecture activity due to a severe illness. Nevertheless, his lectures at the Academy stimulated two of his pupils who went on to leave their mark on 19th century mathematics: Meyer Hamburger and Hermann Amandus Schwarz.

After Weierstraß was appointed ordinary professor of mathematics at Berlin University in 1864, he was succeeded at the Vocational Academy by Siegfried Aronhold. Aronhold had been awarded an honorary doctorate by Königsberg University in 1851, and along with Otto Hesse (1811-1874) he became one of the leading members of the Königsberg school in invariant theory and kinematic geometry. For a detailed discussion of his contribution to the creation of the Clebsch-Aronhold symbolic notation in invariant theory, one should consult Karen Parshall's paper in volume 2 of these proceedings.

The founding of a second chair in mathematics in 1869 was due to the activities of the Academy's Director, Franz Reuleaux (1829-1905). The post went to Elwin B. Christoffel of the Zürich Polytechnical Institute, who was then working mainly on homogeneous differential expressions, partial differential equations, and simply-connected plane surfaces. Christoffel had received his doctoral degree from Berlin University in 1856, and had become a private lecturer there in 1859. Three years later he left Berlin again to assume a mathematics professorship at the then German University in Strassburg.

3. Mathematics at the Technische Hochschule 1879-1945

3.1 Three characteristic features

When the Berlin Technische Hochschule was established in 1879 there were two chairs in geometry and two in mathematics. The chairs in geometry were held by Guido Hauck (Building Academy) and Hugo Ottomar Hertzer (Vocational Training Academy), whereas the other two chairs were occupied by Aronhold and Ernst Kossak, Christoffel's successor, both members of the former Vocational Training Academy. This distinction between geometry (including descriptive geometry) and mathematical analysis, i.e. higher analysis and its applications, remained one of the three characteristic features of the Technische Hochschule. Private lecturers, too, were either geometers like Scholz and Felix Buka (1832-1896) or mathematicians in the above sense, like Hamburger and Otto Reichel (1836-?).[3] Thus we can clearly distinguish between a geometrical and an analytical tradition at the Berlin Technische Hochschule.

The second typical institutional feature of mathematics at the Technische Hochschule was its close ties with applied fields like ballistics, aerodynamics, elasticity theory, electrical engineering, statistics, and last but not least mechanics. One would expect, of course, that the mathematicians and geometers at the TH had a strong interest in applications, but the relationship went deeper than this. Mathematicians like the differential geometer Julius Weingarten and the analyst Fritz Kötter, for example, were appointed as professors of mechanics or rational mechanics. The same was true of István Szabó (1906-1980), who taught these subjects after the Second World War. This circumstance evidently had little effect on their identities as mathematicians, since all three continued to publish mathematical articles. The two most representative examples of this type of mathematician are Georg Hamel (1877-1954), who was a pupil of David Hilbert (1862-1943), and his successor, Wolfgang Haack (*1902), a pupil of Rudolf Haussner who held a joint professorship in mathematics and mechanics at the Berlin Technische Hochschule. These mathematicians strengthened interest in the theoretical background of the engineering sciences, a field in which the Berlin Technische Hochschule enjoyed an excellent reputation. Indeed, since lectures on applied mathematics were rarely ever offered at Berlin University, mathematics students who were interested in learning something about applications were generally advised to attend the courses offered by the Technische Hochschule (Lorey 1916, 289 sq.).

Because of this situation, it is rather difficult to enumerate the mathematical dissertations that were written up to 1945. There were 31 such dissertations if we ignore their subject matter and define a mathematical dissertation at the Berlin Technische Hochschule by stipulating that at least one of the supervisors occupied a professorship of mathematics. This, however, is not a very satisfactory approach. For example, Lichtenstein's dissertation on the capacity of multiple cables would not be considered a mathematical dissertation simply because, as mentioned above, the Technische Hochschule had not yet obtained the right to grant degrees in mathematics. Later on Georg Hamel supervised dissertations that dealt with similar topics, like the one written by Wilhelm Cauer on resistances to alternating current. By the above criterion, this can be classified as a mathematical dissertation since G. Hamel was one of the supervisors.

Even though the mathematicians at the Technische Hochschule belonged to different departments, they could still gather together at the Berlin Mathematical Society. Indeed, a third characteristic feature of mathematics at the Berlin Technische Hochschule was its leading role in the founding and further development of this Society, a role it continues to play up to the present day [Baruch 1922, 34; Tobies 1982, 62]. Ironically enough, the ordinary professors of mathematics at the University only gradually came to accept this new Society.

3.1 The geometrical tradition

In geometry there were two chairs, originally occupied by G. Hauck and H.-O. Hertzer, who had earlier held professorships at the Building and Vocational Training Academy, respectively. In 1907 Hauck was succeeded by Georg Scheffers, a former pupil of Sophus Lie in Leipzig. Scheffers had written his doctoral dissertation on the "Determination of a class of contact transformation groups of three-dimensional space" in 1890 and had passed the habilitation one year later. He was involved to a considerable extent in revising and editing Lie's lectures [Beckert, Schumann 1981, 113-114]. Indeed, his edition of *Die Geometrie der Berührungstransformationen* and other textbooks based on Lie's lectures remain even today one of the best sources for understanding Lie's fundamental mathematical contributions (see the articles by D. Rowe and T. Hawkins in volume 1 of these proceedings). In 1900 he became ordinary professor at the Darmstadt Technische Hochschule, and taught there until 1907 when he came to Berlin. Scheffers published numerous papers on differential geometry, descriptive geometry, and functions of a complex variable, as well as many well-known textbooks. He apparently enjoyed high esteem at the Berlin Technische Hochschule, for he was elected its rector in 1911. Nevertheless, almost to the end of his career, Scheffers had no institutional connection with his mathematics colleagues. The other mathematicians were organized into different departments of the institute for applied mathematics. This also applies to the other mathematician who taught descriptive geometry alongside Scheffers, Erich Salkowski (1881-1943). When Scheffers retired in 1935, no attempt was made to appoint a successor.

Hertzer's chair in geometry, on the other hand, was occupied continuously up till the present day. The three men who held this position after him began their careers as private lecturers at the Technische Hochschule. The first of these was Stanislaus Jolles, who was nearly exclusively interested in synthetic geometry, in particular polar theory and congruences of rays. Jolles received his doctoral degree at Strassburg University in 1882. He then became a private lecturer at the Aachen Technische Hochschule before accepting an appointment at the Berlin Technische Hochschule in 1893.

Jolles' successor, Gerhard Hessenberg, was a far more prominent figure. Mainly influenced by H.A. Schwarz at the Berlin university and G. Hauck at the Technische Hochschule, Hessenberg wrote his dissertation "On the invariants of binary differential forms and their application to the deformation of surfaces" in 1899 under H.A. Schwarz and L. Fuchs. His principal research interests were in the foundations of geometry, set theory and the philosophy of mathematics. It was Hauck who encouraged him to become a private lecturer at the Berlin Technische Hochschule in 1901. Hessenberg was one of the key members of the Berlin Mathematical Society. Before he came to Berlin he had already held chairs in mathematics at Bonn, Breslau, and Tübingen. Unfortunately he died in 1925, the very year he assumed his professorship in Berlin.

Hessenberg was succeeded by Erich Salkowski, who had earned his doctoral degree at Jena university in 1904 before becoming a private lecturer at the Berlin Technische Hochschule in 1907. Most of his scientific papers dealt with the differential geometry of space curves, which highlighted the fruitful interplay of geometry and analysis. He developed a constructive approach to descriptive geometry that proved to be of lasting influence on geometry education in the schools and colleges. After his death in 1943, his position remained vacant for six years, although efforts were made to appoint Wolfgang Haack. But whereas the rector of the Technische Hochschule tried to push for Haack's appointment, the Ministry reacted hesitatingly. On June 19, 1944, the Ministry wrote to the Hochschule: "The definite appointment of a successor to Professor Salkowski can be postponed for the time being." On February 2, 1945—three months before the end of the Second World War—the Ministry demanded a report from Haack and wrote to the administration of the Hochschule: "What says Karlsruhe about this step? The Rector describes the situation too simply" [Ebert Eb 4, pp. 108 and 75].

3.2 The analytical tradition

As with the chairs in geometry, the two chairs in analysis originally held by Siegfried Aronhold and Ernst Kossak were continually occupied by analysts until very recently. In 1883 Aronhold was succeeded by Heinrich Weber (1842-1913) of Königsberg University, a very versatile mathematician, perhaps best known for his famous textbook on algebra written some years later. Weber was a pupil of Otto Hesse in Heidelberg where he wrote his dissertation in 1863 and passed the habilitation in 1866. He remained only one year in Berlin, during which time he mainly studied Galois theory and partial differential equations.

Weber's successor in 1884 was his friend Paul du Bois-Reymond, who previously held a chair at Tübingen University. His dissertation, written in 1859 at Berlin University, dealt with fluids in equilibrium. From his days as a student in Königsberg under F.-J. Richelot (1808-1875) and F.E. Neumann (1798-1895), mathematical physics was his favorite research subject. During his brief tenure at Berlin, he published six papers mainly dealing with partial differential equations and infinite series. He died in 1889.

His successor was Emil Lampe, who like du Bois-Reymond had obtained his doctoral degree from Berlin University under Kummer. His dissertation "On surfaces of 4th order" was completed in 1864. Despite his prolific pen–130 mostly rather short publications—Lampe was better known as a brilliant teacher than an eminent researcher. He was especially interested in the relationship between geometry and algebra, and edited the "Jahrbuch über die Fortschritte der Mathematik" founded by C. Orthmann in 1868. He taught at the Berlin Technische Hochschule from 1889 to 1918.

The man who replaced him in 1919 was Georg Hamel, who influenced mathematics and mechanics at the Berlin Technische Hochschule like no other mathematician before him. A pupil of Hilbert in Göttingen, he wrote his dissertation in 1901 "On the geometries where straight lines are the shortest lines." His habilitation treatise, written three years later, investigated the equations of Lagrange and Euler in mechanics. His best-known achievement was his axiomatization of mechanics, which clearly showed Hilbert's influence as well as his own interest in foundational questions. Hamel's textbook on rational mechanics was published in 1949. Besides articles on geometry, number theory, function theory, calculus of variations, and integral equations, he wrote numerous papers dealing with hydrodynamics, mechanics of solid bodies, and the theory of wires and cables.

During the academic year 1901-02, Hamel was an assistent to Felix Klein (1849-1925). Like Klein, he was deeply interested in all questions of mathematics education. In 1921 he founded the "Reichsverband deutscher mathematischer Gesellschaften und Vereine" (*Reich's* Union of German Mathematical Societies and Associations), and he served as its president up to 1945. Though he was not a National Socialist, Hamel's thinking had a manifest nationalist bent. This may account for his call to establish a "purified German mathematics" in 1933 [Lindner 1980, 106]. At the end of the Second World War he left Berlin because his home had been destroyed by bombs and moved to Landshut in Bavaria. He never took up his lectures again at the Berlin Technische Hochschule, but retired officially no earlier than in 1949.

The other chair in analysis was first occupied by Ernst Kossak, who was succeeded by Wilhelm Stahl (1846-1894) in 1892. Stahl was mainly influenced by the geometer Th. Reye in Zürich and by K. Weierstraß in Berlin, but took his degree in Heidelberg in 1870. At first he worked on physical and mechanical problems dealing with potential surfaces and the bending of beams. Two years later he assumed the chair for synthetic descriptive geometry and graphical statics at the Aachen Technische Hochschule. His research interests then shifted nearly exclusively to synthetic and analytical projective geometry. He taught at the Berlin Technische Hochschule until 1894 when he was succeeded by Georg Hettner, a student of Weierstraß, E.-E. Kummer (1810-1893), and Schwarz. Hettner received his doctoral degree from Berlin University in 1877, and he passed the habilitation at Göttingen two years later. His publications on theta-functions and minimal surfaces were largely influenced by Weierstraß and Schwarz. He also assisted Weierstraß in the publication of the collected works of C.W. Borchardt, C.G. Jacobi (1804-1851), and with Weierstraß's own lectures.

In 1914 Hettner's position went to Rudolf Rothe from the Clausthal Mining Academy. He remained in Berlin for the next 25 years, the last 20 as the colleague of Georg Hamel. Like his predecessor Stahl, Rothe studied at Berlin University, taking his docotrate in 1897. Besides a number of papers on differential geometry, he pursued a variety of physical investigations at the Physikalisch-Technische Reichsanstalt in Berlin. During the First World War, Rothe became heavily involved in ballistic studies. He was well acquainted with the German pope of ballistics, Carl Cranz (1858-1945), who held a professorship in technical physics at the Technische Hochschule since 1920. Being a conservative patriot, Rothe felt obligated to help his country during World War II. Thus despite his refusal to join the Nazi party, he became a member of the Faculty of Defense Technology after his retirement in 1939. This Faculty existed between 1935 and 1945.

During the war years, Rothe's former chair was occupied by Werner Schmeidler (1890-1969), a pupil of E. Landau (1877-1938) from the Breslau Technische Hochschule. Schmeidler had earned his doctoral degree in Göttingen in 1917, where he also passed the habilitation two years later. As a colleague of Emmy Noether during the 1920s, he was one of the many Göttingen mathematicians who took part in shaping the early course of abstract algebra. At the Breslau Technische Hochschule he worked on developing Prandtl's and Glauert's airfoil theory, but in Berlin he turned to the study of singular and algebraic integral equations.

Two years before he came to Berlin Schmeidler joined the National Socialist Party on May 1, 1937 [Ebert Eb 31, p. 61]. His pupil Karl Jaeckel, who became a lecturer in Berlin in 1939, was also a member. [Ebert Eb 31, p. 69]. In addition to these two figures, there were five other pure and applied mathematicians who were Nazi Party members:

Ulrich Graf, joined in April 1937, private lecturer at the Technische Hochschule since 1934 [Ebert Eb 31, p. 9]

Paul Lorenz, joined in 1937, extraordinary professor of mathematical methods of political economy and statistics since 1934 [A Lorenz]

Günther Schulz, joined May 1, 1937, then private lecturer at Berlin University. He became a lecturer at the Technische Hochschule on June 19, 1937 [Ebert Eb 31, p. 48]

Friedrich Tölke, joined April 1, 1933, then private lecturer at the Karlsruhe Technische Hochschule, after 1934 ordinary professor of technical mechanics at the Berlin Technische Hochschule [Ebert Eb 24, pp. 123–125]

Karl Wilhelm Wagner (1883-1953), joined May 1, 1933, ordinary professor of vibration theory

3.3 Timpe's Chair of Mathematics And Economic Mathematics

In 1927 a new professorship of mathematics and mathematical economics was established at the Berlin Technische Hochschule. The first to hold this position was F. Klein's pupil Aloys Timpe (1882-1959). Timpe received his doctoral degree in Göttingen in 1905 and passed the habilitation in Aachen four years later. In 1918 he occupied an ordinary professorship in geodesy at the Berlin Landwirtschaftliche Hochschule. Before he came to the Technische Hochschule he had worked mainly in real analysis and mathematical physics. Afterwards he turned exclusively to finance, mathematical economics, and actuarial theory. This may account for his relatively isolated position within Berlin mathematical circles. Like his colleagues Scheffers, Salkowski, Hamel, and Rothe, Timpe never officially joined the Nazi Party. He was succeeded by Ernst Mohr in 1950.

4. Mathematics at the Technische Universität after the Second World War

The winter semester of 1944/45 was the last before the end of the Second World War during which regular courses were offered at the Berlin Technische Hochschule. At this time there were still six mathematicians on the faculty: the three ordinary professors Hamel, Schmeidler, Timpe, the supernumerary professors Cauer and Schulz, and Schmeidler's assistant István Szabó. Only two of these six, Timpe and Szabó, resumed their former activities when the school reopened as the Technische Universität on April 9, 1946. Thus Timpe was the only ordinary professor of mathematics who resumed lecturing after the Second World War. Salkowski's chair remained unoccupied, and Schmeidler was removed, as were all other members of the National Socialist Party (letter from the preliminary administration of the Technische Hochschule to Schmeidler, dating from December 27, 1945; Ebert Eb 24, p. 270). The Technische Universität wanted to make a new beginning. Along the way, however, these initial good intentions became lost.

Hamel was already 69 years old in 1946, and decided to take a break from his professorial duties before his official retirement in 1949. During the winter semester of 1948/49 he offered a four-hour course on mechanics at the Humboldt-Universität (GDR). In 1949 Wolfgang Haack was appointed to Hamel's professorship. Haack was originally a geometer who habilitated at the Danzig Technische Hochschule in 1929 as mentioned above. In the course of time he studied more and more technical problems, particularly the use of partial differential equations in the study of the dynamics of gases. At the Berlin Technische Hochschule he successfully promoted work on the automatization of air traffic control and electronic computers [Petzold 1979, 396–400].

Schmeidler's vacant professorship was offered to Ernst Mohr, a pupil of Hermann Weyl at Göttingen, where he received his doctoral degree in 1933. In 1938 he habilitated at the Breslau Technische Hochschule in mechanics, and one year later at the University of Breslau in applied mathematics. In 1944, as an extraordinary professor of mathematics at the Germanicized University of Prague, he was denounced as being politically suspect. Thanks to the successful intervention of H. Rohrbach, he was able to survive this frightening episode [Mehrtens 1986, 345].

Mohr was mainly interested in applied mathematics—fluid mechanics, gyroscopic theory and differential equations. He taught at the Technische Universität until his retirement in 1978. When Timpe retired in 1950, Mohr took over his professorship so that Schmeidler could be appointed again to his former chair. In 1959, Schmeidler was even elected as an honorary senator of the Technische Universität.

Mohr was not the only mathematician at the Berlin Technische Hochschule who was threatened by or suffered under the National Socialist system. Others included A. Barneck, R. Fuchs, E.E. Jacobstahl, A. Korn, Hans J. Reissner (1874-1967), and E. Rembs.

Alfred Barneck was a private lecturer whose academic rights were suspended in 1934 because of the "Aryan clause." After the Second World War, he became the headmaster of a *Realgymnasium* and did not resume his activities at the Technische Universität. Richard Fuchs was extraordinary professor of mathematics and had his academic rights suspended in 1938. He died in 1945.

Ernst Erich Jacobsthal was extraordinary professor of mathematics until he lost his position in 1934 and afterward emigrated to Norway. After the war he was made the first honorary member of the Free University in West Berlin. Arthur Korn was ordinary honorary professor until 1933.

Although a mathematician, he mostly lectured on photo-telegraphy. Soon after his suspension he emigrated to the USA.

Another US emigré was Hans J. Reissner, who was ordinary professor of mechanics until his suspension in 1936. Eduard Rembs received his doctoral degree at Bonn University in 1918. When he tried to habilitate at Berlin University in 1938, he was prevented from doing so by Ludwig Bieberbach (1886-1982), the architect of "Deutsche Mathematik," because of political reasons. Rembs later became extraordinary professor of mathematics at the Berlin Technische Universität in 1947 and the official successor to Salkowski in 1949. He resumed earlier geometrical investigations undertaken by Hessenberg and Scheffers.

Other less consequential, but nevertheless characterisitic incidents have been recorded that reveal much about the atmosphere at the Technische Hochschule during this period. One such incident concerned Wilhelm Cauer, who had passed the habilitation at Göttingen University in 1928 and had become a private lecturer at the Berlin Technische Hochschule in 1935. The National Socialist rector v.Arnim wrote on December 18, 1937: "W. Cauer is a half-caste of second degree. Therefore a letter of congratulations is out of the question" [Ebert Eb 10, p. 13].

5. The Mathematicians of the Berlin Technische Hochschule/ Berlin Technische Universität and the Mathematical Society of Berlin

The mathematicans of the Berlin Technische Hochschule played a crucial role when on October 31, 1901, the Mathematical Society of Berlin was founded. Adolf Kneser, then permanent professor at the Mining Academy, and Eugen Jahnke, then a private lecturer at the Technische Hochschule, were the organizers of the founding group which consisted of 38 charter members. Fourteen of these were lecturers or professors at the Technische Hochschule.[5] From the time of its founding up to 1945, the Society was mainly directed by mathematicians from the Technische Hochschule. Julius Weingarten, then professor of mechanics at the Technische Hochschule, was elected its first president [Tobies 1982], and with time it grew to be an accepted centre for mathematics in Berlin. The Society was especially interested in problems concerned with applied mathematics and mathematics education.

As with nearly all aspects of academic life, the Third Reich had a considerable impact on the Berlin Mathematical Society. In 1938, a number of its Jewish members were forced to resign: Alfred Barneck on May 5, Ernst Erich Jacobsthal on October 18, Richard Fuchs on December 11, and Issai Schur (1875-1941), on December 15 [BMG-S1]. These actions, however, were not swift enough to suit all its members. On December 4,

1938, Max Zacharias, a former president during the years 1922/23, asked President Aloys Timpe what the executive board had done or planned to do in order to eliminate non-Aryans from the Society. Timpe answered evasively, referring to an agreement with the Ministry which stipulated that such matters had to be handled confidentially. As a result Zacharias, resigned his membership and wrote Timpe the following letter:

> Dear colleague,
>
> My sincere thanks for your answer to my question. To be sure I have to confess to you that I am astonished and dissatisfied by your answer. I had hoped to be informed by you that the members of the executive board, being impressed by recent events, have taken the necessary steps to eliminate at once the Jewish members. My racial feeling does not allow me to be a member of a society which tolerates Jewish members even today.
>
> Therefore I regretfully declare herewith my withdrawal from the Berlin Mathematical Society.
>
> Heil Hitler!
>
> Max Zacharias

It is interesting to compare this letter with a quotation from the biographical notice written by Dreetz in 1951 on the occasion of Zacharias's 80th birthday [Dreetz 1951]:

> "His life was the prototype of the calm, peaceful life of a scholar. Without much ado he was at hand wherever he could help in any way."

On January 8, 1939, Timpe was forced to circulate a notice informing all Jewish members that they were to resign their memberships. Nevertheless Timpe tried to limit the damage that had been done whenever he could. This is exemplified in the case of Richard Fuchs. Fuchs had been dismissed from his professorship in 1938 because he was not of pure Aryan extraction. The National Socialist rector of the Technische Hochschule, Ernst Storm, prohibited Fuchs from presenting his latest results on airfoil theory at the Technische Hochschule. When Fuchs resigned his membership of the Berlin Mathematical Society, its secretary, Otto-Heinrich Keller, had a conversation with him that led him to retract his notice of withdrawal.

It was not until 1950 that the Berlin Mathematical Society was reestablished through the initiative of thirteen mathematicians.[6] Again, mathematicians at the Technische Universität played a leading role, although the

Society never recovered its former importance. Those who took the initiative this time came from all three Berlin universities, colleges and from the secondary schools. Hermann Ludwig Schmid was the last chairman of the old society during the Second World War (1943-1945). The first official meeting took place on November 20, 1950, when Georg Hamel spoke on the subject "What is geometry?" [BMG-G1]. The chairman of the board invited members of the pre-war Society to become members of the new one, for example Otto-Heinrich Keller, who was full professor of geometry at the Dresden Technische Hochschule since 1947. Keller must have inquired about the former chairman, Max Zacharias, for the secretary, who at the time was Schmeidler, wrote him on January 9, 1952 that Mr. Zacharias was not a member of the new society, adding the strange remark: "nor could we find him in other lists."

Schmeidler, in fact, was one of the leading members of the new society from the very beginning. Already in 1950 he was restored in his former chair, his political background being left out of consideration. On July 26, 1951, the names of the executive board members were sent to the Scientific Research Branch, Economics Group, British Military Government, because the Technische Universität belonged to the British Sector in Berlin. These mebers were:

Chairman: Wolfgang Haack (Technische Universität, British Sector),
Vice-Chairmam: Karl Schröder (Humboldt-Universität, Soviet Sector),
Secretary: Werner Schmeidler (Technische Universität, British Sector),
Treasurer: Alexander Dinghas (1908-1074) (Freie Universität, American Sector).

On November 26, 1951, Ludwig Bieberbach, the most famous National Socialist mathematician, was admitted unanimously to the Berlin Mathematical Society. It was recorded in protocol that with regard to the admission of Mr. Bierberbach the board decided unanimously that it is intolerable for the purposes of the statutes and articles to exclude one of the most important Berlin mathematicians [BMG-Si1]. Because of Bieberbach's admission, E. Rembs withdrew from the Society on April 26, 1953. He did not change his decision, although H. Jonas, a member of the Society's advisory board, had a private conversation with Rembs on behalf of the chairman of the board. On June 12, 1953, A. Timpe also resigned from the Society, as did Ernst Mohr on December 10, 1954. When A. Dinghas nominated Ernst Jacobsthal, who had been vice-chairman from 1922 to 1934, to be an honorary member of the Society, the chairman, Wolfgang Haack, voiced his disapproval. He argued that there had been only two honorary members up to then, Georg Hamel and Erhard Schmidt (1876-1959), whose scientific importance far exceeded that of Jacobsthal. Haack evidently did not take into consideration, at least not officially, that there might have been other grounds for supporting Jacobsthal's nomination.

6. Tables

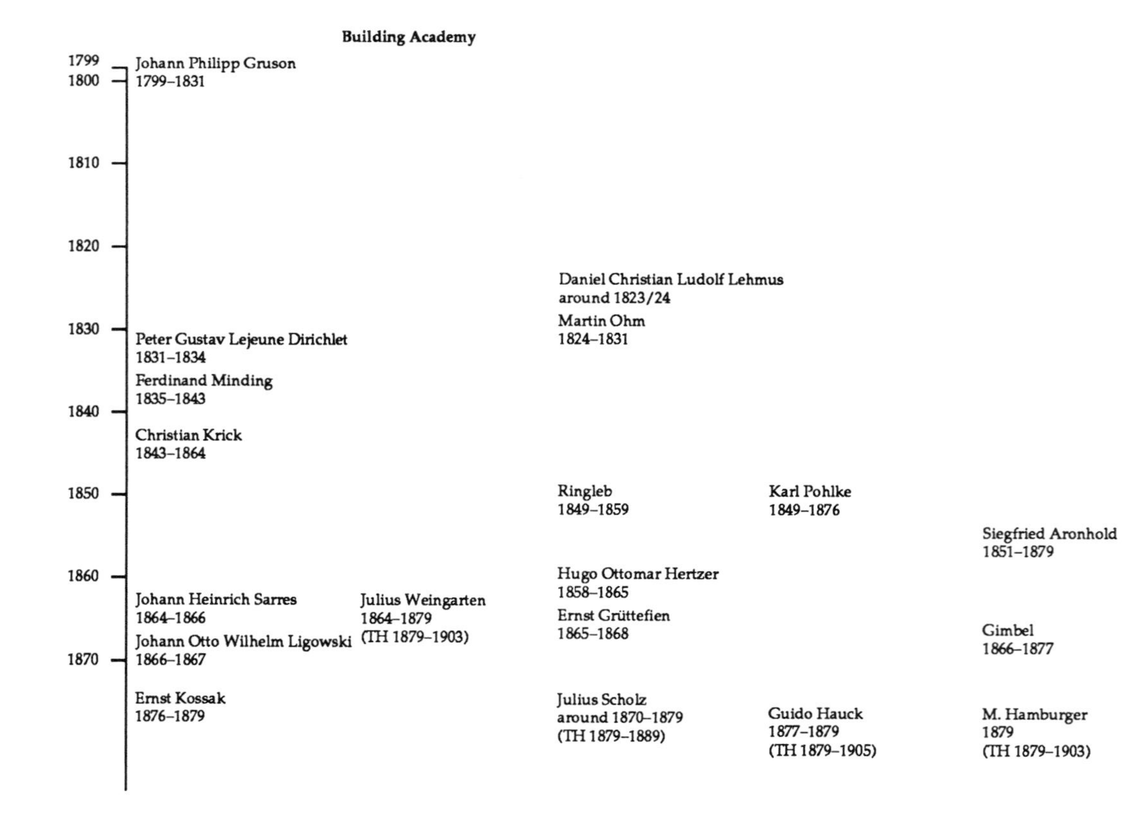
Building Academy
1799
1800
Johann Philipp Gruson
1799–1831
1810
1820
Daniel Christian Ludolf Lehmus
around 1823/24
Martin Ohm
1824–1831
1830
Peter Gustav Lejeune Dirichlet
1831–1834
Ferdinand Minding
1835–1843
1840
Christian Krick
1843–1864
1850
Ringleb
1849–1859
Karl Pohlke
1849–1876
Siegfried Aronhold
1851–1879
1860
Hugo Ottomar Hertzer
1858–1865
Johann Heinrich Sarres
1864–1866
Julius Weingarten
1864–1879
(TH 1879–1903)
Ernst Grüttefien
1865–1868
Gimbel
1866–1877
Johann Otto Wilhelm Ligowski
1866–1867
1870
Ernst Kossak
1876–1879
Julius Scholz
around 1870–1879
(TH 1879–1889)
Guido Hauck
1877–1879
(TH 1879–1905)
M. Hamburger
1879
(TH 1879–1903)

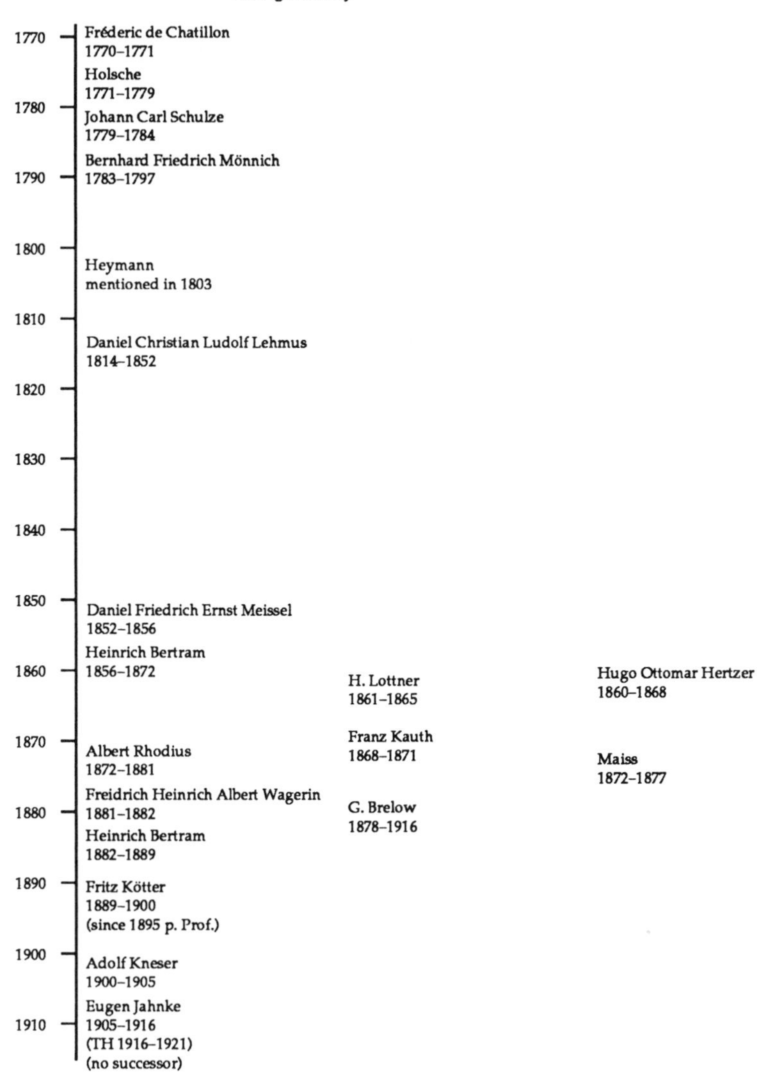
Mining Academy
1770
1780
1790
1800
1810
1820
1830
1840
1850
1860
1870
1880
1890
1900
1910
Fréderic de Chatillon
1770–1771
Holsche
1771–1779
Johann Carl Schulze
1779–1784
Bernhard Friedrich Mönnich
1783–1797
Heymann
mentioned in 1803
Daniel Christian Ludolf Lehmus
1814–1852
Daniel Friedrich Ernst Meissel
1852–1856
Heinrich Bertram
1856–1872
Albert Rhodius
1872–1881
Freidrich Heinrich Albert Wagerin
1881–1882
Heinrich Bertram
1882–1889
Fritz Kötter
1889–1900
(since 1895 p. Prof.)
Adolf Kneser
1900–1905
Eugen Jahnke
1905–1916
(TH 1916–1921)
(no successor)
H. Lottner
1861–1865
Franz Kauth
1868–1871
G. Brelow
1878–1916
Hugo Ottomar Hertzer
1860–1868
Maiss
1872–1877

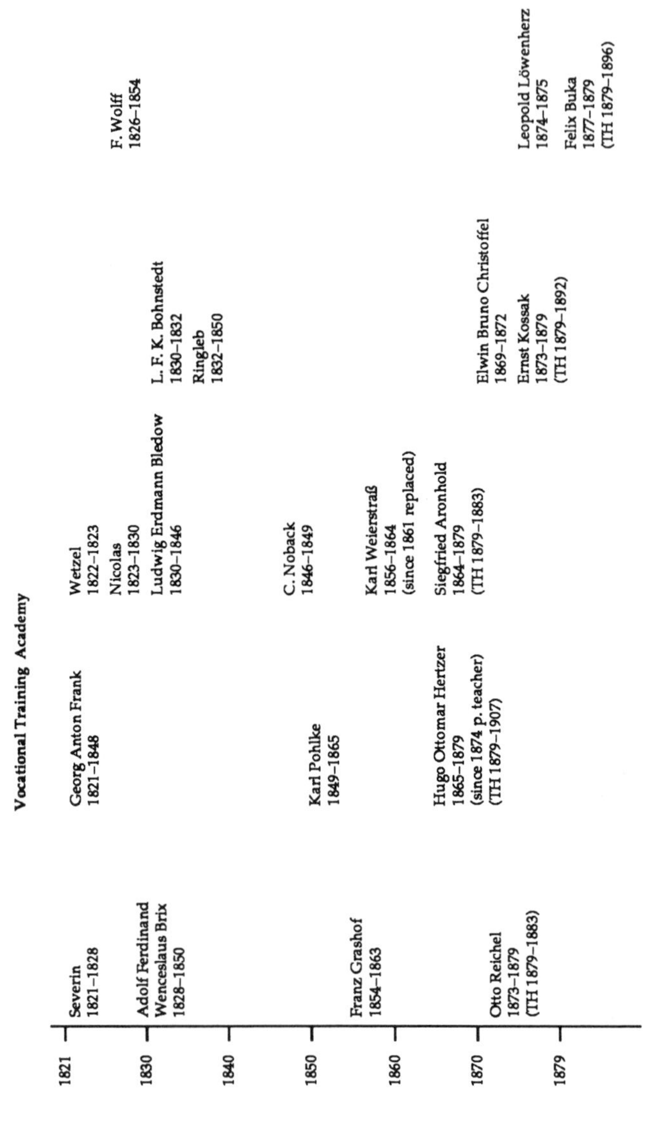
Vocational Training Academy
1821
1830
1840
1850
1860
1870
1879
Severin
1821–1828
Adolf Ferdinand
Wenceslaus Brix
1828–1850
Franz Grashof
1854–1863
Otto Reichel
1873–1879
(TH 1879–1883)
Georg Anton Frank
1821–1848
Karl Pohlke
1849–1865
Hugo Ottomar Hertzer
1865–1879
(since 1874 p. teacher)
(TH 1879–1907)
Wetzel
1822–1823
Nicolas
1823–1830
Ludwig Erdmann Bledow
1830–1846
C. Noback
1846–1849
Karl Weierstraß
1856–1864
(since 1861 replaced)
Siegfried Aronhold
1864–1879
(TH 1879–1883)
L. F. K. Bohnstedt
1830–1832
Ringleb
1832–1850
Elwin Bruno Christoffel
1869–1872
Ernst Kossak
1873–1879
(TH 1879–1892)
F. Wolff
1826–1854
Leopold Löwenherz
1874–1875
Felix Buka
1877–1879
(TH 1879–1896)

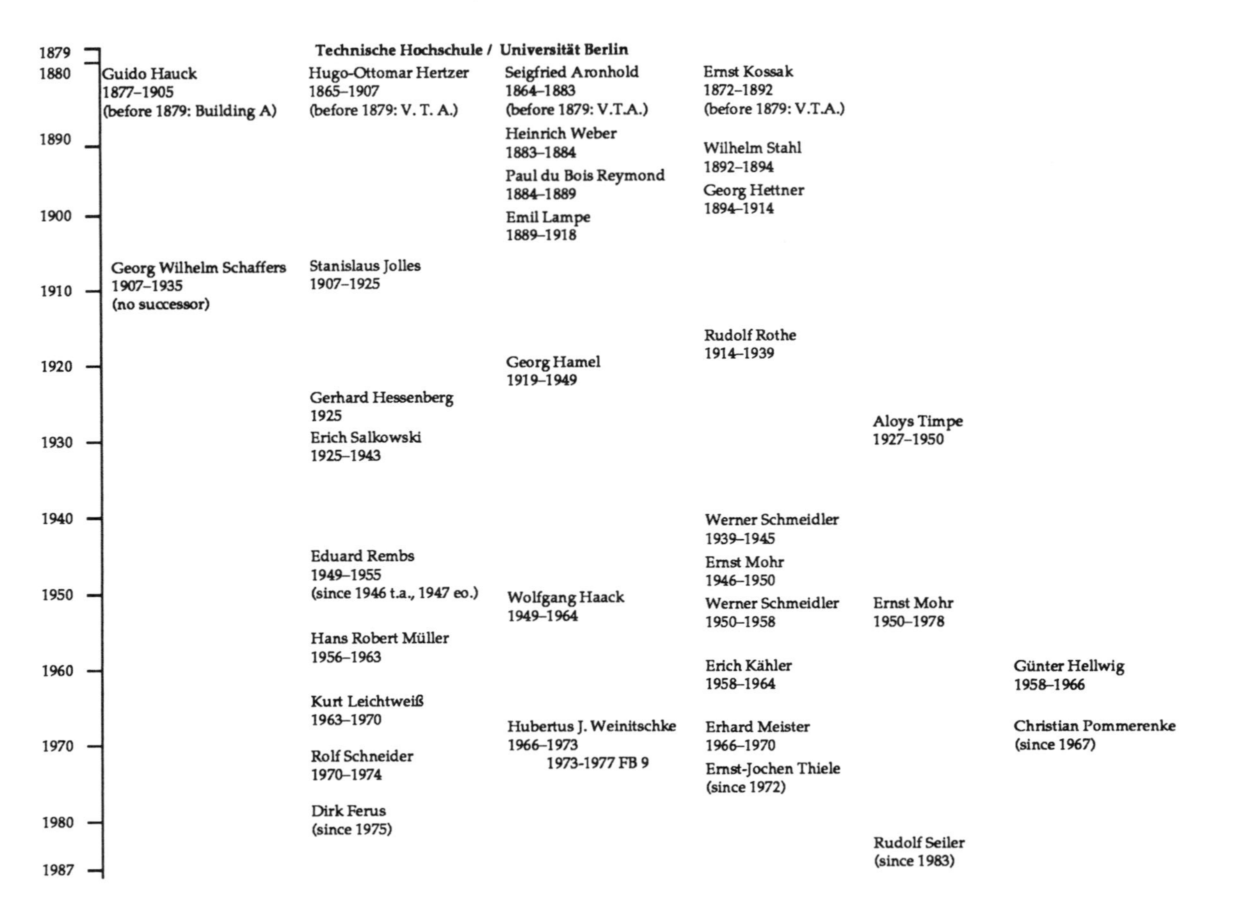
Technische Hochschule / Universität Berlin
1879
1880
1890
1900
1910
1920
1930
1940
1950
1960
1970
1980
1987
Guido Hauck
1877–1905
(before 1879: Building A)
Georg Wilhelm Schaffers
1907–1935
(no successor)
Hugo-Ottomar Hertzer
1865–1907
(before 1879: V. T. A.)
Stanislaus Jolles
1907–1925
Gerhard Hessenberg
1925
Erich Salkowski
1925–1943
Eduard Rembs
1949–1955
(since 1946 t.a., 1947 eo.)
Hans Robert Müller
1956–1963
Kurt Leichtweiß
1963–1970
Rolf Schneider
1970–1974
Dirk Ferus
(since 1975)
Seigfried Aronhold
1864–1883
(before 1879: V.T.A.)
Heinrich Weber
1883–1884
Paul du Bois Reymond
1884–1889
Emil Lampe
1889–1918
Georg Hamel
1919–1949
Wolfgang Haack
1949–1964
Hubertus J. Weinitschke
1966–1973
1973-1977 FB 9
Ernst Kossak
1872–1892
(before 1879: V.T.A.)
Wilhelm Stahl
1892–1894
Georg Hettner
1894–1914
Rudolf Rothe
1914–1939
Werner Schmeidler
1939–1945
Ernst Mohr
1946–1950
Werner Schmeidler
1950–1958
Erich Kähler
1958–1964
Erhard Meister
1966–1970
Ernst-Jochen Thiele
(since 1972)
Aloys Timpe
1927–1950
Ernst Mohr
1950–1978
Rudolf Seiler
(since 1983)
Günter Hellwig
1958–1966
Christian Pommerenke
(since 1967)

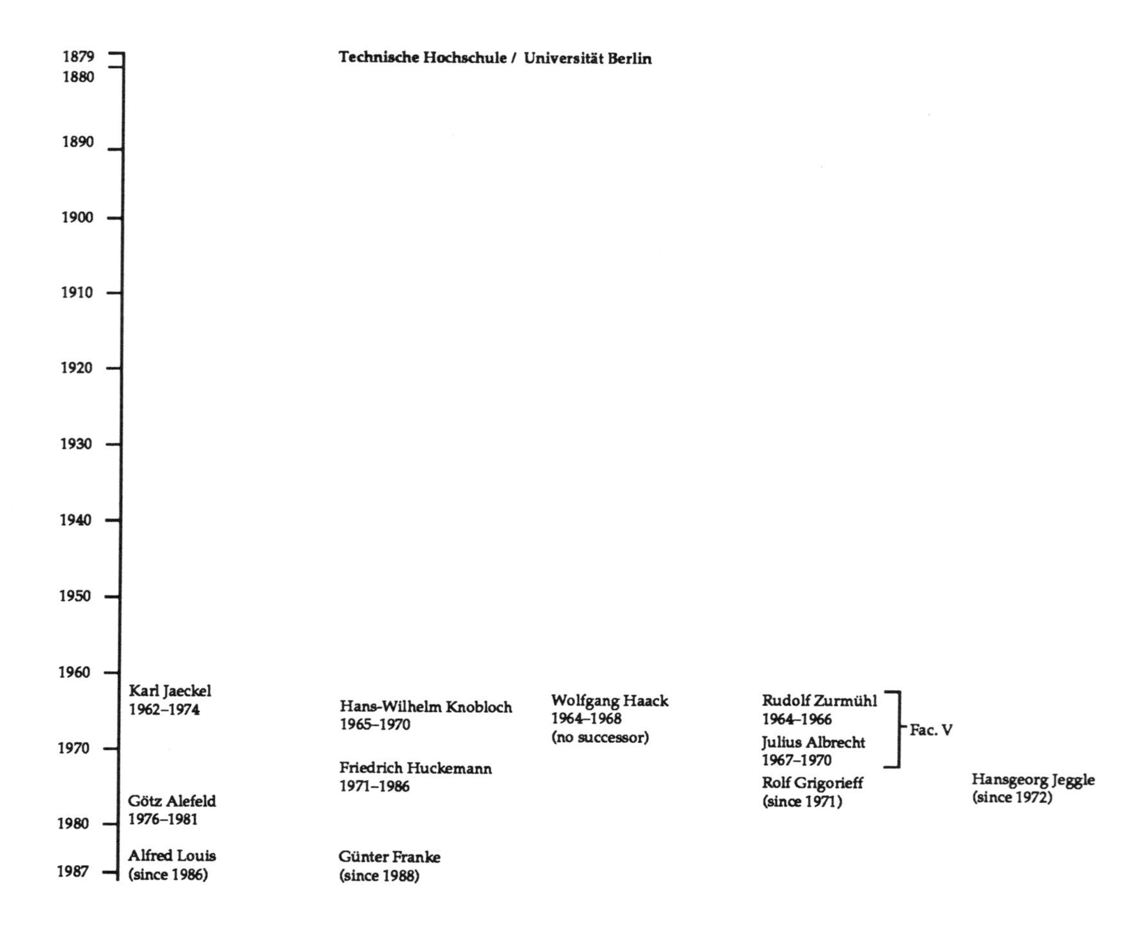

Technische Hochschule / Universität Berlin
1879
1880
1890
1900
1910
1920
1930
1940
1950
1960
1970
1980
1987
Karl Jaeckel
1962–1974
Götz Alefeld
1976–1981
Alfred Louis
(since 1986)
Hans-Wilhelm Knobloch
1965–1970
Friedrich Huckemann
1971–1986
Günter Franke
(since 1988)
Wolfgang Haack
1964–1968
(no successor)
Rudolf Zurmühl
1964–1966
Julius Albrecht
1967–1970
Fac. V
Rolf Grigorieff
(since 1971)
Hansgeorg Jeggle
(since 1972)

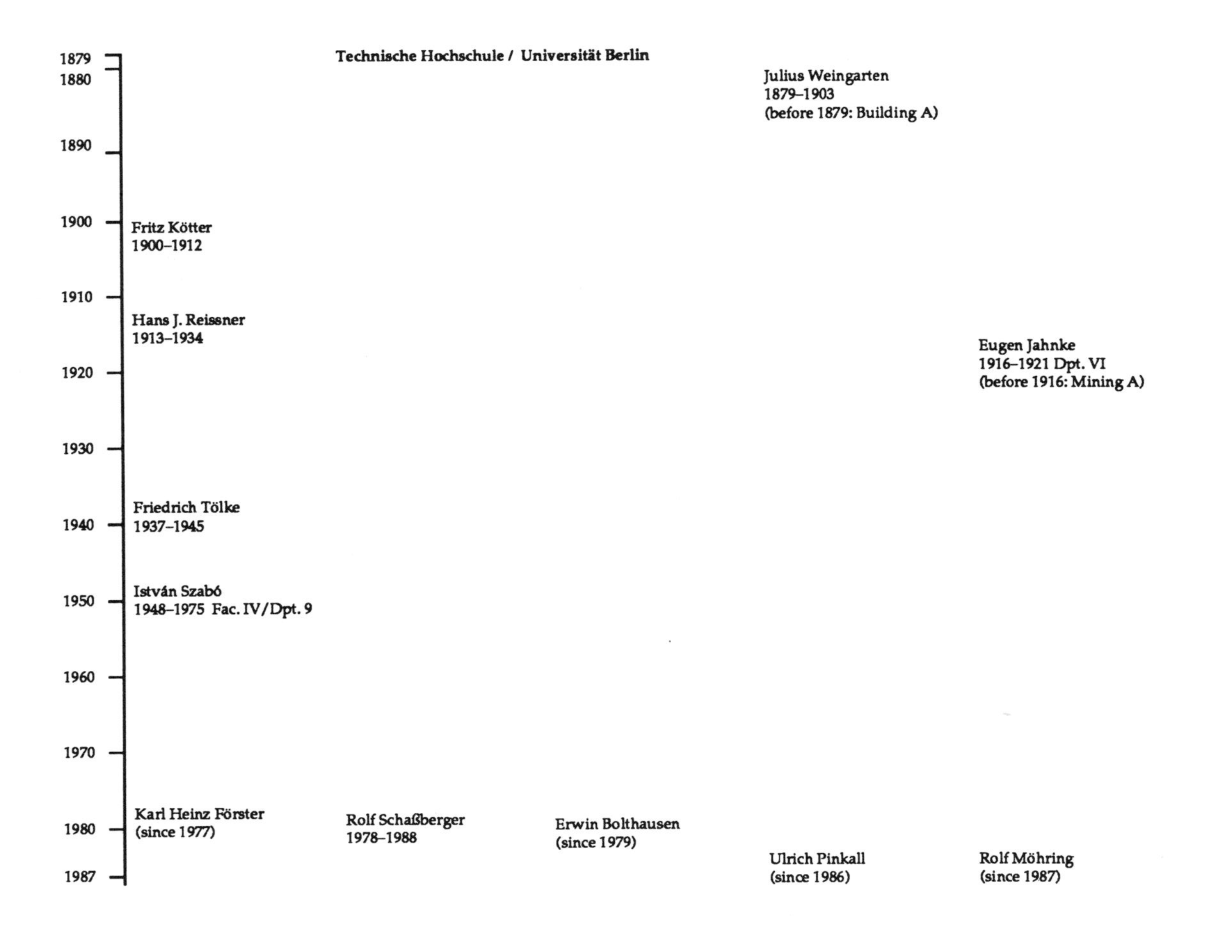
Technische Hochschule / Universität Berlin
Julius Weingarten
1879–1903
(before 1879: Building A)
Fritz Kötter
1900–1912
Hans J. Reissner
1913–1934
Eugen Jahnke
1916–1921 Dpt. VI
(before 1916: Mining A)
Friedrich Tölke
1937–1945
István Szabó
1948–1975 Fac. IV/Dpt. 9
Karl Heinz Förster
(since 1977)
Rolf Schaßberger
1978–1988
Erwin Bolthausen
(since 1979)
Ulrich Pinkall
(since 1986)
Rolf Möhring
(since 1987)
1879
1880
1890
1900
1910
1920
1930
1940
1950
1960
1970
1980
1987

NOTES

The present study is part of a larger research project on mathematics at the Berlin Technische Hochschule and at the older institutions which were combined in order to establish the Technische Hochschule.

1. Graf passed the habilitation in Berlin in 1934 and was professor of mathematics at the Danzig Technische Hochschule from 1938 until 1945; Haack passed the habilitation at the Danzig Technische Hochschule in 1929, was lecturer in Berlin from 1935 until 1937 and professor of mathematics from 1949 until 1968; Jaeckel passed the habilitation at the Breslau Technische Hochschule in 1938 and was professor of mathematics in Berlin from 1959 until 1974; the Breslau Technische Hochschule awarded the honorary degree of engineering sciences to Korn in 1930 who was ordinary honorary professor in Berlin from 1924 until 1933; Mohr passed the habilitation in Breslau in 1934 and was professor of mathematics in Berlin from 1946 until 1978; Steinitz passed the habilitation in Berlin in 1898 and was professor of mathematics in Breslau from 1910 until 1920; Schmeidler, the academic teacher and supervisor of Jaeckel's doctoral dissertation, was professor of mathematics in Breslau from 1921 until 1939, in Berlin from 1939 until 1945 and from 1950 until 1958.

2. This applied to Paul du Bois-Raymond (1831-1889) in 1859, Elwin Bruno Christoffel (1829-1900) in 1856, Richard Fuchs in 1897, Gerhard Hessenberg (1874-1925) in 1899, Georg Hettner (1854-1914) in 1877, Ernst Erich Jacobsthal (1882-1965) in 1906, Adolf Kneser (1862-1930) in 1884, Emil Lampe (1840-1928) in 1864, Leon Lichtenstein (1878-1933) in 1909, Leopold Löwenherz (1847-1892) in 1870, Richard Müller (1802-?) in 1884, Robert Ritter (1905-1959) in 1933, Rudolf Rothe in 1897, Julius Scholtz (1839-?) in 1863, Günther Schulz (1903-1962) in 1928, Amandus Wendt (1855-1939) in 1880, lately to Hans Wilhelm Knobloch (*1927) in 1950 and Ernst-Jochen Thiele (*1928) in 1956.

3. Besides these professors, other lecturers played an important role as mathematics teachers at the Berlin Technische Hochschule. Apart from F. Buka, M. Hamburger, O. Reichel, J. Scholz there were 39 private lecturers in mathematics between 1879 and 1945. In the course of time, eight of them occupied chairs in mathematics at the Berlin Technische Hochschule: W. Haack, G. Hessenberg, K. Jaeckel, E. Jahnke, St. Jolles, F. Kötter, R. Rothe, E. Salkowski. Many others became well-known scientists elsewhere: W. Cauer (control theory), C. Cranz (ballistics), R. Fuchs (analysis, aerodynamics), U. Graf (geometry), J. Horn (1867-1946) (analysis), E.E. Jacobsthal (algebra, number theory), O.-H. Keller (*1906) (geometry, algebra), L. Lichtenstein (analysis, potential theory), E. Steinitz (algebra), F.-A. Willers (practical analysis). The other private lecturers

were: O. Dziobek (1856-1919), H. Grosse, A. Wendt, H. Servus (1858-?), R. Müller, E. Haentzschel, A. Gleichen (1862-1923), G. Wallenberg (1864-1924), E. Meyer (1871-1909), W. v. Ignatowsky (1875-?), A. Barneck (1885-1964), H. Lemke (1874-?), P. Schafheitlin (1861-1924), E. Stübler, G. Grüss (1902-1950), G. Haenzel (1898-1944), M. Sadowsky (1902-1967), P. Lorenz, W. Magnus (*1907), G. Schulz, K. Maruhn (1904-1976).

4. It might suffice to enumerate the names of those mathematicians who held the four traditional chairs of mathematics:

i. Rembs's successors were Hans Robert Müller, Kurt Leichtweiss, Rolf Schneider, Dirk Ferus (since 1975)
ii. Haack's successor was Hubertus J. Weinitschke
iii. Schmeidler's successors were Erich Kähler, Erhard Meister, Ernst-Jochen Thiele (since 1972)
iv. Mohr's successor is Rudolf Seiler (since 1983).

Since 1958 more than 30 new professorships of mathematics were established and 30 private lecturers have habilitated at the Technische Universität Berlin.

5. These fourteen were: E. Dziobek, R. Fuchs, M. Hamburger, E. Haentzschel, G. Hessenberg, E. Jahnke, St. Jolles, A. Kneser, R. Müller, R. Rothe, P. Schafheitlin, E. Steinitz, G. Wallenberg, and J. Weingarten.

6. The thirteen were:

F. Neiss, H.L. Schmid, K. Schröder (Humboldt-Universität, Berlin (GDR))
W. Haack, G. Hellwig, E. Mohr, E. Rembs, A. Timpe (Technische Universität, Berlin (West)),
R. Ritter (Freie Universität, Berlin (West))
R. Sprague (Pädagogische Hochschule, Berlin (West))
Jonas, O. Kempka, Rose (teachers at secondary schools)

References

1. Archival sources

The files of the Berlin Mathematical Society (BMG) are now preserved in the Department of Mathematics of the Technische Universität Berlin. The files have no pagination.

BMG-S1: Berliner Mathematische Gesellschaft, Schriftwechsel file 1 (correspondence)

BMG-G1: Berliner Mathematische Gesellschaft, Angelegenheiten der Gesellschaft, file 1 (affairs of the society)

BMG-Si1: Berliner Mathematische Gesellschaft, Vorstands- und Mitgliederversammlungen ab 1950 (assemblies of the board and of members)

Nachlass Ebert Eb 4: Technische Universität Berlin, Hochschularchiv, Nachlass Hans Ebert, file 4 (for example)

A Lorenz: Technische Universität Berlin, Hochschularchiv, file Ausgeschiedene Lehrbeauftragte, see name Lorenz

2. Publications

Baruch, A. 1922. Nachruf auf Eugen Jahnke. Sitzungsberichte der Berliner Mathematischen Gesellschaft 21, 30–39.

Beckert, H. u. Schumann, H. (Hrsgg.). 1981. 100 Jahre mathematisches Seminar der Karl-Marx-Universität Leipzig. Berlin: Deutscher Verlag der Wissenschaften.

Die deutschen technischen Hochschulen, ihre Gründung und geschichtliche Entwicklung. 1941. München: Verlag der Deutschen Technik.

Dreetz, W. 1952/53. Max Zacharias 80 Jahre alt. Sitzungsberichte der Berliner Mathematischen Gesellschaft, pp. 34–35.

Kneser, A. 1905. Beiträge zur Theorie der Sturm-Liouvilleschen Darstellung willkürlicher Funktionen. Mathematische Annalen 60, 402–423.

Knobloch, E. 1978/79. Die Berliner Gewerbeakademie und ihre Mathematiker. Heimatblätter des Kreises Aachen 34/35, 55–64.

Lindner, H. 1980. "Deutsche" und "gegentypische" Mathematik. Zur Begründung einer "arteigenen Mathematik" im "Dritten Reich" durch Ludwig Bieberbach. In: H. Mehrtens, S. Richter (Eds.) Naturwissenschaft, Technik und NS-Ideologie. Beiträge zur Wissenschaftsgeschichte des Dritten Reiches, pp. 88–115. Frankfurt/M.: Suhrkamp.

Lorey, W. 1916. Das Studium der Mathematik an den deutschen Universitten seit Anfang des 19. Jahrhunderts. Leipzig und Berlin: B.G. Teubner (Abhandlungen über den mathematischen Unterricht in Deutschland, veranlasst durch die Internationale mathematische Unterrichtskommission, ed. by F. Klein, vol. 3, issue 9).

Lundgreen, P. 1988. Technische Bildung in Preussen vom 18. Jahrhundert bis zur Zeit der Reichsgründung. In: Günter Sodan (Ed.). Die Technische Fachschule Berlin im Spektrum Berliner Bildungsgeschichte, pp. 1–44. Berlin: Technische Fachhochschule.

Mehrtens, H. 1986. Wissenschaften im Nationalsozialismus. Geschichte und Gesellschaft 12, 317–347.

Petzold, H. 1979. Konrad Zuse, die Technische Universität Berlin und die Entwicklung der elektronischen Rechenmaschinen. In: R. Rürup (Ed.). Wissenschaft und Gesellschaft. Beiträge zur Geschichte der Technischen Universität Berlin 1879–1979, vol. 1, pp. 389–402. Berlin-Heidelberg-New York: Springer.

Tobies, R. 1982. Zur Geschichte der Berliner Mathematischen Gesellschaft. In: Die Entwicklung Berlins als Wissenschaftszentrum (1870-1930). Zur Entwicklung der Mathematik in Berlin. Hrsg. von der Akademie der Wissenschaften der DDR, Institut für Theorie, Geschichte und Organisation der Wissenschaft. Kolloquium, Heft 30. Berliner wissenschaftshistorische Kolloquien VII, p. 59–92. Berlin.

OBITUARIES AND APPRECIATIONS

"Pogg." means: J. Chr. Poggendorff: Biographisch-literarisches Handwörterbuch zur Geschichte der exakten Wissenschaften. Leipzig-Berlin: since 1863.

Aronhold, Siegfried: Pogg. I, III, IV
M. Cantor, in: Allgemeine Deutsche Biographie 46 (1902), 58–59.
H. Oettel, in: Dictionary of Scientific Biography 1 (1970), 294–295.
anonymous, in: Zentralblatt der Bauverwaltung 4 (1884), 110.
anonymous, in: Schweizerische Bauzeitung 3 (1884), 72.
E. Lampe. 1899. Die reine Mathematik in den Jahren 1884-1899 nebst Actenstükken zum Leben von Siegfried Aronhold. Berlin: Verlag v. Wilhelm Ernst u. Sohn.

Bois–Reymond, Paul du: Pogg. III, IV, V
L. Novy, in: Dictionary of Scientific Biography 4 (1971), 205–206.
L. Kronecker, in: Journal für die reine und angewandte Mathematik 104 (1889), 352–354.
H. Weber, in: Mathematische Annalen 35 (1890), 457–469.
N. Stuloff, in: Neue Deutsche Biographie 4 (1959), 148.
anonymous, in: Centralblatt der Bauverwaltung 9 (1889), 142.

Christoffel, Elwin Bruno: Pogg. III, IV, V
P. L. Butzer, F. Fehér (Eds.). 1981. E.B. Christoffel, The influence of his work on mathematics and the physical sciences. Basel-Boston-Stuttgart: Birkhäuser.

Haack, Wolfgang: Pogg. VI, VIIa
Hans-Joachim Töpfer. 1982. Laudatio zur Ehrenpromotion von Prof. Dr. Dr.h.c. W. Haack am 26. April 1982. In: Fest-Colloquium aus Anlaß der Ehrenpromotion und des 80. Geburtstages von Prof. Dr.Dr.h.c. Wolfgang Haack 26./27. April 1982, pp. 9–12, Berlin 1982.

W. Haack. 20 Jahre Siemens-Rechner im Hahn-Meitner-Institut. Betrachtungen zur Geschichte und Vorgeschichte des Sektors Mathematik [Ms. Archiv des Hahn-Meitner-Instituts für Kernforschung, West-Berlin].

Hamel, Georg Karl Wilhelm: Pogg. V, VI, VIIa
Klaus Mainzer, in: Enzyklopädie Philosophie und Wissenschaftstheorie, J. Mittelstrass, Ed., vol. 2, 1984, pp. 29–30. Mannheim-Wien-Zürich: Bibliographisches Institut.
W. Schmeidler. 1951/52. Bericht über die Hamel-Feier anlässlich des Salzburger Mathematiker-Kongresses 1952. Sitzungsberichte der Berliner Mathematischen Gesellschaft, 27–34.
J. Lense, in: Neue Deutsche Biographie 7 (1966), 583.
G. Faber, in: Jahrbuch Bayerische Akademie der Wissenschaften 1955, 178–180.
O. Haupt, in: Jahrbuch der Akademie der Wissenschaften und Literatur Mainz 1954, 148–154.
W. Schmeidler, in: Jahresbericht der Deutschen Mathematiker-Vereinigung 58 (1955), 1–5.
I. Szabó, in: Jahrbuch der Wissenschaftlichen Gesellschaft für Luftfahrt e.V. (WGL) 1954, 25–28.
J. Willers. 1937. Prof. Dr. Hamel 60 Jahre. Zeitschrift für angewandte Mathematik und Mechanik 17, 311.
W. Kucharski. Georg Hamel zum 70.Geburtstag. Ansprache im Seminar für Mathematik und Mechanik der TU Berlin, am 18. Juni 1948 [Ms. Hochschularchiv der Technischen Universität].
R. Grammel. 1947. Georg Hamel 70 Jahre. Forschungen und Fortschritte 21/23, 281.
W. Kucharski. 1952. Über Hamels Bedeutung für die Mechanik. Ansprache zu seinem 75. Geburtstag im Seminar für Mechanik der Technischen Universität Berlin. Zeitschrift für angewandte Mathematik und Mechanik 32, 293–297.
G. Vogelpohl. 1948. Georg Hamel als Lehrer. Zeitschrift für angewandte Mathematik und Mechanik 28, 131–132.
L. Prandtl. 1948. Georg Hamel siebzig Jahre. Zeitschrift für angewandte Mathematik und Mechanik 28, 129–131.
W. Schmeidler. 1952/53. Über Leben und Werk von Georg Hamel. Sitzungsberichte der Berliner Mathematischen Gesellschaft, 7–9.
W. Haack. 1952/53. Über Hamels Bedeutung in der Mechanik. Sitzungsberichte der Berliner Mathematischen Gesellschaft, 9–16.
Werner Schmeidler, in: Internationale Mathematische Nachrichten 7 (1953), Nr. 27/28, 2–4.
Präsidium der Akademie (Wagner, Jordan, Diepgen, Döblin), in: Jahrbuch der Akademie der Wissenschaften und Literatur Mainz 1952, 127–128.

Hans Martin Klinkenberg, in: Die Rheinisch-Westfälische Technische Hochschule Aachen 1870-1970, Stuttgart 1970, 242.

Hauck, Guido: Pogg. III, IV, V
Vorlesungsverzeichnis Chronik 1904/05, p. 131–133.
G. Hessenberg, in: Zeitschrift für mathematischen und naturwissenschaftlichen Unterricht 37 (1906), 71–76.
anonymous, in: Leopoldina 41 (1905), 38.
Fr. E., in: Deutsche Bauzeitung 39 (1905), 68.
H. Stark, in: Neue Deutsche Biographie 8 (1969), 77.
Die Enthüllungsfeier des Hauck-Denkmals in der Halle des Hauptgebäudes der Königlichen Technischen Hochschule zu Berlin am 14. November 1906. Leipzig 1906.
G. Hessenberg, in: Zentralblatt der Bauverwaltung 25 (1905), 72–73.
E. Lampe, in: Jahresbericht der Deutschen Mathematiker-Vereinigung 14 (1905), 289–311.

Hertzer, Hugo-Ottomar: Pogg. IV, V
anonymous, in: Jahresbericht der Deutschen Mathematiker-Vereinigung 17 (1908), 2. Abt., 170.
E. Lampe, in: Jahresbericht der Deutschen Mathematiker-Vereinigung 18 (1909), 417–422.
Vorlesungsverzeichnis der Technischen Hochschule Berlin, Chronik 1908/1909, 154–157.
anonymous, in: Glasers Annalen für Gewerbe und Bauwesen 64 (1909), 71.
anonymous, in: Vossische Zeitung, 17. November 1908.

Hessenberg, Gerhard: Pogg. V, VI
R. Rothe, in: Sitzungsberichte der Berliner Mathematischen Gesellschaft 25 (1926), 26–44; reprinted in: Jahresbericht der Deutschen Mathematiker-Vereinigung 36 (1927), 312–332.
E. Salkowski. 1926. Hessenberg als darstellender Geometer. In: Sitzungsberichte der Berliner Mathematischen Gesellschaft 25, 45–52.
W. Benz, in: Neue Deutsche Biographie 9 (1972), 24.
W. Lorey, in: Unterrichtsblätter für Mathematik und Naturwissenschaften 32 (1926), 38–39.
anonymous, in: Zeitschrift für angewandte Mathematik und Mechanik 5 (1925), 527.

Hettner, Georg: Pogg. III, IV, V, VI
E. Lampe, in: Sitzungsberichte der Berliner Mathematischen Gesellschaft 14 (1914), 2–7; reprinted in: Jahresbericht der Deutschen Mathematiker-Vereinigung 24 (1915), 51–58.

Jaeckel, Karl Friedrich Wilhelm: Pogg. VIIa
Catalogus Professorum TH Hannover 1831-1956. Hannover 1956, p. 13.
Catalogus Professorum. Festschrift zum 150jährigen Bestehen der Universität Hannover. vol. 2, p. 127. Stuttgart 1981.

Jahnke, Eugen: Pogg. IV, V, VI
A. Baruch, in: Berliner Hochschul-Nachrichten 6. Semester, issue 3 (1921), 36–37.
A. Baruch, in: Sitzungsberichte der Berliner Mathematischen Gesellschaft 6 (1921), 30–39.
E. Knobloch, in: Neue Deutsche Biographie 10 (1974), 307.
St. Jolles, in: Jahresbericht der Deutschen Mathematiker-Vereinigung 31 (1922), 177–184.
anonymous, in: Elektrotechnische Zeitschrift 42 (1921), 1309.

Jolles, Stansislaus: Pogg. IV, V, VI, VIIA

Kneser, Adolf: Pogg. IV, V, VI
Lothar Koschmieder, in: Sitzungsberichte der Berliner Mathematischen Gesellschaft 29 (1930), 78–102.
E. Lohmeyer, R. Hönigswald, C. Schaefer, J. Radon, L. Koschmieder. 1930. Zur Erinnerung an Adolf Kneser. Braunschweig: Vieweg.
Kossak, Ernst, in: Vorlesungsverzeichnis der Technischen Hochschule, Chronik 1891/92, 98–99.
Emil Lampe in: Jahresbericht der Deutschen Mathematiker-Vereinigung 12 (1903), 500–504.

Kötter, Fritz: Pogg. IV, V
anonymous, in: Zeitschrift des Vereines Deutscher Ingenieure 56 (1912), 1422.
anonymous, in: Leopoldina 48 (1912), Nr. 9, 86.

Lampe, Emil: Pogg. III, IV, V, VI
anonymous, in: Leopoldina 54 (1918), 67–68.
K. Strubecker, in: Neue Deutsche Biographie 13 (1982), 459–460.
E. Jahnke, B. G. Teubner, in: Archiv der Mathematik und Physik 28 (1919/20), 1–16.
A. Korn, in: Sitzungsberichte der Berliner Mathematischen Gesellschaft 18 (1919), 4–12.
anonymous, in: Elektrotechnische Zeitschrift 39 (1918), 879.
E. Jahnke, in: Verhandlungen der deutschen Physikalischen Gesellschaft 21 (1919).

Mohr, Max Ernst: Pogg. VIIa
Mathematica, Ad diem natalem septuagesimum quintum data, Festschrift Ernst Mohr zum 75. Geburtstag am 20. April 1985. Berlin 1985.

Rembs, Eduard: Pogg. VI, VIIa
Max Pinl. 1969. Kollegen in einer dunklen Zeit. Jahresbericht der Deutschen Mathematiker-Vereinigung 71, 167–228, espec. pp. 192–195.

Rothe, Rudolf: Pogg. IV, V, VI
G. Richter.1920/21. Rektoratswechsel. Die Technische Hochschule 3 , 205.
W. Schmeidler, in: Elektrotechnische Zeitschrift 63 (1942), 550.
G. Chr. Grüß, in: Zeitschrift für angewandte Mathematik und Mechanik 22 (1942), 302–303.
W. Dreetz, in: Zeitschrift für mathematischen und naturwissenschaftlichen Unterricht 74 (1943), 29–31.
anonymous: Telegraphen-Fernsprech-Funk- und Fernseh-Technik 31(1942), 334.
G. Grüß, 1938. R. Rothe zum 65. Geburtstag. Zeitschrift für angewandte Mathematik und Mechanik 18, 317–318.
anonymous, in: Elektrotechnische Zeitschrift 35 (1914), 1017.

Salkowski, Erich: Pogg. V, VI, VIIa
U. Graf. 1941. Zum 60. Geburtstag von E. Salkowski. Zeitschrift für angewandte Mathematik und Mechanik 21, 62.
Der Lehrkörper der TH Hannover 1831–1956, p. 10. Hannover 1956.

Scheffers, Georg Wilhelm: Pogg. IV, V, VI, VIIa
Technischer Literaturkalender, 3rd edition 1929, col. 561.

Schmeidler, Johannes Werner: Pogg. V, VI, VIIa
J. Schönbeck. 1968. Mathematik. In: Geschichte der Christian-Albrechts-Universität Kiel 1665-1965, vol. 6 Geschichte der Mathematik, der Naturwissenschaften und der Landwirtschaftswissenschaften, K. Jordan, Ed., pp. 9–58, espec. p. 53 Neumünster.
G. Hellwig u. K. Jaeckel. 1960. Werner Schmeidler 70 Jahre. Zeitschrift für Flugwissenschaft 8, 177.
Technische Universität Berlin, Akademische Reden, vol. 2, Berlin 1959. Lecture by H. R. Müller (W. Schmeidler), pp. 12–17.

Stahl, Wilhelm: Pogg. III, IV
Vorlesungsverzeichnis der Technischen Hochschule, Chronik 1893/94, 99–100.
Th. Reye, in: Journal für die reine und angewandte Mathematik 114 (1895), 45–46.
Th. Reye, A. Brill, in: Jahresbericht der Deutschen Mathematiker-Vereinigung 4 (1894/95), 36–45.
H. M. Klinkenberg. Die Rheinisch-Westfälische Technische Hochschule Aachen 1870-1970, p. 256. Stuttgart 1970.

Timpe, Anton Aloys: Pogg. V, VI, VIIa

anonymous, in: Prof. Dr. Timpe 75 Jahre. Petrusblatt 13 (1957), Nr. 12, S. 7.
anonymous, in: Prof. Dr. Aloys Timpe 75 Jahre. Unitas 97 (1957), 49–50.

Weber, Heinrich: Pogg. III, IV, V, VI
A. Voss, in: Jahresbericht der Deutschen Mathematiker-Vereinigung 23 (1914), 431–444.

Weierstraß, Karl: Pogg. II, III, IV, V, VI, VIIa
L. Kiepert, in: Jahresbericht der Deutschen Mathematiker-Vereinigung 35 (1926), 56–65.
C. Runge, in: Jahresbericht der Deutschen Mathematiker-Vereinigung 35 (1926), 175–179.
Biermann, K.-R. 1973. Die Mathematik und ihre Dozenten an der Berliner Universität 1810-1920. Stationen auf dem Wege eines mathematischen Zentrums von Weltgeltung, pp. 61–63 and more often. Berlin: Akademie-Verlag.
The voluminous literature is mentioned in Pogg.

Weingarten, Julius: Pogg. III, IV, V
St. Jolles, in: Vorlesungsverzeichnis der TH Berlin 1910/11, 157–161; French translation in: Bulletin des Sciences mathématiques 2e Série 35 (1911), 142–148.
St. Jolles, in: Sitzungsberichte der Berliner Mathematischen Gesellschaft 10 (1911), 8–14.
anonymous, in: Leopoldina 46 (1911), 21.
L. Bianchi, in: Atti della Reale Accademia dei Lincei, Serie quinta, Rendiconti, Classe di scienze fisiche, mathematiche e naturali 19 (1910), 470–477.

Applied Mathematics
In the United States during World War II

Warren Weaver (1894–1978), left, and Richard Courant (1888–1972), with shovel in hand, at the Groundbreaking Ceremony for Warren Weaver Hall, Courant Institute for Mathematical Sciences, November 20, 1962
Courtesy of Springer-Verlag Archives

Mathematicians at War: Warren Weaver and the Applied Mathematics Panel, 1942–1945

Larry Owens

The Second World War was a time of testing for American scientists. Not only did the war challenge their courage as individuals, it tried as well the mettle of their professions. This was an opportunity not to be squandered, a lesson scientists had learned well after the First World War: proving the temper of wartime science, they would be able, they hoped, to lay claim to public support in the peace to follow.[1] Though less famous than the atomic physicists, for example, who successfully converted wartime achievements into postwar influence, the country's mathematicians likewise sought to make the most of their opportunities.[2] Partway through the war, it appeared things were well in hand.

Indeed, military and scientific leaders both had come to realize that the development and deployment of increasingly complex weapons systems demanded considerable mathematical expertise. In November 1942, Vannevar Bush, the head of the Office of Scientific Research and Development (OSRD) and the figure who spearheaded the mobilization of science early in the war, created the Applied Mathematics Panel to coordinate the services of mathematicians and to serve as a clearinghouse for mathematical information pertinent to the war. To direct the panel, Bush chose Warren Weaver, an applied mathematician from the University of Wisconsin and the director of the Rockefeller Foundation's Natural Sciences Division since 1934. Weaver was also the chief of the fire control analysis section of OSRD's National Defense Research Committee (NDRC).[3] Over the next two and a half years, Weaver's panel supervised an effort that employed close to three hundred people, including such mathematicians as John von Neumann, Richard Courant, Jerzy Neyman, Garrett Birkhoff, Harold Hotelling, and Oswald Veblen; wrote several hundred technical reports; and spent nearly three million dollars.[3a] (See table 1.) The panel encouraged new developments in statistics, numerical analysis and computation, the theory of shock waves, and operational research, and served as a training ground for mathematically-minded workers in fields like economics, one of the more famous being the eventual Nobel Prize winner Milton Friedman. The panel also promoted the institutionalization of applied mathematics through its support, e.g., of Brown University's

ISBN 0-12-599662-4.

Program in Applied Mechanics, Jerzy Neyman's Statistical Laboratory at Berkeley, and Richard Courant's group in applied mathematics at NYU. When ground was broken for the Courant Institute at New York University in 1962, Warren Weaver was there to wield a shovel for the building that would bear his name.[4]

Table 1
Major Contractors for the Applied Mathematics Panel

Major Group	Scientific Personnel	Contract Number	Number of Personnel	Amount
AMG-C	E.J. Moulton S. Mac Lane	sr1007	57	$656,000
SRG-C	H. Hotelling A. Wald M. Friedman W.A. Wallis	sr618	48	$563,000
BRG-C	J. Schilt	sr818	15	$157,000
			120	$1,376,500
SRG-P	S.S. Wilks	sr860	15	$240,000
Princeton	M. Flood	sr1365	20	$80,000
AMG-IAS	J. von Neumann	sr1111	2	$46,500
AMG-NYU	R. Courant	sr945	22	$223,650
AMG-N	E.J. Moulton	sr1379	14*	$131,000
AMG-B	R.G.D. Richardson	sr1066	12	$98,000
AMG-H	G. Birkhoff	sr1384	7	$50,200
NBS	A.N. Lowan		36	
U. New Mex.	E.J. Workman	sr1390	25	$185,000
BRG-Cal	J. Neyman	sr817	—	$110,400
CIW	W. Adams	sr1381	10	$60,000
		TOTALS:	283	$2,601,250

(C-Columbia, P-Princeton, N-Northwestern, B-Brown, H-Harvard NBS-National Bureau of Standards; AMG-Applied Math. Group, SRG-Statistical Res. Group, BRG-Bombing Research Group; * As of 2-12-45; ** As of 7-19-44—figure includes ancillary staff)

In retrospect, the panel might seem an unalloyed success, for it managed to insinuate mathematicians inton a variety of wartime projects and, in particular, gave applied mathematics a boost at a time when, by all accounts, it was paid little heed. Yet judged in terms of its larger ambitions—the central coordination of wartime mathematics—the panel failed. Furthermore, the success it did achieve split the nation's mathematicians into angry factions. This essay concentrates on the panel's failures, for while its successes have been well memorialized, its forgotten trials and tribulations illuminate both the uneven development of American mathematics at the outbreak of World War II as well as the imperial ambitions of those who, like Vannevar Bush, James Conant, and Warren Weaver, took the lead in the mobilization of wartime science.

As with OSRD generally, the activities of the mathematics panel were distributed across the country. At NYU, the group led by Courant studied underwater sound transmission and the behavior of shock waves central to the understanding of shaped charges and jet and rocket nozzles. R.G.D. Richardson and William Prager led the Brown group in studying problems involving engineering and applied mechanics. Princeton's Samuel Wilks analyzed fire effect diagrams and weather forecasting. At the National Bureau of Standards, the panel assumed coordination of the Mathematical Tables Project begun under the New Deal's Works Projects Administration in 1938.[5] Smaller contracts were maintained with John von Neumann at the Institute for Advanced Study and with George and Garrett Birkhoff at Harvard. A group of three contracts with Merrill Flood at Princeton, Walter Adams and the Carnegie Institution of Washington, and E.J. Workman at the University of New Mexico involved the tactical employment of the new B-29 long-range bomber.[6]

The bulk of the panel's work, however, was concentrated in three groups located at Columbia University (see table 1). Established early, the Applied Mathematics Group (AMG), the Statistical Research Group (SRG), and the Bombing Research Group (BRG) drew on substantial statistical and mathematical talent; concentrated primarily on fire control, bombing, and statistical analysis; and, by all accounts, were enormously successful thanks largely to a critical mass of talented individuals and previous experience accumulated in Weaver's NDRC fire control division.

The statistical group was the largest and, arguably, most successful of the panel's efforts to organize mathematics. Established and guided early in its career by Weaver himself, the group benefitted from close and generally cordial relationships with Admiral Julius Furer, the Navy's Coordinator of Research and Development. The Navy, indeed, set the group their first problem: a study of the comparative effectiveness of different gun configurations in air-to-air combat involving fighter aircraft.

Table 2
Distribution of Active AMP Studies as of July 19, 1944

1.	Aircraft	9 studies
2.	Fire Control Equipment	6
3.	Antiaircraft Fire Control	4
4.	Damage Probabilities	6
5.	Bombing	6
6.	Ballistics	3
7.	Gas Kinetics and Explosion Theory	6
8.	Naval Warfare	9
9.	Statistical Problems	9
10.	Computations and Tables	3
11.	Miscellaneous	8

This initial work led to studies of the geometry of combat, pursuit curves and aerial probabilities, and aircraft vulnerability. Columbia's statisticians designed automatic bomb sights, homing torpedoes, and guided missiles, and worried about search strategies for enemy submarines. In addition to such problems of fire control analysis, they dealt with the very different questions of quality control, a concern undramatic in comparison but economically essential for military inspectors obliged to test the reliability of munitions quickly and with little waste.[6a] Both problem areas inspired theoretical developments that were summarized in highly influential postwar treatises—notably in Abraham Wald's *Sequential Analysis* (1947) and the group's *Techniques of Statistical Analysis*. Moreover, this wartime enterprise proved formative in the early careers of men who went on to become powerful figures in the statistical and economic fields. Milton Friedman, for instance, claimed that concepts central to his economic theory were derived from his wartime analysis of the proximity fuse done under the auspices of the Columbia SRG.[7]

Weaver's panel was plagued by all those organizational irritations common to programs grown large quickly in times of war. The Columbia AMG exemplified the administrative difficulties that could afflict even a relatively successful unit. Conceived in the beginning as a central base of operations for mathematical consultants, the group eventually assumed responsibility for a large number of inhouse studies that demanded both competent administration and informed scientific leadership. By January 1944, it had become apparent to Weaver that E.J. Moulton, the group's director, was comfortable with neither responsibility and that activities at Columbia were poorly coordinated and badly run. Conditions had worsened so much by that summer that the panel's executive committee decreed that Moulton change jobs. An applied mathematics group was established at Northwestern to enable his tactful reassignment.[8]

Other problems hinged not so much on lack of administrative ability, as on the eccentricities of genius. For instance, Weaver bent over backwards to apply the much respected talents of Norbert Wiener as a critical check on the quality of mathematical studies—to no avail. Julian Bigelow, an MIT engineer and Wiener coworker, reported that Wiener was just not interested in problem-solving, whether for himself or others, "simply on the basis of utility, particularly if it lacks the qualities suggestive of an elegant, general, formal solution—as do so many of the problems confronting the bolts-and-nuts realist. He simply will not hammer out an inelegant though adequate solution with the cheap-and-nasty tools of the everyday applied mathematician. . . ." If Weaver insisted on getting Wiener's opinion of the work of others, then he should "get a stack of paper, pencils and erasers, hire a hotel room in N.Y., Wash. D.C., or wherever else convenient, send him an emergency telegram preferably to reach him at his house at 2:00 A.M. stressing an emergency of catastrophic consequences requiring his decision immediately; when he arrives rush him to the room and lock him in, with request for a written report to be printed and published immediately, then come around again in about 1/10th the time you think it will take him to read it, and your report will be ready for you."[9] As a matter of fact, Wiener did work for Weaver under Section D-2 that was of fundamental importance for the statistical theory of prediction; but he was of little use within the framework of Weaver's hopes to establish centralized supervision of wartime mathematical work. Indeed, he distrusted "all plans that might depend on a high degree of subordination of individuals to a completely authoritative setup from above, which would assign each man the narrow frame within which he was to work." Wiener had enjoyed an early collaboration with Bush in the 1920's, but his wartime experiences caused him to view Bush and, to some degree, his organizational collaborators in OSRD as fanatical gadgeteers turned authoritarian bureaucrats who preferred the predictability of machines to the eccentricities of human creativity.[10]

Wiener was not the only mathematician Weaver had trouble keeping in harness; he had similar difficulties with von Neumann and Jerzy Neyman. In the summer of 1943, the panel contracted with von Neumann for work on the theory of shock waves; after months of silence, Weaver wrote in January to find out what had been done. Von Neumann excused his lack of communication, confessing that he was frequently out of town, consulting on a wide range of subjects for the Army, Navy, and the NDRC itself. He estimated, for Weaver's sake, that in the previous four months he had probably spent 40% of his time in Princeton and, while there, had worked largely on his panel assignment. Von Neumann's silence frustrated Weaver's sense of administrative accountability; nevertheless, although he found the arrangement somewhat "irregular," von Neumann's association was too valuable to be discontinued. At Berkeley, Jerzy Neyman was engaged in statistical analyses of train bombing that had originated in a request from the Army Air Forces that Weaver's NDRC Section D-2 design a bombardier's calculator. Here also, Neyman's association with the panel became increasingly difficult for a number of reasons: a pronounced dislike for "selling" his expertise to his military patrons and a tendency to postpone the computational chores assigned him by the panel for the sake of highly general theoretical studies of great interest to statisticians but little use to practical-minded generals. Although Weaver made an effort to separate the "pure" from the "applied" portions of the tasks assigned Neyman, the panel eventually decided to terminate the contract.[11]

Neither the eccentricities of genius nor incompetent administration were the cause of the panel's frequently sour relationships with the military. A case in point was the major tactical study of the B-29 that the panel was invited to undertake by the Army Air Forces in the summer of 1944.[12] Weaver established a steering committee and contracted with Merrill Flood at Princeton for analytical studies, with Walter Adams of the Carnegie Institution's Mount Wilson Observatory for model studies of flight formations, and with E.J. Workman of the University of New Mexico Physics Department for experimental flight testing at Alamagordo. Over the next seven months, the panel spent over $300,000 on its B-29 study, conducted 1500 flights, shot 130,000 feet of film, and kept 350 people at work. But here also the project soon became a major headache. To simplify a complex story, the Army Air Forces were in the midst of their wartime identity crisis and the new bomber was valuable and hotly contested property with reputations and influence riding on its success. The Army Air Proving Ground at Eglin Field, the Board of Bombardment in Orlando, Florida, the Second and the Twentieth Air Forces, as well as charismatic leaders like Hap Arnold, all fought for jurisdiction and the authority to define the character of this new potent weapon. Moreover, several of the military men involved in the project protested loudly that NDRC civilians had little business intruding in matters of military tactics. As one of them put it—How long would the Air Force let itself be pushed around by civilians?[13] Workman, on the other hand, turned out to be deeply protective of a project he had begun locally and resented intrusions by Weaver's panel that would produce, he was convinced, "a beautiful theoretical solution" and little else.[14] Weaver himself became extraordinarily frustrated by what seemed to be petty jurisdictional disputes within the Air Force, angered by Workman's habit of bypassing the panel in order to report directly to the military men on the project, and suspicious of the quality of Workman's experiments. Before the project ground to a halt, Workman had both cursed the panel's director and threatened to throw a colonel bodily from his office.[15] Weaver's hopes that the panel might coordinate a major study of air warfare never materialized.

More significant than these piecemeal tactical defeats was a larger strategic failure that compromised the panel's chances of success from the very beginning. In 1943, *Time Magazine* reviewed *Science at War*, a recent book by the popular science writer George Gray that enumerated the achievements of wartime science. In a chapter titled "The Next Decimal Place," Gray celebrated the assistance mathematics was giving ballistics, aerodynamics, optics, acoustics, and electronics—contributions that were being made, he suggested, in spite of the tendency of powerful academicians to denigrate applications as "impure" mathematics. Since modern war was so largely based on mathematics, *Time* quoted Gray, "the U.S. has been severely handicapped by its shortage of topflight mathematicians."[16]

Gray's accusation provoked an angry rejoinder from Marston Morse, the president of the American Mathematical Society (AMS). "The actual fact," he wrote, "is that the deficiency lies... in the failure of the civilian authorities to use mathematics at an early time, in adequate numbers and in the proper way."[17] *Time*'s editors only aggravated matters when they thanked Morse for his "stout reminder of men too easily forgotten." Morse's oblique indictment of Bush and the Office of Scientific Research and Development was quickly echoed by Marshall Stone, a Harvard mathematician and another important figure in the AMS. Stone wrote to Weaver that such articles as that appearing in *Time* could only harm the public standing of mathematicians and "cause future damage to the development of mathematics in the United States and perhaps in the rest of the world as well."[18] Morse and Stone had reason to be angry; the notoriety attracted by the OSRD and the new panel ignored early and strenuous efforts by the mathematical profession to become involved in the war. As early as 1940, the AMS and the Mathematical Association of America had established a War Preparedness Committee involving such men as John von Neumann, Norbert Wiener, Thornton Fry, and Samuel Wilks in order to bring problems relevant to the war effort to the attention of mathematicians.[19]

Bush and Conant had, indeed, underestimated the potential role of mathematics and ignored the possible assistance of the AMS; the NDRC was dominated by physicists, chemists, and engineers and had no mathematical representative. They continued to snub the AMS even when Morse and Stone, in the winter of 1941, sought to alert them to the Society's initiatives. In March 1942, following discussions at the AMS December meeting in Chicago over the unhappy state of affairs, Morse and Stone arranged a meeting with Bush, Conant, and Frank Jewett (who was both the president of the National Academy of Sciences and a member of NDRC) to present them a memo on "Mathematics in War" and to seek arrangements to improve its utilization.[20] Recognizing that the assistance of mathematicians was crucial in the struggle against an enemy that carried his calculations to "the fourth significant figure," the authors of the memorandum expressed their disappointment that current provisions for aid were inadequate and shortsighted and urged "the immediate alteration of prevailing arrangements so that the mathematical intelligence of the country" could be applied to war. It was imperative that NDRC promptly grant "a suitably qualified and recognized mathematician" the authority to inspect defense work in progress and to select competent colleagues for work yet to be undertaken. Furthermore, the authors made plain what, to their minds, constituted "suitable qualification"—history had demonstrated, they wrote, not only that the cultivation of mathematics led to the discovery of natural law and the mastery of nature, but that in many instances mastery had been achieved "only through the skillful application of pure mathematics developed without reference to the immediate needs of physics or engineering."[21]

In an attempt to soothe ruffled feathers, Jewett established a Joint Committee on Mathematics under the sponsorship of the National Academy of Sciences and the National Research Council; its members consisted of Stone, Morse, Griffith Evans, George Birkhoff, Oswald Veblen, Warren Weaver, H.P. Robertson, Walter Bartky, Harry Bateman, and Dunham Jackson, with Stone, Morse, and Evans constituting an executive committee and Morse serving as chairman and mathematical liaison to NDRC.[22] Morse's committee failed to satisfy anyone. It certainly disillusioned the mathematicians and Jewett surmised that Stone was disappointed by not being named chairman. Weaver, excluded from the executive inner sanctum, only discovered through secondhand reports that the committee was, in fact, doing anything at all. The committee was not entirely quiescent—largely through the individual initiative of certain members, it did lend aid to a small number of projects but as Morse once admitted to Weaver "Our Committee had a very quiet existence for a long time."[23] The leaders of OSRD (siding, one must imagine, with those physicists and engineers maligned by the mathematical memo) seem to have expected little from the committee and doubted that a comfortable working relationship with the AMS was generally possible. After the formation of the joint committee, Ward Davidson, an NDRC staff assistant, met with Veblen, Morse, Stone, and Weaver to arrange liaison and reported afterwards that he had "just about no success in any of these conversations." He was unable, he said, "to bridge the wide gap between the view of an engineer and those of a 'pure' mathematician. My imagination just didn't go far enough to understand how problems that seem to me to be rather practical could be handled effectively in the quite rare atmosphere of abstract mathematical thinking."[24] When Weaver was eventually named chair of NDRC's own mathematics committee, the NAS-NRC group became a moot issue.

Should it be surprising, then, that when *Time,* almost a year and half after the establishment of the joint committee, announced that a shortage of able mathematicians had damaged the war effort, insult compounded injury? Stone quickly entered the lists on behalf of the mathematical profession and assaulted Weaver with a barrage of letters that raised a wide range of serious issues. OSRD had, in effect, given mathematicians the "brushoff," he wrote. Weaver responded that he had protested vigorously his appointment as the panel's chair when he learned that neither the AMS nor the NAS-NRC committee had been consulted. He was told that "this was war"; "the reorganization had to be carried out quickly and by a small group." He had his assignment and could accept or refuse it.[25] Why wasn't the panel able, Stone inquired, to discover and employ many of the country's "undeniably gifted mathematicians"? Weaver answered with a primer on panel procedures, noting the difficulties under which it worked. Besides, in spite of all the government red-tape, the burdens of security, and, he pointedly remarked, "the waste of time and energy caused by the self-centered few who are more interested in prestige and their own reputation than they are in helping to win the war," the panel was doing its job. He was well aware of the fact that he was criticized because there was "no stronger representation of the profession" on the panel's Executive Committee; but an organization that included Richard Courant and Samuel Wilks and depended on Jerzy Neyman and Harold Hotelling had to be pretty good.

Moreover, not all mathematicians were likely candidates for the panel. The work demanded a particular kind of person, Weaver surmised, one familiar with the NDRC's manifold activities, comfortable with military personnel, current with weapons development, and, personally, tolerant, unselfish, and cooperative—a team-player who was not convinced that his own ideas were "transmitted to him by Almighty God." Given these necessary qualities, the more mathematics he knew, the better. "It is unfortunately true that these conditions exclude a good many mathematicians, the dreamy moonchildren, the prima donnas, the a-social geniuses. Many of them are ornaments of a peaceful civilization; some of them are very good or even great mathematicians, but they are certainly a severe pain in the neck in this kind of situation." Stone shot back that he thought Weaver had his priorities mixed up. Were organizational qualities really more important than good mathematics? Granted that good mathematicians were often eccentric, would not a competent organization adapt to the best individuals? Weaver answered, several months before Bigelow's advice to lock Wiener in a room, that it was attention to good administration that, in the unpleasant conditions of war, made mathematical contributions possible at all.[26]

There was more at stake in this dispute than hurt feelings or the comparative worth of the brilliant eccentric versus the mathematical team-player. Beneath superficial irritations ran a deeper and more profound disagreement over who, in fact, was best suited to apply the mathematics everybody agreed the war needed. Echoing the feelings of Ward Davidson, Weaver wrote that he was not one of those who felt that mathematics was indivisible or that pure and applied mathematicians were easily convertible. There was a difference between the two disciplines and it was generally the applied mathematician who had the qualities of character and training to be useful in the current crisis. Certainly there were pure mathematicians who could do applied work, he admitted, and applied mathematicians who could do abstract and theoretical work—but their number was small. There was something in the training and discipline of applied mathematicians, Weaver mused, that instilled an attitude of service, while it was "part of the beauty and power of pure mathematics to be abstract, general, imaginative, and independent." Stone agreed that if one were faced with clearly defined problems, then it made sense to put to work the appropriately trained "technicians." In many cases, however, it was the formulation of the problem that was crucial, and here "broader mathematical experience and fuller knowledge of the resources of mathematics" favored the pure mathematician.[27]

That summer, in an address to the Wellesley meeting of the American Mathematical Society, Stone recounted the Society's struggles to get competent mathematicians into the war effort and blamed Weaver's panel for disappointing results. Mathematical work would be better carried out without the mediation of the panel and "there were many mathematicians thus engaged."[28]

Several points should be made about this exchange of letters. In the first instance, from a vantage point outside the OSRD circle, it was certainly reasonable to interpret Weaver's efforts to mobilize mathematics as a powergrab on the part of a bureaucratic junta that effectively excluded the legitimate representatives of the mathematical profession from Washington's corridors of power.[29] To a degree, this perception was the consequence of egos bruised in the political maneuverings that accompanied the coordination of wartime science. But there was more to it than that. What Morse and Stone sensed in this affair was a larger struggle for the future that hinged on the relation between mathematics and its applications—and on who was best suited to make them.

Mathematics had experienced dramatic though uneven growth in the United States since Daniel Gilman lured the British mathematician J.J. Sylvester to Baltimore in 1876 to establish a program in pure mathematics at the new Johns Hopkins University. Over the next half century, pure mathematics made great strides at a number of universities, in particular at Chicago, Harvard, and Princeton. Ironically, however, progress in pure mathematics came at the expense of the practical interests that had long characterized American mathematics.[30] A committee assembled to report on Brown University's new program in applied mechanics reported in 1941: "there has been... since 1900 a marked tendency to emphasize pure mathematics. The success of this development is a great source of national strength and should be a cause for national pride. But it is highly unfortunate that in our enthusiasm for pure mathematics, we have foolishly assumed that applied mathematics is something less attractive and worthy."[31] R.G.D. Richardson noted in 1943 that "of those mathematicians whose names are starred in American Men of Science, the number who are now working in the field of applications is almost negligible. The percentage has decreased with each of the six issues and the new list... will not change this picture. Only one man (E.W. Brown) interested primarily in applied mathematics has been elected to the presidency of the American Mathematical Society since 1900, though before that time nearly all the presidents were in that field." There were few doctorates in applied mathematics, he continued, few graduate programs, and a small number of articles published in journals. "Steps should be taken," Richardson argued, "to strengthen one of the weakest links in the American scientific chain."[32] Richardson's complaints were shared by G.D. Birkhoff, Stone's Harvard colleague: "American mathematicians have shown in the last fifty years a disregard for this most authentically justified field of all."[33] In spite of Birkhoff's sentiments, E.B. Wilson remembered even Harvard as having abandoned earlier American commitments to applied mathematics.[34] In the 1920's, Princeton's Oswald Veblen failed in an attempt to establish an institute dedicated to applied mathematics in part, as William Aspray speculates, because of suspicions about the longterm productivity of research mathematicians.[35]

Nevertheless, there was applied mathematics to be found in the United States, as indeed in any industrializing country, if one looked for it in the right places—and those places were not so much departments of mathematics as industrial laboratories, engineering schools, and departments of physics. Ever since Edison's introduction of the world's first central power station on Pearl Street in Manhattan in 1882, the electrical industry had generated a continuing series of problems that stretched the often rudimentary mathematical talents of inventors and engineers. Francis Upton showed Edison how to use Ohm's Law to understand the "subdivision of the electric light" and thereby calculate the economics of central power. Charles Steinmetz wrestled complex variables into a mathematical tool that allowed engineers at General Electric to design more powerful and competitive dynamos.[36] John Carson mastered the arcane operational calculus and George Campbell studied the mathematical physics of cables to help AT&T complete its longdistance communications network. Vannevar Bush himself earned his reputation as a mathematically-talented electrical engineer who authored an introduction to mathematical methods, collaborated with Norbert Wiener, and pioneered the field of mechanical analysis at MIT in the late 20's and 30's. Most engineering schools echoed MIT's appreciation for applied mathematics if not always its willingness to cooperate with its mathematics department.[37]

Beyond engineering, American physicists felt the need for sophisticated mathematical tools as they began to import developments in quantum mechanics and relativity. Statistics was beginning to find audiences in the fields of insurance, economic forecasting, and industrial quality control. Harold Hotelling at Columbia, Samuel Wilks at Princeton, and Jerzy Neyman at Berkeley were creating small but important centers for statistical research. Moreover, the thirties witnessed the immigration of several highly competent Europeans who quickly took the lead in the institutionalization of applied mathematics notably, Richard Courant, William Prager, John von Neumann, and Theodore von Karman. All in all, the situation in applied mathematics in the United States at the outbreak of the war was probably best summed up by Thornton Fry, Bell Lab's director of mathematical research, in a widely-read article on industrial mathematics. Despite the expanding market and obvious necessity for applications, applied mathematicians were hard to find, Fry noted;

there were throughout U.S. industry only some 150 at work, and most of those had received their training as physicists and engineers.[38]

Thus, when the leaders of the AMS inventoried the strengths and weaknesses of the mathematical profession at the beginning of the war, there was cause to be disturbed. Pure mathematics was strong, with a well-established institutional base; applied mathematics, however, was undernourished and institutionally weak. Worse, at least for the professional aspirations of the AMS, it was not mathematics departments that were supplying market demands for applications, but physics departments and schools of engineering. On the verge of a war in which the status and public appreciation of the mathematical profession would depend on its contributions to mobilization, American mathematicians were dismayed to discover that engineers and physicists, on the one flank, and immigrants, on the other, had stolen a march on them. Given Bush's tilt towards engineering, a belief that an overemphasis on applications would set a "test of immediate social utility" in postwar education, and a nagging fear that wartime organizations like OSRD and the AMP were harbingers of "federal-political control,"[39] it is small wonder that Stone and others found much to worry about in Weaver's efforts to coordinate wartime mathematics.

This litany of failure and compromised hopes could be continued indefinitely, but enough has been said to make the point that the ambitions shared by Bush, Conant, and Weaver to see the Applied Mathematics Panel play a key coordinating role in wartime mathematics failed badly. At the very beginning, in 1940 with the establishment of the National Defense Research Committee, Bush had made a serious mistake, as Oswald Veblen noted, in assuming "that mathematics would not play a fundamental role" in the war—the committee had recruited physicists, chemists, and engineers, but not mathematicians.[40] Worse, Bush alienated the leadership of the mathematical profession when he rebuffed their initial overtures and denigrated their ability to contribute effectively in the wartime environment. Given the potential rewards attached to successful wartime contributions and the fears of many mathematicians that engineers and applied scientists might coopt the direction of mathematical work after the war, their ambivalence if not antagonism towards the panel was only to be expected. Furthermore, even if Bush had persisted in seeking out a working relationship with the leaders of the American Mathematical Society, it seems improbable that the bureaucratic frame-of-mind so characteristic of OSRD, prizing dependable team-play and administrative deference, would have set well with many mathematicians. A further complication was the panel's late creation: by the time it was established in 1943, those components of the NDRC that required mathematical skills had often made their own arrangements; when NDRC, the Army, or the Navy did turn to the panel for help, it was often to take advantage of experience in fire control analysis Weaver had acquired earlier—thus its opportunities for creative mathematical contributions were, to some extent, predetermined, and limited, by the happenstance of its creation.

In other ways, however, the panel must be accounted a success. It did publicize the role of mathematics in the war and thereby promote its continuing importance afterwards. It accelerated the institutionalization of applied mathematics through its support of groups like Courant's at NYU and Richardson's at Brown, schools that were able to translate wartime associations into postwar support.[41] Even with the military, or maybe especially with the military, the panel succeeded. Despite military suspicions of interfering civilians and doubts about the relevance of professorial talents, the panel's labors facilitated the slow awakening of the military to the importance of mathematics for war.[42]

By most accounts, the mobilization of science during the Second World War was an enormous success. The experience of the Applied Mathematics Panel suggests that these achievements were neither complete nor without conflict and were conditioned, at least in some cases, by professional disputes that transcended the immediate limits of the war.

NOTES

I would like to thank the archivists of the National Archives for their expert assistance; especially Lee Johnson, Marjorie Ciarlante, and Janet Hargett, as well as the staff of the Central Research Room, who do their best in frequently trying conditions. Alex Roland and Judith Goodstein also have my thanks for their advice and criticism.

1. Daniel Kevles, "George Ellery Hale, the First World War, and the Advancement of Science in America," *Isis* 59 (1968):427–437.
2. Michael Sherry (see below) has offered the most perceptive sociological account of wartime scientific workers in his *Rise of American Air Power*, in which he sees them as members of a professionalizing elite anxious to put a bad thing to good and self-serving use.
3. See Warren Weaver's slender autobiography, —iбcene of Change A Lifetime in American Science (New York: Charles Scribner's Sons, 1970); for Weaver's work at the Rockefeller Foundation, see Robert Kohler, "The Management of Science: Warren Weaver and the Rockefeller Foundation Programme in Molecular Biology," *Minerva* 14 (19761977): 279–306.

3a. As originally established, the membership of the panel consisted of representatives of divisions having considerable mathematical work, in addition to an executive committee that included Weaver, Thorton Fry, Richard Courant, Samuel Wilks, and E.J. Moulton.

4. For reminiscences of the Panel's wartime work, see W. Allen Wallis, "The Statistical Research Group, 1942-1945, *Journal of the American Statistical Association* 75 (June 1980): 320–335, and Mina Rees, "The Mathematical Sciences and World War II," *American Mathematical Monthly* 87 (1980): 607–621.
5. R.C. Archibald, "The New York Mathematics Tables Project," *Science*, September 25, 1942, pp.294–296.
6. See Wallis, Rees, *op.cit.*; most of the information concerning the Panel's activities is drawn from the AMP files in the records of the OSRD located at the National Archives.

6a. For a summary of the Columbia SRG and its wartime involvements, see the Final Report of the SRG, Columbia University, 9-29-45, National Archives, OSRD Contractors' Reports, AMP.

7. For Friedman's claim, see Wallis, p.329.
8. Weaver to Moulton, January 6, 1944; and July 26, 1944, AMP.
9. Weaver to Stone, January 19, 1944; Bigelow to Weaver, April 22, 1944, AMP.
10. Norbert Wiener, *I Am A Mathematician* (Cambridge: MIT Press, 1954 edition), esp. chapter 12; the Wiener quote is on p.231.
11. Von Neumann to Weaver, January 15, 1944; AMP Executive Committee Minutes, April 3, 1944; Weaver to Neyman, November 17, 1943; Weaver to Conant, October 28, 1942; Weaver to Stone, January 19, 1944; see also, Constance Reid, *Neyman From Life* (New York: Springer, 1982).
12. For the best study of the tactical challenges confronting the AAF and the new bomber towards the end of the war, see Michael Sherry, *The Rise of American Air Power. The Creation of Armageddon* (New Haven: Yale University Press, 1987). Sherry's treatment of both the Army Air Forces' commitment to complex technologies and the role played by civilian "experts" during the war is first-rate (though the experts highlighted in his study are neither hard scientists nor mathematicians). See also his *Preparing for the Next War: American Plans for Postwar Defense, 1941-1945* (New Haven: Yale University Press, 1977), esp. chapter 5, "Soldiers and Scientists."
13. Weaver Diary, AMP, January 21, 1945.
14. Mina Rees Diary, AMP, June 19, 1945.

15. Weaver to Conant, December 8, 1944; Weaver to Davidson, November 25, 1944; Mina Rees Diary, AMP Panel, December 18, and June 26, 1944; Warren Weaver Diary, AMP, February 10, 1945, and June 27-28, 1944; Weaver to Hayward, July 3, 1944.
16. *Time Magazine*, November 29, 1943.
17. *Time Magazine*, December 20, 1943.
18. Marshall Stone to Warren Weaver, November 28, 1943; the Weaver-Stone correspondence is found in the records of the OSRD, Record Group 227, General Records, Organization-Applied Mathematics, National Archives, (abbreviated as AMP). Gray, who, by the way, worked for the Rockefeller Foundation, probably got his information on American mathematics from, among others, Bush and, as Stone suggests, Weaver himself.
19. Marston Morse and William Hart, "Mathematics in the Defense Program," *American Mathematical Monthly* 48 (1941): 293–302.
20. The memorandum had been coauthored by Dunham Jackson, G.A. Bliss, G.C. Evans, and Stone himself. Stone to Weaver, 11-28-43, AMP.
21. Frank Jewett to Marston Morse, March 30, 1942, and memorandum "Mathematics in War," National Academy of Science Archives, Administration, Organization, NAS-Committee on Mathematics, Joint with NRC 1942: General.
22. Jewett to Conant, April 22, 1942, NAS Archives, Admin., Org., NAS-Committee on Mathematics, Joint with NRC 1942: General.
23. Morse to Jewett, February 22, 1943, and Morse to Weaver, March 1, 1943, NAS Archives, Admin., Org., NAS: Comm. on Mathematics, Joint with NRC 1943.
24. Weaver to Stone, December 6, 1943, AMP; Davidson to Conant, August 23, 1944, AMP.
25. Stone to Weaver, November 28, 1943; Weaver to Stone, December 6, 1943, AMP. Weaver protests too much, it seems; the files of the AMP certainly indicate that Weaver had more input than he suggests into Bush's plans to create the Panel.
26. Weaver to Stone, December 6, 1943; Stone to Weaver, December 14, 1943; December 29, 1943, AMP.
27. Weaver to Stone, December 29, 1943; Stone to Weaver, January 12, 1944, AMP.
28. Mina Rees, one of Weaver's technical aides, reported on Stone's talk: "Memo to WW and TCF," August 16, 1944, AMP. One important area of mathematical work, hinted at in Stone's remarks, has not been

mentioned in this essay, and certainly demands consideration in its own right and that is operational research (or operations analysis). Virtually all of the services had, by the end of the war, established independent groups of mathematicians and other experts, sometimes with and sometimes without connections to the Mathematics Panel, to help them solve tactical problems relevant to the deployment of weapons. Some of these groups had great success working with their military patrons, and several had important consequences in the postwar world. See, e.g., Philip Morse, In *At the Beginning*, for an account of the Navy's operational research group, and Fred Kaplan, *The Wizards of Armaggedon*, on the origins of the Rand Corporation in operational research for the Air Force. There is substantial material on operational research in the records of the AMP; while OR lay outside the Panel's field of responsibility as originally envisioned, Weaver found himself drawn perforce into the tangled web of operational research. Needless to say, the Panel had great difficulty sorting out its working relationship with such groups, although it did come to provide valuable assistance in serving as a kind of employment bureau when mathematicians were needed by various military groups in the field. Some of the best work on OR is being done by Robin Rider: see her talk on "OR during the war in the U.S. and Great Britain," Society for the History of Technology, annual meeting, the Hagley Museum, October 23, 1988.

29. Weaver confessed to Stone that the Panel was never set up to cover all of mathematics, only to advise NDRC. While that may be true, OSRD's occupation of the institutional "highground" lent its activities enormous and, to men on the outside like Morse and Stone, worrisome influence.

30. "In historical perspective, what is most interesting about mathematics in the United States from 1900 to the outbreak of World War II is its growth and its decided orientation to pure mathematics," Nathan and Ida Reingold, eds., *Science in America, A Documentary History 1900-1939* (Chicago: University of Chicago Press, 1981), p. 110. Substantial material on the history of American mathematics is scarce and still largely commemorative. See *The Bicentennial Tribute to American Mathematics 1776-1976*, ed. by Dalton Tarwater, and the aging but still valuable account by D.E. Smith and J. Ginsburg, *A History of Mathematics in America Before 1900*. There is, however, evidence of awakening interest: see, e.g., John Servos, "Mathematics and Physical Sciences in America, 1880-1930," *Isis* 77 (1986): 611–629; Nathan Reingold, "Refugee Mathematicians in the U.S.A.," *Annals of Science* 38 (1981): 313–338; John L. Greenberg and Judith Goodstein,

"Theodore von Karman and Applied Mathematics in America," *Science* 222 (December 23, 1983): 1300–1304. See also the introduction to *History and Philosophy of Modern Mathematics,* edited by William Aspray and Philip Kitcher as Volume XI of the *Minnesota Studies in the Philosophy of Science* (1988); as well as the essays in this volume by Karen Parshall, Helena Pycior, and William Aspray.

31. "A Report on Advance Training in Applied Mechanics, With Special Reference to the School of Applied Mechanics at Brown University, by Marston Morse, Mervin Kelly, George Pegram, Theodore von Karman, and Warren Weaver, November 1941," Vannevar Bush Papers, W.A. Jessup file, Library of Congress.
32. R.G.D. Richardson, "Applied Mathematics and the Present Crisis," *American Mathematical Monthly* 50 (1943): 415–423.
33. G.D. Birkhoff, "Fifty Years of American Mathematics," in *Semicentennial Addresses of the American Mathematical Society,* (New York: American Mathematical Society, 1938), Volume 2, p. 313.
34. Discussing the balance of pure and applied mathematics at Harvard, Wilson wrote to MIT's J.A. Stratton in 1957: "You are probably familiar with the discussion of this matter which has been going on at Harvard since before the beginning of this century. The older professors—W.E. Byerly and B. Peirce—who were dominant in the department in the 1890's were sure that the department should cover both pure and applied mathematics, but the German-trained young mathematicians, Osgood and Bôcher, were opposed to anything but pure mathematics being in the department of mathematics and after the death of Peirce and the retirement of Byerly, they dominated the department; they did not want to cover geometry and algebra in adequate fashion, but to specialize on analysis. In the main, this has been the policy of the Harvard department ever since. (And you know well enough that it is the policy of the section of mathematics of the National Academy of Sciences.) George Birkhoff did considerable personally with mathematical applied mathematics which did not get down to numerical physical applications, but he did not reform the department. His son, Garrett, a very good pure mathematician, does by preference concern himself now with hydromechanics even getting down to numerical results; but whether this makes any real dent on the thinking of the department, I do not know. I have some evidence that there are members who believe the department should stay pure and leave to Dean Van Vleck the job of developing applied mathematics at Harvard which he seems to be doing very

effectively." Wilson to Stratton, 10 January, 1957, MIT Archives, AC4, Box 214, Folder 14.

35. William Aspray, "The Emergence of Princeton as a World Center for Mathematical Research, 1896-1939," in Aspray and Kitcher, *op. cit.*, pp.346–366.
36. On Steinmetz, see Ronald Kline, "Science and Engineering Theory in the Invention and Development of the Induction Motor, 1880-1900." *Technology and Culture* 28 (1987): 283–313.
37. For engineers and the sophisticated demands of electrical work, see Thomas Hughes, *Networks of Power*; for the MIT example, see Karl Wildes and Nilo Lindgren, *A Century of Electrical Engineering and Computer Science at MIT, 1882-1982* (Cambridge, MA: The MIT Press, 1985); Larry Owens offers insights into the origins and circumstances of the mathematical work of another electrical engineer in "Vannevar Bush and the Differential Analyzer: The Text and Context of an Early Computer," *Technology and Culture* (1986).
38. Thornton C. Fry, "Industrial Mathematics," *American Mathematical Monthly* 48(6) (1941): 139.
39. Stone to Weaver, December 14, 1943, AMP.
40. Mina Rees Diary, November 13, 1944; Courant memo to Weaver, January 6, 1944, in Courant file, AMP.
41. See Constance Reid, *Courant in Göttingen and New York: The Story of an Improbable Mathematician* (New York: Springer, 1976), for an account of how Courant parlayed his wartime contacts into contributions to his long-sought institute for applied mathematics at Columbia.
42. The Army Air Force, according to its official historians, had opposed the creation of an umbrella agency like OSRD, and its reaction "to the new agency was mixed." "The scientists who served with OSRD helped soon to make the AAF a willing, even eager, collaborator. . . ." Wesley Craven and James Cate, eds. *The Army Air Forces in World War II* (Chicago: University of Chicago Press, 1955), Volume 6, p. 236.

John von Neumann (1903–1957) and J. Robert Oppenheimer (1904–1967)
standing before the Institute for Advanced Study Computer
Courtesy of Springer-Verlag Archives

The Transformation of Numerical Analysis by the Computer: an Example from the Work of John von Neumann

William Aspray

Although numerical methods for determining approximate solutions to mathematical problems had existed from the time of Newton, the classical discipline of numerical analysis is generally regarded as beginning with the work of Gauss. Building on his results, mathematicians of the nineteenth and early twentieth centuries developed numerical methods for practical hand calculation, to solve small systems of linear equations, invert small matrices, approximate integrals, and solve ordinary differential equations.[1] These methods remained relatively unchanged until the introduction of the computer after the second world war. Keying on the computer's great power to carry out long sequences of calculations rapidly and without human intervention, John von Neumann and a few other visionary mathematicians saw the opportunity to extend numerical techniques to a much wider domain of mathematics and mathematical applications, e.g. to large linear systems and partial differential equations, and so to their applications in classical and quantum physics, economics, biology, and the earth sciences. Von Neumann recognized that certain adjustments needed to be made to numerical analysis in order to employ the computer effectively and reliably in these pursuits. The computer differed in several important respects from the desk calculator or punched-card equipment that had previously driven numerical analyses. First, the computer was much faster—on the order of 10^3 to 10^5 times faster—than hand methods in its ability to carry out multiplications.[2] However, its ability to centrally store data, instructions, and intermediate results was more severely limited. Practically unlimited amounts of these types of information could be stored on punched cards and made effectively available in a punched-card system, while only 20 to 150 numbers could be stored in the first of the new high-speed calculating devices.[3] As von Neumann and Herman Goldstine described the situation:

> ... in an automatic computing establishment there will be a "lower price" on arithmetical operations, but a "higher price" on storage of data, intermediate results,

ISBN 0-12-599662-4.

> etc. Consequently, the "inner economy" of such an establishment will be very different from what we are used to now, and what we were uniformly used to since the days of Gauss. Therefore, new computing methods, or, to speak more fundamentally, new criteria for "practicality" and "elegance" will have to be developed.

We are actually now engaged in various mathematical and mathematical-logical efforts aiming towards these objectives. We consider them to be absolutely essential parts of any well-rounded program which aims to develop the new possibilities of very high speed, automatic computing.[4]

An analysis of the complexity of various types of numerical problems and the speed of various calculating devices convinced von Neumann and his collaborator Herman Goldstine that new ground could be broken in the areas of partial differential equations, integral equations, and systems of linear equations. However, they also concluded that problems with numerical stability, i.e. with the accumulation and amplification of round-off errors, would require substantial changes in the classical methods of numerical analysis. For example, their analysis suggested that in the solution of systems of linear equations it may be necessary to abandon classical elimination methods in favor of successive approximation methods, which, though they may require more multiplications, have an intrinsic numerical stability and hence are less demanding in the number of digits that must be carried through the calculation. They began an investigation of computer-oriented numerical analysis in 1946, at almost the same time they began construction of the Institute for Advanced Study computer. Their research continued over the next decade, almost until the time of von Neumann's death in 1957.

The remainder of this article describes two of von Neumann's forays into computer-oriented numerical analysis, as an indication both of his research contributions to this area and of the changes that took place in numerical analysis by the introduction of the computer.[5]

Partial Differential Equations and the Question of Numerical Stability

Numerical methods for the solution of total differential equations were well established prior to the second world war, but only a few contributions had been made toward numerical methods for the solution of partial differential equations. The most important prewar results were the work of Courant, Friedrichs, and Lewy on numerical stability and of Southwell on relaxation methods. In the immediate postwar years problems of fluid dynamics—including important applications to atomic energy, petroleum exploration, and weather forecasting—came to the fore. Von Neumann's involvement in all of these applications, through his consultancies to government and industry, led to his interest in numerical methods for solving the partial differential equations governing fluid dynamics. His chief contribution to this area of numerical analysis was his investigation of numerical stability.

This work built directly on that of Courant, Friedrichs, and Lewy. In 1928 these authors published a paper introducing the vexing problem of numerical stability. They demonstrated that there need not be an absolute relationship between the solution to a partial differential equation and the solution to the conventionally given difference equations approximating it. Using as an example an equation describing compressible, non-viscous flow, they showed that the solutions to the difference equations would converge to the solution of the exact equation just in case the difference equation grid met certain rigid conditions. More specifically, they proved that the difference equation solution was reliable only if the ratio $\Delta t/\Delta x$ of the discrete time and space intervals was less than a fixed number c. Otherwise, the approximation errors would be amplified beyond control. This was known as the Courant condition and became a standard concern of numerical analysts.

The Courant, Friedrichs, Lewy paper rigorously presented stability criteria for PDEs of the general hyperbolic and parabolic type, with one dependent and two independent variables. Von Neumann extended their results to more complicated systems of equations: those with more than one dependent variable, more than two independent variables, higher order than two, etc.[6] But he gained his generality at the sacrifice of rigor. He recognized this powerful method he introduced as heuristic, and he claimed its value on empirical grounds, arguing "as far as any evidence that is known to me goes ... it has not led to false results so far."[7]

Von Neumann never published a general account of his stability analysis, though he lectured on his results on several occasions.[8] Perhaps most representative of his approach is a consultant's report he wrote in July 1948 for Standard Oil Development Company.[9] In it he applies his method to a practical problem, a model of gas-liquid flow around an oil well or along a long oil field being produced at one end.[10]

The problem assumes there is a single, thin oil-bearing layer, which may be on an incline. The layer is regarded as two-dimensional, but is described in terms of a single spatial variable because of symmetry considerations. The independent variables are the distance x from the well and the time t. The dependent variables are the pressure p and the liquid saturation s of the semipermeable medium in the oil-bearing layer. This formulation yields a set of partial differential equations of the parabolic, heat conduction type.

Von Neumann's problem is to determine whether there is a numerical method for the solution of these equations that can be solved using desk calculators or time rented on the IBM SSEC.[11] The problem involves a two- dimensional grid. Using the specifications agreed upon with the company, the range of the oil field is five miles, with $\Delta x = 1,000$ feet; thus, requiring approximately 25 x-netpoints. The time span is ten years, with $\Delta t = 30$ days; thus, requiring approximately 120 t-netpoints. This meant that there were $25 \times 120 = 3000$ netpoints x 50 multiplications/netpoint = 150,000 multiplications. Based on this consideration alone, von Neumann ruled out the use of desk calculators.

However, he pointed out that there were a number of additional mathematical factors to consider before one could be confident that a numerical solution was possible on a faster calculating machine. He dealt straightforwardly with the discontinuities that arose in the system from a conflict between the initial and boundary conditions. He also showed how to solve a special system of total differential equations to provide initial values. Most of the remainder of the report addresses the more difficult problem of error amplification (numerical stability), which he writes is "of paramount importance" and "is altogether controlling with respect to the stepwise integration methods that can be used."[12]

For this reason, von Neumann devotes several sections of the report to a general discussion of error amplification, before returning to the problem at hand. He identifies four sources of error in the stepwise numerical solution of total or partial differential equations:[13]

1. Each individual step of the integration process introduces an error by replacing a differential quotient by an approximating finite difference quotient.
2. Within each integration step, arithmetic operations are carried out inexactly because of round-off errors.
3. These elementary round-off errors accumulate, since there is one integration for each netpoint in the integration net.
4. In the course of the integration these elementary errors are not only accumulated but also amplified, as they become multiplied by various factors.

It is this amplification of errors that is the insidious problem in the solution of partial differential equations, and so the one to which von Neumann devotes his attention.

He begins his stability analysis by considering a simpler PDE, but one which contains the "decisive" terms from the oil-gas flow equations. The equation he chooses is

$$\frac{\partial p}{\partial t} = m\frac{\partial^2 p}{\partial x^2}.$$

Arguing analogously to Courants, Friedrich, Lewy (1928), von Neumann shows that each Fourier component in the error term of the conventional difference equation solution to this PDE is amplified by

$$A = [1 - \frac{4m\Delta t}{(\Delta x)^2}\sin^2\left(\frac{h\Delta x}{2}\right)]^{\frac{(t-t_0)}{\Delta t}}$$

where h is the Fourier frequency. He can then show that the maximum value of A is less than or equal to one (so that the error term is not amplified) just in case $\Delta t \leq \frac{(\Delta x)^2}{2m}$. He also notes that there is no relationship between the computational stability and the stability of the PDE, which is the heat conduction equation and known to be stable.[14]

With this example as guide, von Neumann repeats his stability analysis for the full oil-gas flow equation. He shows that it is impossible, with existing computing equipment like ENIAC or SSEC, to meet the stability conditions if the problem is approximated by difference equations in the conventional way. To meet the stability conditions and hold $\Delta x = 1000$ as specified by the company, he determines that Δt might have to be less than .1 for the LP case and 1000 times smaller for the CS case—as compared with the value $\Delta t = 30$ specified by the company. This requirement increased the number of multiplications in the numerical solution by factors of 300 and 300,000 for the two cases, respectively—placing them outside the capacity of the fastest computing equipment of the time.

These results led von Neumann to seek a method of stepwise integration for the oil-gas problem more stable than the conventional one. He suggested a method first introduced at Los Alamos, which damped out the unstable element in the conventional method and resulted in full stability.[15] The price paid for this stability was having to solve an additional system of simultaneous linear equations, which represented the implicit form of the solution, for every value of t. In the LP (long oil field) case this amounted to a system of 27 linear equations in 27 unknowns for each of 120 values of t. Von Neumann pointed out that this price was at first appearance "formidable," but in practice was much lower because each of the equations contained only 3 of the 27 unknowns. Using a slightly improved version of this implicit method, together with a method for solving systems of linear equations that was optimal for his particular circumstances, he estimated the size of the oil-gas flow problem: 170,000 multiplications per problem, requiring 200 man-weeks at a desk calculator, seven hours at the SSEC if it were working at peak efficiency, or 18 hours if SSEC were working at its customary 40% efficiency.

The method von Neumann developed turned out not to have practical utility for Standard Oil, but both the stability analysis and the implicit solution method became working tools in numerical analysis.

Random Numbers and Monte Carlo Methods

A rather different pattern of impact of the compact on numerical methods is illustrated by the history of the Monte Carlo method. It is a variety of model sampling, similar to that which had been employed in a limited number of cases by statisticians for many years prior to the appearance of the computer. The method involves establishing a relatively simple stochastic model of a complex problem. Random numbers with an appropriate statistical distribution replace the seemingly random physical events under consideration, and a solution is inferred by calculating the behavior of the random numbers in the stochastic model. For example, the problem of neutron diffusion, in which at each instance of time millions of neutrons might undergo fission, scattering, or absorption, is infeasible to solve by the traditional approaches of theoretical or experimental physics. However, one can replace the neutrons, with their seemingly random behavior, by an appropriately chosen set of random numbers that describe the histories of the neutrons over time, and a solution can be calculated without need of any experimental study.

Von Neumann's friend and associate, the mathematician and theoretical physicist Stanislaw Ulam, has explained how the Monte Carlo idea came to him while playing solitaire during an illness in 1946.[16] He noticed that it was much simpler to gain an idea of the outcome of such a complex phenomenon as solitaire by making various trials with the cards and counting proportions of outcomes, than it was to formally compute all of the combinatorial possibilities. It occurred to him that this same observation should apply to his work on neutron diffusion at Los Alamos, where it was practically impossible to solve the integro-differential equations governing scattering, absorption, and fission of particles.

Computing machines, which were just becoming available, could readily handle the trial calculations and replace the experimental apparatus of the physicist. During one of von Neumann's visits to Los Alamos in 1946, Ulam mentioned his idea. After a few hours of skepticism, von Neumann was won over and began rapidly to develop the possibilities into a systematic procedure. As Ulam describes it, Monte Carlo "came into concrete form and started developing with all its attendant tricks rudiments of a theory after I proposed the scheme to Johnny."[17]

What seems to be the first written account of the method was given by von Neumann in a letter to Robert Richtmyer of Los Alamos in early 1947. Von Neumann's missive was bound with Richtmyer's reply into a Los Alamos report and distributed accordingly.[18] Von Neumann suggested the possibility of using Monte Carlo to trace the isotropic generation of neutrons from a variable composition of active material along the radius of a sphere. He believed the problem was well-suited to the ENIAC, which he estimated would require five hours to calculate the actions of 100 neutrons through the course of 100 collisions apiece.[19] He also suggested that a more general formulation of the problem, taking into account the energy and momentum exchanges, was possible using Monte Carlo, provided a more powerful calculating device like the one being built at the Institute for Advanced Study were available.

The first actual Monte Carlo calculations by computer were carried out on the ENIAC, by a team from Los Alamos headed by Nicholas Metropolis.[20] The first calculations simulated chain reactions in critical and supercritical systems, starting from an assumed initial space and velocity distribution of neutrons. Within a year Monte Carlo was being used for a wide variety of applications. The method attracted considerable interest at a mathematics conference held at the Institute for Numerical Analysis (INA) in Los Angeles in July 1948; and this led to a formal symposium on the topic at INA in June 1949.[21] The symposium gives some indication of the rapid growth and application of the method: to nuclear physics, statistical mechanics, materials science (in the aircraft industry), and partial differential equations.[22]

The method was applied originally to probabilistic problems, like neutron diffusion, but soon ways were found to solve deterministic problems, like differential equations or linear operator equations, by this fundamentally probabilistic method. One of the first such applications was made in 1948 by Enrico Fermi, Ulam, and von Neumann to estimate the eigenvalues of the Schrödinger equation. Von Neumann and Ulam found one of the first methods for inverting matrices using Monte Carlo methods.[23] Von Neumann also discovered a Monte Carlo method for solving a typical hyperbolic differential equation.[24]

The Monte Carlo method consumed large quantities of random numbers. Methods for generating random numbers of sufficient quantity and quality already existed for hand computation, but there was the belief that with the increased use of the computer there would be a need for many more random numbers and that they would have to satisfy much more stringent randomness conditions. Von Neumann considered the general question of how to produce a sequence of random decimal digits.[25] He mentioned two methods, physical and arithmetical, for producing random sequences. Physical methods involve using some sensing device to count a presumably random physical process, e.g. the clicks of a Geiger counter, from which the random digits are fed into the computer. He found this method unsatisfactory, unless the random sequence was recorded, otherwise there was no way of later checking computational results using the same random sequence; but he noted that recording these sequences taxed the weakest link of the existing computer systems, the memory.

Thus von Neumann turned to arithmetic methods. He developed one of the first arithmetic methods for producing pseudorandom numbers—the middle-square method.[26] It was a practical solution: it took only 3 milliseconds on the ENIAC to produce a random number using the middle-square method, but 600 milliseconds to read a random number from a punched card. Von Neumann went further and developed a set of statistical tests for checking sequences for random properties.[27] He also developed methods, e.g. the rejection technique, for producing sets of numbers with particular multivariate distributions to use in applications where the standard rectangular distribution was not suitable.[28]

This interest in Monte Carlo methods seems to have been the motivation for several empirical studies von Neumann undertook on the computer to investigate the properties of randomness. He became interested in calculating long strings of decimal digits of certain real numbers to determine their statistical deviations from randomness.[29] The question under investigation was whether these decimal expansions would remain closer to their expectation values than a random sequence of the particular length normally would. In July 1949 ENIAC was used to calculate 2000 digits in the sequences of e and π. The former, but not the latter, was found to deviate significantly from a random sequence of digits.[30] A similar investigation was carried out in 1955 using the Institute for Advanced Study computer to calculate decimal expansions of real algebraic numbers. In collaboration with Bryant Tuckerman, von Neumann generated 2000 partial quotients in the continued fraction expansion of $\sqrt[3]{2}$. The results were inconclusive. But these studies encouraged other empirical and theoretical investigations of randomness; and by this time Monte Carlo techniques were established and were being employed by hundreds of researchers in a wide range of application areas.

GENERALIZATIONS

The introduction of the computer invigorated and redirected the study of numerical analysis. This change is evident in the immediate postwar decade in the growth of publications and practitioners, the emergence of specialty professional societies and journals like The Society for Industrial and Applied Mathematics and SIAM Review, and the spin-off of disciplines like operations research and management science.

The computer stimulated many new avenues of research in numerical analysis, as are suggested by the two examples we have described from the work of John von Neumann. In fact, his contributions to numerical analysis extend beyond partial differential equations and Monte Carlo methods, and include methods for linear systems, eigenvalues, shocks, and linear programming. Some of his contributions introduced quantitative change. For example, once von Neumann and other numerical analysts identified appropriate algorithms, the computer could solve much larger systems of linear equations than were practical to solve by hand. Other changes were of a qualitative nature. For the first time there existed means to numerically solve partial differential equations, study non-linear phenomena, and investigate stochastic models of complex stochastic and deterministic phenomena.

The computer made the numerical analysts reinvestigate the time-honored methods of their discipline. Truncation errors became less significant when using a machine able to handle massive numbers of calculations, but round-off errors and the statistical properties of random numbers became more significant. Mathematicians had to reconsider the methods they were using, even for such classical problems as inverting a matrix or finding its eigenvalues, to see how well they could be handled by a computer. Could the methods be entirely automated? Did they have numerical stability? What was the relationship between the number of iterations required and the precision that must be carried through the calculation? How much initial data and intermediate results must be retained in storage? These and other questions were addressed for the first time upon the introduction of the computer.

Von Neumann was aware from its inception that to fashion the computer as an effective scientific tool required the development of numerical methods as well as computing machinery. He was also cognizant that the different "inner economy" of the computer from past calculating devices meant that numerical analysis would have to undergo fundamental revision. Thus he began an investigation of numerical methods at the same time he began work on the design of the Institute for Advanced Study computer, and, as such, was among the first scientists to study computer-oriented numerical analysis. Some of his results have endured over the past forty years; the methods of Monte Carlo and linear programming are examples. In other cases his work was superseded by later work: his methods for numerically solving partial differential equations and inverting large matrices, methods for accommodating shocks, and numerical stability analysis. But even in these cases his work brought into focus the nature of the problems and offered directions for further study.

Under von Neumann's direction, the Electronics Computer Project at the Institute for Advanced Study became an important training center for computer-oriented numerical analysis. Herman Goldstine has identified almost fifty scientists who participated in its study there.[31] The Institute was by no means the only center actively studying numerical analysis during this time. There were major centers at the National Bureau of Standards in Washington, the Institute for Numerical Analysis at UCLA, and New York University in the United States; and other centers in Britain and continental Europe. Von Neumann was a regular visitor to many of these centers until such visits were precluded by his commitments as an AEC commissioner. It is quite likely that his elevated status in the mathematical and government communities helped to legitimize the new discipline of numerical analysis and improve the flow of research funds. For all of these reasons, von Neumann should be regarded as one of the originators of the modern discipline of numerical analysis—a discipline renewed and refocused by the introduction of the computer.

NOTES*

1. Hand calculation refers not only to calculations done long-hand, with pencil and paper, but also to the use of logarithmic and other tables, slide rules, desk calculators, punched-card tabulators, and a few other aids to doing the calculation without assistance from a programmable, automatic calculating machine.
2. Von Neumann argued that multiplication required approximately the same amount of time as division, and that it dominated (was substantially greater than) the time required for addition or subtraction; so that multiplication time was the appropriate measure of speed of arithmetic operation for a calculating device.
3. Even the new computers under construction, like that at the Institute for Advanced Study, could store only a few thousand instructions in primary memory. Of course, these high-speed calculation devices could store practically unlimited amounts of information in secondary storage (e.g. on punched cards), but the time required to

* This work was supported in part by National Science Foundation grant NSF/SES–8609543.

access this data was sufficiently slow that it compromised the overall performance of the machine. Memory design involved not only how much could be stored, but how rapidly it could be accessed. Information on punched cards could be accessed no faster by punched-card systems than by computer systems, but because the punched-card systems carried out their calculations at a relatively slower rate the time to access information from the cards was not out of balance with the multiplication time, as it was in the new computing systems.

4. Goldstine and von Neumann (1946, p. 6).
5. An extended account, covering all of von Neumann's contributions to numerical analysis and setting them in the context of his other work in computing science, will appear in my book, *John von Neumann and the Origins of Modern Computing.*
6. Letter, von Neumann to Werner Leutert, 15 October 1950. (Von Neumann Papers, Library of Congress).

   One of von Neumann's earliest extensions of the CFL stability analysis was given in a 12 page letter written from Los Alamos to Goldstine, dated 21 February 1947. In this letter von Neumann describes how his interest in meteorology primarily had led him to seek integration methods for hyperbolic differential equations where the CFL condition $c\Delta t \leq \Delta x$ does not hold. (Goldstine Papers, American Philosophical Society Library)
7. Letter, von Neumann to Werner Leutert, 10 November, 1950 (Von Neumann Papers, Library of Congress). This correspondence is part of a three-sided argument, which runs from July 1950 to August 1951 (all in the von Neumann Papers), over the value of von Neumann's method as reported in O'Brien, Hyman, and Kaplan(1950). Leutert, a mathematics professor at the University of Maryland, sent von Neumann a copy of his manuscript "Eine neue Methode zur numerischen Lösung linearer partieller Differentialgleichungen mit konstanten Koeffizienten" and published a paper, Leutert (1951), which claimed that von Neumann's method does not adequately decide numerical stability, using the heat equation to illustrate his point. In the course of the correspondence Hyman and Kaplan communicate to von Neumann, and in a different way von Neumann to Leutert, errors in Leutert's reasoning; but Leutert remained resolute. Von Neumann wrote in exasperation to Kaplan, 15 August 1951:

   "I would like to restate, that I think Leutert misses the point completely. As far as I can see, he treats the heat equation by a different numerical method from the one for which the particular stability condition in question was derived. I have been trying to make it clear at every occasion, including in my correspondence with Leutert, that a stability criterium, in the sense in which I use this word, always relates to a specific numerical method for solving a (partial) differential equation, and not to the differential equation itself."

8. Lectures he presented at Los Alamos Laboratories in February 1947 are cited and summarized in von Neumann and Richtmyer (1947). A lecture given in August 1947 to the Mechanics Division of the Naval Ordnance Laboratory at White Oak, Silver Spring, Maryland is reported in Hyman (1947). Von Neumann authorized an account of his White Oak results to be presented in a paper by O'Brien, Hyman, and Kaplan (1950).

   In the White Oak lecture von Neumann considered the wave-equation

$$\frac{\partial^2 u}{\partial x^2} = \frac{\partial^2 u}{\partial t^2}$$

   He showed how the Courant, Friedrichs, Lewy difference representation is numerically stable whenever $(\Delta x)/(\Delta t) \geq 1$, and how under another difference representation involving a parameter a stability is assured when $(\Delta x)/(\Delta t) \geq \sqrt{1-4a}$ Von Neumann claimed that, although he had restricted himself to the simple problem of the wave equation, his method would apply equally well to more complicated equations.

9. Von Neumann (1948a).

10. The oil well and oil field have similar equations governing their flow. Von Neumann treats the two cases, which he calls *circular symmetry* (CS) and *linear parallelity* (PL), in the same report.

11. Von Neumann had already determined there were no (known) analytic solutions to the problem. He had further shown that if one assumes that the flow is horizontal, the equations can be simplified greatly, to a total differential equation that can be solved straight forwardly by step-wise numerical integration. However, he was dubious whether such an idealized result could be of any practical value.

    The SSEC was one of the fastest calculators of its day, using a hybrid of relay and vacuum tube technology. For more information about the SSEC, see Bashe et al. (1986).

12. Von Neumann (1948a, p. 689).

13. A similar account is given in Goldstine and von Neumann (1946, pp. 15–17).

14. To summarize von Neumann's stability analysis: give a Fourier expansion of the error expression and trace the values assumed by the general term of the expansion to see whether it exceeds one. One might also include as part of his heuristic method the use of implicit difference representation to achieve numerical stability. O'Brien, Hyman, and Kaplan (1950) provide a systematic discussion of von Neumann's method, and indicate that he had applied it to a wide range of problems during the second world war.
15. The method is discussed in von Neumann and Richtmyer (1947).
16. Ulam's most complete account of these origins is given in an unpublished manuscript, entitled "The Origin of Monte Carlo" (Ulam Papers, American Philosophical Society Library).
17. Ibid.
18. The Los Alamos report is Richtmyer and von Neumann (1947). In an unpublished manuscript entitled "Introduction to Papers on the Monte Carlo Method," dated 12 April 1983, Ulam claims that von Neumann's letter is probably the first written account of the method.
19. He also suggested that several "man-days" using manual-graphical methods might also suffice, and planned to ask Princeton University statistician S.S. Wilks about his experience using such techniques on a bombing problem during the secong world war. There is no record of the outcome.
20. A talk given by R.D. Richtmyer on "Monte Carlo Methods" at a meeting of the American Mathematical Society 24 April 1959 (Ulam Papers, American Philosophical Society Library) describes these early Monte Carlo calculations and other points about the early development of the method.
21. These meetings are described in Hurd (1985). Ulam and von Neumann participated in both of these conferences. Hurd also discusses another early meeting on Monte Carlo, sponsored by IBM's Department of Education in November 1949. See Hurd (1950).
22. A copy of the program for the Monte Carlo symposium resides in the Ulam Papers (American Philosophical Society Library).
23. This method is described in Forsythe and Leibler (1950).
24. This method is described in a 13–page letter von Neumann wrote to Abraham Taub, 5 July 1949 (Ulam Papers, American Philosophical Society Library). The equation is one that describes one-dimensional

Lagrangian hydrodynamics, with an equation of state where p depends on v alone, namely:

$$\frac{\partial^2 X}{\partial t^2} = \frac{-v}{\frac{\partial X}{\partial x}} \frac{\partial p}{\partial x}, \quad v = \frac{\partial X}{\partial x}, \quad p = F(v).$$

Von Neumann was excited about the promise of the method, but apparently could not find a way to adapt the method to higher-dimensional problems, for there is no further mention of it.

25. Von Neumann (1951).
26. This method involves taking an arbitrary eight-digit number as the first number in the random number sequence. The $(n+1)$-st number in the sequence is formed by squaring the nth number and taking the eight middle digits of the result.
27. See von Neumann's letter of 3 February 1948 to Alston Householder and of 3 December 1948 to Cuthbert Hurd (von Neumann Papers, Library of Congress).
28. See von Neumann (1951). Also, Hammersley and Handscomb (1964, pp. 36–37).
29. Goldstine (1972, p. 293) suggests Ulam as the inspiration for this research program.
30. The results of these investigations are reported in Metropolis, Reitwiesner, and von Neumann (1950). The details of the calculation are reported in Reitwiesner (1950). The calculation for e required 11 hours of machine time for computing and checking, and an equal amount of time for punched-card handling; that for π took approximately 70 hours.

    In a letter of 8 September 1949 von Neumann presented to Reitwiesner his analysis of the ENIAC calculations. He summarized his findings by stating: "The first 2000 digits of e *do not look like a random sample* of decimal digits. Oddly enough, they seem to deviate from randomness by being *too uniformly distributed!* This is very remarkable, and it would seem to deserve further study." (Emphasis in original. Goldstine Papers, American Philosophical Society Library)
31. Goldstine (1972, p. 292).

BIBLIOGRAPHY

Bashe, Charles J., Lyle R. Johnson, John H. Palmer, and Emerson W. Pugh. 1986. *IBM's Early Computers*. Cambridge, MA: MIT Press.

Courant, R., K. Friedrichs, and H. Lewy. 1928. *Über die partiellen Differenzengleichungen der Mathematischen Physik,* Mathematische Annalen, 100, pp. 32–74.

Forsythe, George E. and Richard A. Leibler. 1950. *Matrix Inversion by a Monte Carlo Method,* Mathematical Tables and Other Aids to Computation, 4, pp. 127–129.

Goldstine, Herman H. 1972. *The Computer From Pascal to von Neumann.* Princeton: Princeton University Press.

Goldstine, Herman H. and John von Neumann. 1946. *On the Principles of Large Scale Computing Machines.* Manuscript of lecture given at the 15 May 1946 meeting of the Mathematical Computing Advisory Panel, Office of Research and Inventions, Navy Department, Washington, DC. Published in A. Taub, ed. *John von Neumann Collected Works,* V, pp. 1–32.

Hammersley, John M. and D.C. Handscomb. 1964. *Monte Carlo Methods.* London: Methuen.

Hurd, Cuthberd C., ed. 1950. *Seminar on Scientific Computation.* New York, NY: International Business Machines Corporation.

Hurd, Cuthbert C. 1985. *A Note on Early Monte Carlo Computations and Scientific Meetings,* Annals of the History of Computing, 7, pp. 141–155.

Hyman, M.A. 1947. *Stability of Finite Difference Representation.* Mechanics Division Technical Note SB/Tm-18.2–11. Naval Ordnance Laboratory, White Oak, Silver Spring, MD. (October 16).

Kendall, M.G. and B. Babbington Smith. 1939. *Tables of Random Sampling Numbers,* Tracts for Computers, 24. Cambridge University Press.

Leutert, Werner. 1951. *On the Convergence of Approximate Solutions of the Heat Equation to the Exact Solution,* Proceedings of the American Mathematical Society, 2, pp. 433–439.

Metropolis, N.C., G. Reitwiesner, and John von Neumann. 1950. *Statistical Treatment of Values of First 2000 Decimal Digits of e and of π Calculated on the ENIAC,* Mathematical Tables and Other Aids to Computation, 4, pp. 109–111. Reprinted in A. Taub, ed. *John von Neumann Collected Works,* V, pp. 765–767.

O'Brien, George C., Mortan A. Hyman, and Sidney Kaplan. 1950. *A Study of the Numerical Solution of Partial Differential Equations,* Journal of Mathematics and Physics, 29, pp. 223–239.

Reitwiesner, George W. 1950. *An ENIAC Determination of π and e to more than 2000 Decimal Places*. Mathematical Tables and Other Aids to Computation, 4, pp. 11–15.

Richtmyer, R.D. and J. von Neumann. 1947. *Statistical Methods in Neutron Diffusion*. Los Alamos Scientific Laboratory Report LAMS-551 (April 9).

Von Neumann, John. 1948a. *First Report of the Numerical Calculation of Flow Problems. Consultant's report to Standard Oil Development Company. Prepared 22 June to 6 July 1948*. Published in A. Taub, ed. *John von Neumann Collected Works*, V, pp. 664–712.

Von Neumann, John. 1948b. *Second Report on the Numerical Calculation of Flow Problems*. Consultant's report to Standard Oil Development Company. Prepared 25 July to 22 August 1948. Published in A. Taub, ed. *John von Neumann Collected Works*, V, pp. 713–750.

Von Neumann, John. 1951. *Various Techniques Used in Connection with Random Digits*. Chapter 13 of Proceedings of Symposium on "Monte Carlo Method" held June-July 1949 in Los Angeles. Journal of Research of the National Bureau of Standards, Applied Mathematics Series, 12, pp. 36–38. Summary written by George E. Forsythe. Reprinted in A. Taub, ed. *John von Neumann Collected Works*, V, pp. 768–770.

Von Neumann, J. and R. D. Richtmyer. 1947. *On the Numerical Solution of Partial Differential Equations of Parabolic Type*. Los Alamos Scientific Laboratory, Technical Report LA-657. Reprinted in A. Taub, ed. *John von Neumann Collected Works*, V, pp. 652–663.

Von Neumann, John and Bryant Tuckerman. 1955. *Continued Fraction Expansion of* $2^{\frac{1}{3}}$. Mathematical Tables and Other Aids to Computation 9, pp. 23–34. Reprinted in A. Taub, ed. *John von Neumann Collected Works V, pp. 773–774.*

Notes on the Contributors

Thomas Archibald is Assistant Professor of Mathematics at Acadia University, Wolfville, Nova Scotia, Canada. He took his Ph.D. in the history of science and technology from the University of Toronto. His research has centered on mathematical physics, particularly the development of potential theory, in 19th-century Germany. He is the author of "Carl Neumann versus Rudolf Clausius on the propogation of electrodynamic potentials," *American Journal of Physics,* 1986.

William Aspray is Associate Director of the Charles Babbage Institute for the History of Information Processing at the University of Minnesota. His research centers on the history of mathematics and computing during the 20th century. His recent publications include: "The Emergence of Princeton as a World Center of Mathematical Research," in W. Aspray and P. Kitcher, eds., *History and Philosophy of Modern Mathematics,* (Univ. of Minnesota Press, 1987); (co-editor with Arthur Burks), *John von Neumann's Papers on Computing and Computer Science,* (MIT Press/Tomash Publishers, 1987); "The Mathematical Reception of the Computer," in E. Phillips, ed., *Studies in the History of Mathematics* (Mathematical Association of America, 1987).

Amy Dahan-Dalmedico is affiliated with the Centre de la Recherche Scientifique (C.N.R.S), Paris. Her research has concentrated on Cauchy's contributions to algebra and group theory and mathematical physics during the early decades of the 19th century. She is the co-author of *Une histoire des mathématiques, routes et dédales,* (Le Seuil, 1986) and *Mathématiques au fil des ages,* (Gauthier-Villars, 1987) and author of: "La mathematisation des theories de l'elasticite par A. L. Cauchy et les debats dans la physique mathématique francaise (1800-1840)," *Sciences et Techniques en perspective,* 1984/85; "Les travaux de Cauchy sur les substitutions – Etude de son approche du concept de groupe," *Archive for History of Exact Sciences,* 1980.

Ivor Grattan-Guinness is Reader in Mathematics at Middlesex Polytechnic, England. From 1974 to 1981 he was editor of *Annals of Science,* and he is the founder-editor of *History and Philosophy of Logic,* which was launched in 1980. He is a member of the Executive Committee of the International

Commission on the History of Mathematics and President of the British Society for the History of Mathematics. He recently edited *History in Mathematics Education* (Paris, 1987), the proceedings of a workshop held in Toronto in 1983, and his lengthy study of the mathematical sciences in early 19th-century France is scheduled to appear in 1989 as volumes 2, 3, and 4 of the series *Science Networks*.

Eberhard Knobloch is Professor of History of Exact Sciences and Technology at Berlin Technical University and editor of *Historia Mathematica*. He is an expert on illustrated technical manuscripts of the Middle Ages and he has also published extensively on the work of Leibniz and Euler. He is a collaborator on the Academy edition of Leibniz' complete works and letters, and a co-editor of *Humanis* and *Technik*. He edited the 1983 volume *Zum Werk Leonhard Eulers* and authored *Die Mathematische Studien von G. W. Leibniz zur Kombinatorik*, Studia Leibnitiana Supplementa, vols. 13, 16, and "Unbekannte Studien von Leibniz zur Eliminations- und Explikationstheorie," *Archive for History of Exact Sciences*, 1974.

Jesper Lützen is Professor of the History of Exact Sciences in the mathematics department at the University of Copenhagen. His recent publications include: *The Prehistory of the Theory of Distributions*, (Springer, 1982); "Sturm and Liouville's Work on Ordinary Linear Differential Equations. The Emergence of Sturm–Liouville Theory," *Archive for History of Exact Sciences*, 1984; "Joseph Liouville's Work on the Figures of Equilibrium of a Rotating Mass of Fluid," *Archive for History of Exact Sciences*, 1984. His comprehensive scientific biography of Liouville is scheduled to be published by Springer-Verlag in 1989.

Larry Owens is Assistant Professor of History at the University of Massachusetts, Amherst. The holder of two doctorates, one in biochemistry and one in history, he is the author of "Pure and Sound Government: Laboratories, Playing Fields, and Gymnasia in the 19th-Century Search for Order," *Isis*, 1985, and "Vannevar Bush and the Differential Analyzer: The Text and Context of an Early Computer," *Technology and Culture*, 1986. He is presently revising his dissertation, *Straight-thinking Vannevar Bush and the Culture of American Engineering* (Princeton University, 1987), for publication.

Erhard Scholz is Professor of Mathematics at Wuppertal University. His publications include: *Geschichte des Mannigfaltigkeitsbegriffs von Riemann bis Poincaré*, (Birkhäuser, 1980); "Riemann's frühe Notizen zum Mannigfaltigkeitsbegriff und zu den Grundlagen der Geometrie," *Archive for History of Exact Sciences*, 1982; "Projektive und vektorielle Methoden in Culmanns graphischer Statik," *NTM*, 1984; *Symmetrie – Gruppe – Dualität. Zur*

Beziehung zwischen theoretischer Mechanik und Anwendungen in Kristallographie und Baustatik des 19. Jahrhunderts, Science Networks, vol. 1, 1989.

Gert Schubring is affiliated with the Institut für Didaktik der Mathematik at Bielefeld University; his speciality is institutional and educational issues in 19th century mathematics. Among his publications are: *Die Entstehung des Mathematiklehrerberufs im 19. Jahrhundert. Studien und Materialen zum Prozess der Professionalisierung in Preussen (1810–1870)*, (Beltz, 1983); "The Conception of Pure Mathematics in the Professionalization of Mathematics," in H. Mehrtens, H. Bos, I. Schneider, eds., *The Social History of Nineteenth Century Mathematics*, (Birkhäuser, 1981); "The Rise and Decline of the Bonn Naturwissenschaften Seminar," *Osiris*, vol. 5, 1989.

Dirk Jan Struik, the dean of American historians of mathematics, is Professor Emeritus at the Massachusetts Institute of Technology. A student of J. A. Schouten and Tullio Levi-Civita during the 1920s, he was a participant in and eye-witness to the emergence of modern tensor analysis at that time. One of his early forays into historical writing was his two-part article "Outline of a History of Differential Geometry," *Isis*, 1933/1934. His *Yankee Science in the Making*, a detailed account of the scientific culture in colonial New England, was first published by Little & Brown in 1948. The same year saw the appearance of Professor Struik's *Concise History of Mathematics*, his most well-known book, which has since gone through numerous editions and reprintings, and has been translated into at least 18 different languages. Presently he is working on a study of the early development of tensor analysis and preparing to write his autobiography.

Renate Tobies is affiliated with the Karl-Sudhoff-Institut at Leipzig University and specializes in educational and institutional developments in modern mathematics. Her publications include *Felix Klein*, Biographien hervorragender Naturwissenschaftler, Techniker und Mediziner, vol. 50, 1981, and the 2-part article "Zu Veränderungen im deutschen mathematischen Zeitschriftenwesen um die Wende vom 19. zum 20. Jahrhundert," *NTM*, 1986/87. She is presently revising her Dissertation B, "Die gesellschaftliche Stellung deutscher mathematischer Organisationen und ihre Funktion bei der Veränderung ger gesellschaftlichen Wirksamkeit der Mathematik (1871-1933)," for publication.